A Practical Guide to Ecological Modelling

A Practical Guide to Ecological Modelling

Using R as a Simulation Platform

Karline Soetaert and Peter M.J. Herman

Netherlands Institute of Ecology, Yerseke, The Netherlands

 Springer

Dr. Karline Soetaert
Netherlands Institute of Ecology
Centre for Estuarine & Marine Ecology
(NIOO-CEME)
4400 AC Yerseke
PO Box 140
The Netherlands
k.soetaert@nioo.knaw.nl

Dr. Peter M.J. Herman
Netherlands Institute of Ecology
Centre for Estuarine & Marine Ecology
(NIOO-CEME)
PO Box 140
4400 AC Yerseke
The Netherlands
p.herman@nioo.knaw.nl

Additional material, the R-examples and the R-code of all figures, is available as an R-package (ecolMod),
which can be found on the official R-website (http://cran.r-project.org/).
The R-example files are also available on the website of this book at www.springer.com

ISBN: 978-90-481-7936-7   e-ISBN: 978-1-4020-8624-3

Cover illustration: View at the Oosterschelde from the Netherlands Institute of Ecology, Yerseke,
The Netherlands. Photograph taken by Thomas Haverkamp.

Printed on acid-free paper

9 8 7 6 5 4 3 2 1

Preface

Why Another Ecological Modelling Book?

For several years we have taught courses on ecological modelling at the level of graduates or starting PhD students. The audience typically consists of students from biology, geology, bio-engineering, and, less frequently, from sciences such as physics and chemistry. For most of these students our course was a first acquaintance with the field of ecological mathematical modelling. Although often difficult, it was an intellectual adventure both for them and for us. The course was set up as an initiation to the subject, starting from the most basic of principles but nevertheless leading to quite advanced applications.

This book is based on lecture notes that were written in 2001 and that accompany this modelling course. We prepared the original notes because we felt there was shortage of a book covering all the material that we wanted to address in our lectures. The notes also served as a readable text that the students can consult while preparing for the exam. This allowed us to teach the theory in rather short lessons, leaving more time to cover the practical exercises, during which we directly interacted with the students individually.

This book is written for young researchers who want to get more out of their data than just description. Even when they see the possibilities of modelling to help them gaining insight in the processes they study, two factors might frighten them away from this path. One is mathematical formalism, theorems and detailed proofs, the bread and butter of the applied mathematician. The other is complicated, semantic and near-philosophical ecological theory. We have steered away from these two extremes. We have tried to write a book that is readable for an audience with a basic formation in ecology and a basic knowledge of mathematics. Although enough material is presented that may also interest the more experienced ecologic modeller, it is not a book only readable for either full-blown mathematicians or ecological theoreticians.

Throughout the book we have tried to be practical, emphasizing the diversity that exists in mathematical models and techniques. We discuss only the essential aspects of mathematical methods, without pretension to mathematical rigour: often one does not need to understand the fine details of a technique to correctly apply it, but it helps greatly to have an intuitive understanding of its foundations. This is where our emphasis has been placed.

Despite our preference for practical and simple approaches, fact is that the basic methods of solution are often adequate only for the most simple of models. As application of more efficient and complex mathematics may considerably speed up solution, and thus avoid frustration, we do not neglect to mention some more advanced techniques that can make the life of a modeller so much more pleasant.

Ecological modelling has multiple roots. Many theoretical ecological models go back in some way to the pioneering work of Lotka and Volterra. A new approach, aiming at environmental modelling, was based on engineering principles from about the 1960's onwards. We have taken this approach as a starting point for the book, because it is much more directly based on conservation laws and therefore an easier vehicle to explain principles underlying modelling. However, we have tried to bridge, from there, to the more classical ecological approach in later sections of the book.

Scope and Content

A brief mention of the book's contents reveals its scope.

We start by giving arguments as to why models are useful, from the scientific point of view as well as with respect to management. This sets the scene and explains why we are doing what we are doing. Here we introduce some model applications, both simple and complex, that will be expanded on further in the book.

After having introduced the semantics of models, we then proceed with the basic principles of transferring ecology into equations. This is where our book differs most from other books, which generally assume that such knowledge is already available, or can be deduced from the rather complex examples that these books generally contain. During our lectures we became aware that the mental switch from descriptive to process-based thinking is the largest leap for most of our students, it is NOT the maths. Therefore we spend much effort to explain and detail the formulation of ecological interactions.

Next we deal with how space can be incorporated into the mass equations. We concentrate on one-dimensional problems with various geometries, with a short excursion to three dimensions.

We then continue with the mathematical solution of the models, mentioning where applicable possible sources of difficulty and error. This section deals with differential and numerical calculus, the basic mathematical concepts are introduced as they are needed, and compiled in an appendix. This is definitely the most demanding part of the book, but necessary to put theory into practice.

In a next chapter, the derivation of the steady-state solution and subsequent analysis of its properties introduces concepts such as stability, domains of attraction, multiple stable states and bifurcation.

Thus far, the models that were discussed fall into the category of deterministic differential equations. In the remaining part of the book, some other types of models are dealt with. They include difference (discrete time) equations, dynamic matrix models, and sequential decision models, also known as dynamic programming models.

As it is essential for making robust modelling applications, we generally spend a lot of time during our practical courses on designing, testing, validating and improving ecological equations. This is the topic of the final book chapter, which discusses various techniques for analysing model behaviour.

Each chapter is organised as follows: an introduction sets the scene of what is to follow, and, if relevant, puts the chapter in perspective with respect to previous chapters. The first sections give the basics and theory, if appropriate illustrated by (simple) examples. Certain sections (starred) probe beyond the elementary level, and may be skipped at first reading. We also find it important to actually show how to implement models, such that the reader may acquire hands-on experience. Thus, each chapter includes case studies that illustrate (nearly) all methods discussed in the main text, and put the theory into practice. Where possible, we chose published models that are simple enough and were amongst the first in their kind, to illustrate concepts. The code to run these examples, implemented in the R computer language, is included and discussed in the book and can be found on the accompanying website or on the official R-website (see below).

R, the Modelling Platform Used in this Book

For those who are being initiated in the field, the learning of a new (programming) language, on top of the new sets of principles that surround mathematical modelling may be very demanding. Therefore, during our practical courses, the problems have been kept simple, such that the students can implement them in a spreadsheet, a software package that most of them know or should know. These exercises, and their solutions, can be found on the accompanying website.

For this book, we have taken a different approach and we use R for our examples, mainly because it is free software, it is rapidly gaining popularity, R-code is highly readable, and... we simply like it.

Although R was not originally developed to be used as a modelling tool, it is very well suited for this task. In our day-to-day work, we use R mainly to develop simple models or to visualise model output. We also use R to interface with compiled models written in Fortran. R is then used for post-processing the model output (making graphs, creating summaries, performing tests...).

As the use of the R-language is growing rapidly, students are now becoming acquainted with R during their statistical courses. We expect (or hope) that it is only a matter of time before the use of spreadsheets in introductory modelling courses can be avoided.

In an appendix, we give a -very short- introduction to R. A more extensive introduction can be found on the book's website.

The Books Website

The book comes with some additional material, which can be downloaded from the book's website on www.springer.com.

The files that contain the example codes are in a subdirectory named after the chapter. As the code is generally small, we have printed almost all of it in the book.

We make liberal use of diagrams and figures in our book. This has a purpose: diagrams visualise concepts and relationships, while figures are a critically important tool for analyzing model output. R has been used for making these diagrams and figures. The source code of all book figures has been bundled in an official R-package (ecolMod – Soetaert and Herman, 2008) and put on the R-website (CRAN). It can simply be installed as a regular R-package, after which the figures of each chapter can be generated by running a demo, named after the chapter.

Acknowledgements

Many people have provided valuable input/feedback or reviewed parts of this book. We would like to thank especially our colleagues Jack Middelburg, who commented on the overall concept and on most chapters but also stimulated us to start writing this book, Filip Meysman, Johan van de Koppel, Matthijs Vos, Dick van Oevelen, Andreas Hofmann, Wolf Mooij and Marcel Klaassen. Whereas all these persons provided great help, of course all remaining errors are our own responsibility. Also thanks to our post-docs (Marilaure Grégoire, Sophie Rabouille, Caroline Ulses, Jim Greenwood), our (former) PhD students (Dick van Oevelen, Jeroen Wijsman, Filip Meysman, Henrik Andersson, Karel Van den Meersche, Andreas Hofmann, Pieter Provoost, Tom van Engeland, Julius Kones, Paul Obade, Tom Cox) and all MARELAC and ECOMAMA students, for challenging us to explain the modelling process from the most basic up to the highest level.

The Royal Dutch Academy of Science (KNAW) supports our research, in which we make frequent use of mathematical models. The University of Ghent, the Free University of Brussels, and the University of Nijmegen allowed us to teach ecological modelling to undergraduate and graduate students. We express our gratitude to the Netherlands Institute of Ecology, Centre for Estuarine and Marine Ecology, our base institution for providing us with the opportunity to finalise this book. Finally, we dedicate the book to those that are near to us: our spouses, Carlo Heip (KS) and Rosette Mortier (PH), and our children, Maarten and Eva (KS), Eva, Gerard and Judith (PH).

Yerseke, The Netherlands   K. Soetaert
Yerseke, The Netherlands   P.M.J. Herman

Contents

Chapter 1
Introduction

The natural environment, with all the physical, chemical and biological processes going on, is of an overwhelming complexity. Scientists try to make sense of this complexity, so as to, among other things, have a certain predictive capability. This activity only progresses by the construction of models that make suitable abstractions of reality. The purpose of these abstractions is to highlight the relevant aspects of complex phenomena, select the most important processes that drive their dynamics in time and space, and relate the descriptions of state and process to each other.

There is a reward for this difficult work. Models can be used for various purposes: for testing our understanding of a certain phenomenon, for extracting quantitative information, for making time- and spatially averaged budgets and for prediction.

In this chapter we will review basic concepts of modelling, their building blocks as well as their purpose and use.

1.1 What is a Model?

A model is a simple representation of a complex phenomenon. It is an *abstraction*, and therefore does not contain all the features of the real system. However, a model does comprise all the characteristic ones, those essential to the problem to be solved or described.

Consider the example of someone about to buy a car. He is comparing two very different kinds of car: a white van with a diesel engine, and a red sports car. They differ in many respects: colour, height, weight, power of the engine, price, fuel consumption, glamour, type of seats, electronics etc. etc. When this person is buying a car mainly to transport goods, he will probably only consider the characteristic 'volume', to arrive (very quickly) at a decision. When he is buying the car only for leisure trips, he might be most concerned with the probability of seducing someone to go with him, the maximum speed, the impression made when cruising along the sea coast. Then aspects such as price, glamour, colour, power divided by weight may come into play. Depending on his question, very different aspects of the car should be considered, and there are few questions for which all aspects are really relevant. Conscious or not, the car buyer makes abstractions in his mind that suit his purpose,

K. Soetaert, P.M.J. Herman, *A Practical Guide to Ecological Modelling*,
© Springer Science+Business Media B.V. 2009

and uses these to reach a decision. Basically, this is also what scientists do when constructing models.

Mankind has always used models to solve certain problems or describe certain complex phenomena. Consider a number of examples (Fig. 1.1).

- A map is a model in which the position, names, etc... of countries are emphasised, whilst irrelevant details such as the position of houses, agricultural information... are ignored. It is a model, because it is simpler than the real world, and it is in a form that can be easily understood and interpreted (Fig. 1.1 A).
- A miniature of a ship will have the shape of a real ship, but the position of furniture, instrumentation, etc... will be ignored (Fig. 1.1 B).
- In a conceptual model of a sediment ecosystem, the complexity of the food web may have been collapsed into a few functional groups. It is a model, because it synthesises our knowledge in a simplified way (Fig. 1.1C).

What characteristics are essential to the problem? Obviously, this depends on the *aims* of the model and on the definition of the problem. It also depends on the *scale window* of the model. It is not possible to describe everything, from molecules and elementary physics up to the functioning of the whole earth or universe in a single model. The modeller will choose a *typical temporal and spatial scale* to describe the phenomenon of interest, e.g. a scale of seconds, days, years, centuries, a spatial scale of millimetres, centimetres or kilometres. Processes at much smaller scales (sub-scale) are not explicitly modelled; processes at much larger scales (supra-scale) are imposed onto the model. A eutrophication model for a lake will typically have a one-year temporal scale, and a spatial scale of the whole lake. It will resolve processes with a typical rate between once per day to once per year, but it will simplify processes that are completed in seconds, minutes or hours. It will impose multi-year fluctuations in external conditions without modelling all the relevant processes determining these conditions. With respect to its aim, it will focus on nutrients and on processes that consume or produce nutrients, as well as those processes that are

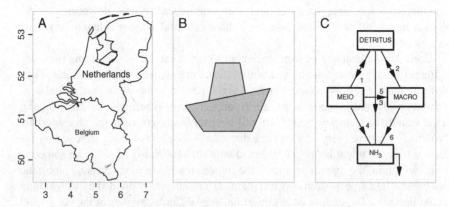

Fig. 1.1 Three examples of models. They have in common that they focus only on the object of interest, ignoring the irrelevant details. What is irrelevant depends on the aim of the model

directly affected by the consumers and producers (such as grazers of phytoplankton). It will likely not consider all the details of the life history of one species of snails in the lake.

In this ecological modelling book, we discuss mechanistic *mathematical* models, i.e. models where the ecological processes (the mechanisms) are described in a mathematical sense. Mathematics should be considered as a 'language', a formal framework suitable for conveying quantitative information and for systematic quantitative deduction. Mathematical treatment of a problem has a number of advantages: mathematical models are formally precise: any inconsistency in the formulation of processes will be easily detected. They are also numerically precise: by making quantitative *predictions* that can be tested against observations, the fit or non-fit of the model to data is inevitably shown. This will, amongst others, demonstrate whether our concept of the natural phenomena was valid or needs refinement.

1.1.1 A Simple Example: Zooplankton Energy Balance

Most ecological interactions involve the exchange of mass or energy or both. For instance, growth of algae involves the uptake of nutrients from the environment; feeding of a zooplankton individual on these algae involves the transfer of biomass from the algae to the animal. In addition, the animal will produce faeces and respire, which transfers mass to detritus and CO_2; predation on the zooplankton will remove animal biomass.

As a *verbal* model of the interaction between the zooplankton population and the algae we note that the biomass of the zooplankton will increase by the ingested food, whilst defecation, predation on the zooplankton (e.g. by fish) and respiratory costs will decrease the biomass (Fig. 1.2).

We can put this model into more *quantitative* form by writing a simple mass balance equation for the zooplankton biomass:

Zooplankton biomass storage = food ingested − faeces production
−respiration − biomass removed by predators

(1.1)

From this equation we learn that the biomass storage term will be positive, i.e. there will be growth, if the source (ingestion) is bigger than the sinks (faeces+respiration+predation). If the source is smaller than the sinks, storage will be negative and biomass will decline.

As we will see in this book, it is only a small step to rewrite this verbal model in a mathematical form, known as a *differential equation*:

$$\frac{dZOO}{dt} = \text{Ingestion–Defaecation–Respiration–Predation} \quad (1.2)$$

where $dZOO/dt$ is now much more precisely defined than the somewhat loose term 'storage': it is the rate of change of zooplankton biomass through time, that is, the

Fig. 1.2 Based on a
conceptual diagram of the
energy balance of a
zooplankton organism, it is
but a small step to write a
mathematical model

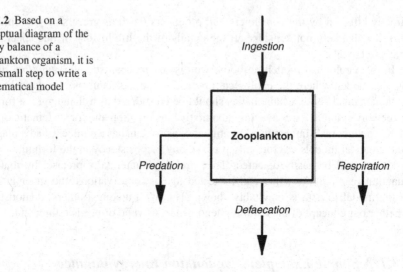

speed at which the zooplankton biomass will change. Obviously, the more positive
is the balance on the right hand side of the equation, the faster the biomass of zooplankton will increase.

In a next step, we rewrite all the terms on the right hand side in a more quantitative way as a function of zooplankton (ZOO) and algal (ALG) biomass. There exist
many possible formulations, for instance:

$$\frac{d\text{ZOO}}{dt} = imax \cdot \frac{\text{ALG}}{\text{ALG} + ks} \cdot (1 - \gamma) \cdot \text{ZOO} - r \cdot \text{ZOO} - m \cdot \text{ZOO} \qquad (1.3)$$

The first term on the right hand side represents ingestion minus faeces production; the second term denotes respiration and the last term is the mortality; *imax*, *ks*,
γ, *r* and *m* are constants (so-called parameters). This equation may look impressive,
but it obeys a number of simple rules that will be explained in detail in this book.

Most of the ecological, but also physical and chemical models are written as one
or as a set of coupled differential equations. In a similar way as in the example, these
equations are the mathematical translations of our conceptual understanding how a
system works, i.e. how certain quantities change through time.

One consequence of the simplification process is that we (must) make a number
of *assumptions* about reality. In the zooplankton example, we assumed that algae
and zooplankton can each effectively be described by one quantity. Thus, we assume that any spatial differences in their concentrations do not affect the model
dynamics. Also, differences in zooplankton species composition or in edibility of
the different algae over the model time scale are considered unimportant. By assigning only grazing, faeces production, respiration and mortality as the governing
factors for zooplankton dynamics, behavioural responses to, for instance, predator
presence are excluded. It is virtually impossible to list all the assumptions implicitly

or explicitly made, but finding the most relevant assumptions is very crucial during the modelling process.

Once formulated, the model needs to be *solved* in order to express not the rate of change of the zooplankton biomass as a function of time, but the actual biomass itself. As scientists have worked on differential calculus for a long time, there exist powerful techniques for solving these differential equations. For example,

- By integration (the inverse operation of differentiation) of the example model, we may predict how the zooplankton biomass will evolve in time.
- Stability analysis allows investigating the model equilibrium conditions, conditions that will lead to chaotic behaviour, etc...

These aspects will receive attention in separate chapters of the book.

1.2 Why Do We Need Models?

Traditionally, engineers and physicists have depended greatly on mathematical models and modelling has been an integrated part of their scientific method for a long time.

In the ecological sciences, the Lotka-Volterra model (Lotka, 1925; Volterra, 1926) was one of the first models based on sound mathematical principles, and since then the importance of mathematical models as tools in many ecological disciplines has become widely accepted.

Nevertheless, the interest for mathematical modelling in biological sciences has been modest, compared to other disciplines.

There are plenty of ecological problems or applications that can benefit from or cannot be solved without mathematical modelling.

1.2.1 Models as Analysing Tools

Science proceeds through systematic recording of observations and experimental results and then interpreting these results in terms of probable causal factors, or generating *hypotheses* that are consistent with these data (Fig. 1.3). Usually it is not very difficult to generate a hypothesis (or even ten hypotheses) when we are confronted with experimental data. The real problem is to ensure that the hypothesis is really consistent with the data at hand, and to generate hypotheses that are of wider validity than the specific experimental system. It is here that models can play a crucial role. In our quest to determine whether our understanding of the system is valid, and to convince our colleague scientists, we often go beyond the qualitative description and use a model to demonstrate our right. If your colleagues are easily convinced, this model can be some rough (back-of-the-envelope) calculation; else it should be quite sophisticated.

Fig. 1.3 The scientific
method: observations and
models are used for
interpretation and hypothesis
generation, and thus explain
real-world phenomena

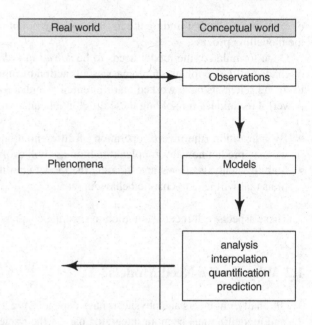

Experimentalists might argue that every mechanism can be elucidated by carefully designed experiments, and one does not need to resort to models. However, experiments are most efficiently performed if we have some idea about what we are looking for, i.e. if we have some model telling us what to expect. In fact, one can only devise a reasonable experiment if one has at least a mental or conceptual model of the possible outcomes in mind. The difference between a conceptual and a mathematical model is in the rigour and the formalism, not in the question whether a model is used or not. Therefore, some form of model, whether mathematical or conceptual, will always *guide the experimental part* of the scientific method, and there is no reason to logically separate 'modellers' from 'experimentalists'.

Besides, some experiments are impossible to carry out (think of astronomy!). This occurs, when the spatial or temporal scales are too large, or when there stand, between dream and deed, regulations and practical considerations (to paraphrase a famous Dutch poem). When this is the case, one generally applies a model first, followed by careful model verification, and this is often the only way to proceed in understanding the system.

Some measurements are possible but may be too expensive, unpractical or may take too long a time. Here models can be devised as a (cheap) alternative to performing measurements and to distinguish between competing explanations that cannot be elucidated by observations. Once a mathematical model has been devised, it can be tested in a variety of conditions. This for instance may reveal hidden properties and functions of ecosystems that can be tested to data.

As analysing tools, *mathematical* models have a certain edge over conceptual models, as during their development scientists are forced to think more clearly. This is because the modeller has to specify every component relevant to the problem, every term that enters a model equation, and to be explicit about the assumptions. These assumptions are a necessary by-product of the simplification. We may make them deliberately, because we think many aspects do not really matter. Sometimes, however, we have to rely on assumptions because the necessary knowledge about the process details are lacking.

The resulting model then can be thought of as a logical proposition that, given a set of assumptions, produces certain consequences, or model output. It suffices to note that a certain model is unable to reproduce natural phenomena in order to realise that we have insufficient conceptual knowledge and refinement of the model is due. Refinement can be a simple reformulation of the same processes, or an extension of the model to include more processes, or a different choice of most important variables and processes to consider. The process leads to identifying gaps in our knowledge that may give direction to research priorities and new experiments to be performed.

1.2.2 Models as Interpolation, Extrapolation, and Budgeting Tools

In addition, some measurements may not be easily *interpolated or extrapolated in space and time*.

Combining point samplings with averaging routines (e.g. by means of a Geographic Information System) is one way to obtain spatial (or temporal) averages.

Often though, interpolation by means of a model that is based on our conceptual understanding of the underlying mechanisms, and that is driven by appropriate external conditions (e.g. currents, atmospheric data) may be more accurate than simply taking the mean of all measurements, especially as the data set becomes sparser. Figure 1.4 schematizes such a condition. We used ocean primary production in spring as the mental example for this figure. When measurements are plentiful (filled circles and full lines) simple averaging will give a good impression of the real process. If, however, for some reason measurements lack in the first half of the spring peak (e.g. because the ship had to take another crew, it was too stormy to go out, an instrument broke down, or it was simply too expensive...), then interpolation using process understanding, may provide a much better estimate of the real process than simple averaging.

This ability to make accurate spatial and temporal averages is one reason why mathematical models are often used for making mass and energy budgets of ecosystems. As a mechanistic model is a mimic of how the system works, it can, to a certain degree, also be used for extrapolating in space, and in time.

Other, 'black box' techniques (such as neural network models) may faithfully reproduce existing data sets and allow some forecasting, but here too modelling may provide better alternatives. Not necessarily because they give better r^2 (they usually don't) but rather because they will give more confidence when extrapolating. Most importantly though, mechanistic models have the benefit of increasing our understanding of the mechanisms, whilst statistical descriptions give only shallow understanding.

The basic difference between statistical (black-box) and mechanistic models is that the black box routines consider ecosystems as something to be measured, not to

Fig. 1.4 Purely statistical
interpolation introduces large
errors if based on a too small
data set (*dashed line*).
Model-based interpolation
(*grey area*) may be better
suited in these cases

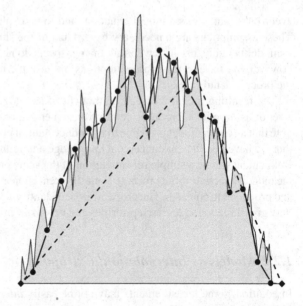

be understood. In such (statistical) models, it is assumed for instance that all areas
of similar type behave in the same way. Discrepancies are adjusted using correction
factors, adding more structure, etc... These adjustments often make it possible to
generate good predictions. However, by adding more and more detail, there may
come a time where the black-box method will start to have the complexity of a truly
mechanistic model. In the end then, both the statistical and mechanistic model may
give similar r^2, but the black box method will lack the added value of increased
knowledge into the biological, chemical and/or physical interactions and the ability
to extrapolate (Fig 1.5).

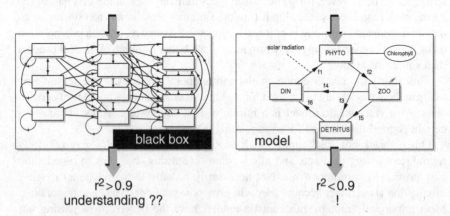

Fig. 1.5 Black-box methods may give hight r^2 but they do not increase our understanding of
the system. Mechanistic models, even if they perform less well in predicting, have the benefit of
increased knowledge of system functioning

1.2.3 Models to Quantify Immeasurable Processes

All too often ecologists cannot measure the processes in which they are truly interested. Manipulative experiments have greatly expanded our capability to study environmental systems, but they also introduce artefacts that are difficult to avoid.

It is generally much easier to measure *concentrations* or *densities*, rather than process rates but often the *rates* are the quantities that we want to determine. Fortunately, both are related: concentration gradients arise due to the processes which consume or produce the substances. Mathematical models describe how these processes affect the substances. Thus, if the output of these models is fitted to the concentration data, it may be possible to obtain quantitative estimates of the process rates. In this way, something that is easily measurable (the concentration) is translated into something hard to quantify (the rate) by means of a (set of) mathematical equation(s).

For instance, based on profiles of oxygen vertically into the sediment, it is possible to derive oxygen fluxes across the sediment-water interface, by fitting a so-called 'diagenetic' model to the data. Thus the measurements of the oxygen concentration profiles performed by micro-electrodes are converted into the rates that have generated the profiles (Fig. 1.6).

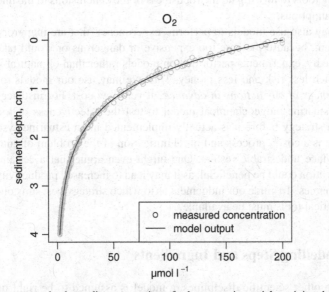

Fig. 1.6 At great water depths, sediment organisms feed on organic particles raining down from the upper water layers. The more food available, the more organisms survive and the higher the total biomass. But how can we measure the amount of food deposition? Using micro-electrodes, mounted on sophisticated landers, biogeochemists measure high-resolution profiles of oxygen concentration as a function of sediment depth. Fitting these oxygen profiles with a model estimates oxygen fluxes, from which organic matter deposition rates are calculated

Generally, quantification makes use of a process called 'data assimilation' or 'calibration' where a model is fitted to the data by fine-tuning certain key parameters. We will discuss model calibration in Chapter 4.

1.2.4 Model Prediction as a Management Tool

Once we have acquired an understanding of phenomena and have formalized this as a model, we can use it to make *predictions*. Such prediction can be both quantitative, where we forecast the level of a response, or qualitative, where we simply predict the presence or absence or the sign of a response.

Such predictive capability may then be used to examine certain socio-economic problems (productivity and sustainability of ecosystems; the effects of global warming). Managers in most countries are required to ensure that the communities and ecosystems are managed on an ecologically sustainable basis. Any attempt to ensure sustainability requires the prediction of the system in the future.

These managers are paid to decide, and they will decide, if need be, based on a gut feeling of what they think might happen. Through mathematical modelling scientists take up their responsibility and try to provide some rigour to these decisions. The model may not be perfect (it never is), but a model-guided decision has the advantage that the assumptions on which it is based are transparent and there exist modelling tools to investigate the robustness of the conclusions to the uncertainty in these assumptions.

We may also use models to perform 'experiments' that are inappropriate in the real system, because they are too expensive or dangerous or would take too long. Undoubtedly, experiments performed on models rather than on natural systems involve much less risk and less money. Thus we may use our models to assess the *consequences of our actions in advance*, at very low cost. For instance, we might use an estuarine biogeochemical model to test the effectiveness of a certain purification strategy before it is actually implemented in an estuarine system. Waste treatment is a costly process and blind mitigation of the problem of eutrophication may produce undesirable results. One might even argue that a certain degree of eutrophication could be beneficial, as it may lead to increased productivity and more food resources. To guide our judgement as to which strategy best suits our purposes, mathematical tools must be available.

1.3 Modelling Steps and Ingredients

Just as in other scientific disciplines, a model is assumed to be right until proven wrong. It represents our best opinion of the situation at one stage of the investigation, and this opinion must be changed, i.e. the model reconsidered, if we find at a later stage that the facts are against it.

This indicates that the interaction between models and *data* is very important. Not only do we need data to test the model and if necessary to refute it, equally important are data for driving (forcing) the model or providing good parameter values.

Successful modelling in ecology is almost always iterative: the initial versions of a model highlight areas of uncertainty and weakness, which can then be addressed in later versions. Although one may discern several successive phases in model making, it is possible that these phases need to be cycled through (iterated) several times, before we can be content with the result (see Fig. 1.7).

In this and in following sections, we will discuss each of the modelling phases and introduce some of the 'slang' used by modellers.

Initiating a model starts by defining the **problem**, and the objectives for the modelling.

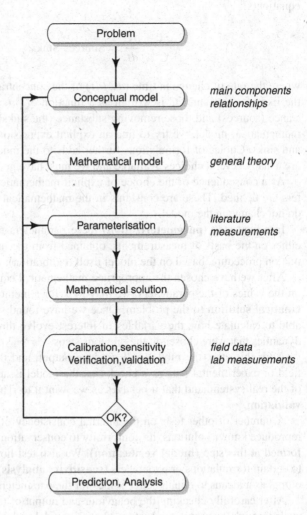

Fig. 1.7 Modelling steps and ingredients

The creation of a new model generally takes off with simplifying the system, i.e. the writing of the **conceptual model**. In this book, we deal with models that express exchanges of energy, mass or momentum between certain quantities, usually concentrations, biomasses, densities. In the conceptualization step we select the basic components in a model (the substances) and the relationships (flows) that exist between them, we decide upon the units that we will use in the model, and the 'scale windows' for time and, if appropriate, space. At this point, we make the most fundamental **assumptions** of the model, by deciding what is important for the problem and what is not.

The interactions and flows defined in the conceptual model must be formulated explicitly as mathematical expressions in the next step (**formulating the mathematical model**). Generally, the model will be written as a (set of) differential equation(s):

$$\frac{d\mathrm{C}}{dt} = \text{Sources-Sinks} \tag{1.4}$$

where the rate of change in time ($d\mathrm{C}/dt$) of the concentration, density, biomass or the like of a substance C is written as a function of processes adding to the substance (sources) and those removing substance (the sinks). Whilst formulating the mathematical model, we try to find an explicit expression for each of the sources and sinks. Choice of formulations further adds to the model assumptions, as there are usually several choices available and we must have good reason to pick one.

As a consequence of the choice of explicit mathematical formulations, parameters are defined. These are constants in the mathematical equations, and therefore do not change in the model.

In a next step (**parameterization**) these parameters must be assigned values, either on the basis of measurements, obtained from the literature or using an estimation procedure, based on the model itself (calibration).

After we have chosen the appropriate mathematical equations and have decided on the values of the parameters, we may look for a method that will give a **mathematical solution** to the problem. Once we have found such solution, we will be able to calculate how the variables of interest evolve through time, based on the dynamics that were chosen in the previous steps.

After this step, the model will produce output, and this may be compared to field or experimental data as a check that the model is an accurate representation of the real system, and that it behaves as we want it to. This will be done in **model validation.**

A number of other tests on the internal consistency of the model, its ability to reproduce known solutions, its conformity to conservation laws etc. . . are also performed at this step (**model verification**). We also test how sensitive the model is to certain formulations or parameters (**sensitivity analysis**), and we may use model output as a means to obtain better values for the parameters (**model calibration**).

After carefully checking the behaviour and output of the model, we then either cycle through the process and reformulate the model or decide that it is good enough for **model application**. Here we may use it to predict a system's behaviour under different scenarios, to analyse the dynamics of certain interactions, to make budgets, or for quantification of unmeasured processes.

1.4 The Modeller's Toolkit

Although 'modern' modelling is computationally intensive, the initial stages only require pen-and-paper and contemplation. The most simple models may be fully executed with these basic tools, perhaps aided by hand calculators.

For the more complex applications, we resort to computers. Here, using appropriate software is one of the cornerstones of successful modelling.

Spreadsheets are not particularly suited for modelling, but are easy-to-use tools for making graphs. Thus, they mainly serve at visualizing mathematical equations, although they can also be used for implementing the simplest models. Using spreadsheets may have some educational benefit, but it cannot be recommended for any 'serious' modelling. To do that, there are two possibilities.

1. Some software environments allow interactive model development. These typically provide various software tools that perform difficult tasks such as root finding, integration, and automatic differentiation. Most commonly cited examples of such environments are MATHEMATICA, MAPLE, MATLAB, which are commercially available products. A valid alternative is the R software, which is freely available (http://www.r-project.org/). This is the toolkit that we use in this book.

All these languages allow writing concise programs, lead quickly to results, and are quite flexible. However, they have a steep learning curve (not to be underestimated!), and they are generally slower than lower-level languages.

Most of these software packages also allow interfacing with software written in lower-level languages, such as C or Fortran. How the interaction between R and Fortran can be achieved is explained in Appendix A.3.

2. Lower-level languages give rise to compiled code that is much faster and are therefore preferred for models that have to be applied many times or for computationally demanding models. Moreover, as their syntax comprises but a fraction of the syntax of the environmental languages discussed above, they are easier to learn. However, one may spend quite some time implementing basic routines, or trying to find these routines on the web. This contrasts to the environmental languages where these are triggered by a simple statement. Fortran is the language most frequently utilized for scientific computing, and it is the language that we are most familiar with, but C and C^{++} are also widely used. These languages can interface with, or simply be converted into one another. Useful mathematical routines in low-level languages are available from the web, commercial packages or the literature.

Chapter 2
Model Formulation

Model formulation is the step where our knowledge of a natural system is translated in mathematical form. It involves two steps: the construction of a conceptual model and the formulation of this conceptual model into mathematical equations.

We start by choosing the main components (state variables) and the flows that describe exchange of matter, energy or momentum, between them. We show that, based on the principle of *conservation*, the conceptual model equations simply express the rate of change of the state variables as the sum of all the flows that enter minus all flows that leave the compartment. This is conveniently depicted as a conceptual diagram or a flow chart, where the state variables are depicted by boxes, which are connected by arrows symbolising the flows.

In the next step, the flows are formulated explicitly as mathematical expressions. Although ecological systems are complex and cross many different levels (from cell to organism to an entire ecosystem), the processes share many features and can therefore be expressed with similar mathematical expressions. A number of basic and very simple rules apply. The most important one is that an ecological interaction (flow) can be written as the product of a maximal rate times the compartment performing the work, and, if appropriate, one or more limitation and inhibition terms.

Although the distinction is somewhat artificial, we discuss separately how to formulate ecological and chemical interactions, and how inhibition can be mathematically represented. We demonstrate that there exist several ways in which model equations can be coupled. Finally we discuss the forcing functions (driving variables) that are commonly used in ecological models.

2.1 Conceptual Model

At the stage of conceptual model formulation, we determine the model complexity. For practical as well as aesthetic reasons, a model should be as simple as possible, but not simpler (this quote has been ascribed to Einstein). The largest intellectual challenge of modelling consists in the creative simplification of a scientific problem, in such a way that no great injustice is done to realism.

The choice of modelled components and processes involves a subtle balance between realism and complexity. If knowledge is poor, then the model will be unable to contain many details. The more we know about a system, the more complex the model can be. However, the most complex model is not necessarily the most effective! Ecologists who have spent half their career studying all the details of a particular process, have a hard time accepting that neglecting, rather than incorporating, most of this knowledge will lead to better models. Yet, the will to make simplifications is the first step to becoming a modeller.

Note also that model complexity must depend on the problem to be solved. Thus, for one system, many different models may exist, all addressing different questions.

A conceptual model can be represented in a conceptual diagram or flow chart (Fig. 2.1) which contains the following components:

The **state variables** (or dynamic variables) are the components that we are interested in. In ecological models they are often the biomasses or densities of organisms, the concentrations of nutrients and so on. State variables are those variables that appear on the left-hand side of the model equations as time derivatives: we specify their rate of change in the model (see later), and after model solution will determine their values in time.

Time derivatives of a state variable represent the 'speed' at which the value of the state variable will change over an infinitesimal time period (dt) and are represented as: $\dfrac{dStateVariable}{dt}$.

In Appendix B we give a more elaborate definition of the concept.

In a flow chart, the state variables are often denoted by rectangular boxes connected by arrows. The arrows are the **flows or interactions** amongst the various state variables; they represent the external sources and sinks to a state variable, (where source arrows point towards and sinks leave the state variable boxes).

Forcing functions, or external variables, are important external factors that drive or regulate the system. As they are not calculated in the model and are not constant in time, they are generally imposed to ('forced upon') the model as a data series. Frequently used forcing functions in ecological models are: light intensity, temperature, flow rates, wind, inputs of toxic substances to ecosystems, rate of harvesting (effort) by humans...

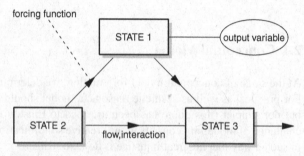

Fig. 2.1 Conceptual model diagram, with the main model components; STATE1, STATE2 and STATE 3 are state variables, connected by flows (*arrows*)

In addition to state variables, a model generally contains **output variables**. These often relate to measured quantities, and allow comparison of the model to reality. In addition, output variables can be instantaneous rates that one wants to inspect, or long-term averaged rates. They are computed based on other components of the model (forcing functions, state variables, parameters). For instance, algal biomass is a state variable in many aquatic models, but it is usually not measured as such. Chlorophyll, a related and more easily measurable quantity is often used as a proxy to algal biomass instead. In these models, chlorophyll will be an output variable that is calculated based on the state variable algal biomass (using a conversion factor) and that can be compared to the data. Other output variables in these models might be (instantaneous) net algal growth, or yearly averaged primary production, because these are interesting quantities.

Choosing an appropriate **spatial and temporal scale** is equally important at the stage of conceptual model formulation. For instance, in astronomy, one might choose a length scale that allows expressing, in not too large numbers, the distance between the sun and planets, whilst in molecular dynamics, the spatial scale should allow representing the distance separation between the atoms. Our choice of temporal and spatial scale of the model has great repercussion on the choice of processes to represent or not in the model. We will generally only crudely represent processes that fall outside of the model's scale window.

Together with the model components, and the temporal and spatial scales, we also decide about the **model currency**, the elemental composition in which we want to express the state variables. There are many different ways of expressing biomass in models. One can use dry weight or (especially for animals containing large calcareous structures) ash-free dry weight, carbon content, nitrogen content, phosphorus content, protein content, energy equivalents (i.e. energy released upon oxidation) or other measures. Important is to choose a currency that is both accurate and convenient. For instance, in models that describe phosphate, but not nitrate or ammonium, the best choice for biomass of living organisms will be in terms of phosphorus.

2.1.1 The Balance Equation of a State Variable

Most models that we will deal with in this book are concerned with the exchange of mass, energy or momentum. For instance, environmental models may describe:

- Uptake of dissolved inorganic nitrogen by phytoplankton (ecosystem models).
- Conversion of carbohydrates and nutrients into more complex proteins (physiological models).
- Heat transfer from the air to the water column (physical models)
- Transfer of momentum from air to sea by the action of the wind on the water (turbulence models)

One of the most powerful laws when studying such interactions is the **conservation law**. This law states that neither mass, energy nor momentum can be created or lost by ordinary means, i.e. flows of these quantities have to come from known sources, or go to known sinks. Stated otherwise, any flux entering some state variable should either accumulate in this state variable, or lead to another state variable or to a known sink.

When this principle applies, i.e. the state variable is expressed in mass, energy or momentum, we can state the **balance equation** for the state variable as:

$$\frac{d\,\text{StateVariable}}{dt} = \text{sources} - \text{sinks} \qquad (2.1)$$

It expresses that a state variable can only change in time when there is either an external input (source) or an external output (sink), i.e. material flows into or out of the state variable.

Consider the following simple model (Fig. 2.2): two state variables (SV1 and SV2) are connected with one flow (F1); F0 is a flow from the external world to state variable 1 (SV1), whilst F2 leaves state variable 2 to the external world. The conservation law states that any increase in mass (or energy, momentum) of state variable 2, due to F1 induces an equally large decrease in the magnitude of state variable 1. The total mass in the model (sum of SV1 and SV2) changes due to F0 and F2.

We may write the balance equations, representing the rate of change of the state variables as:

$$\frac{d\text{SV1}}{dt} = \text{F0-F1}$$
$$\frac{d\text{SV2}}{dt} = \text{F1-F2} \qquad (2.2)$$

We now use the conservation of mass principle to test whether the model makes sense. The easiest test is by making total mass budget calculations. If the model conserves mass, then the total mass in the model should change only due to flows

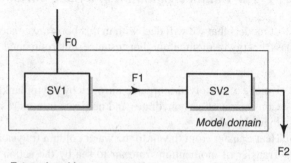

Fig. 2.2 Illustration of the balance equation of state variables. See text for details

from and to the external world. In the simple model example total mass changes as:

$$\frac{dSV1}{dt} + \frac{dSV2}{dt} = \text{F0-F1+F1-F2}$$
$$= \text{F0-F2} \tag{2.3}$$

Total mass changes only due to flows connecting the model domain with the external world, therefore, the model is mass conserving.

Note: the conservation principle does not apply to all conceivable state variables. Many ecological models use individuals as their currency. In many models (e.g. predator-prey models), individuals is a proxy for mass and in these cases the conservation principle still holds. However, in other models transfer of mass is not considered. For instance, in population models, newborns may be generated in the youngest age class, depending on numbers in the reproductive age classes, without conservation: the numbers in the reproductive age classes do not decrease because they have given rise to newborns.

In this book, we mainly deal with models for which the conservation principle does hold. Yet, you may find examples of non-conservative models in Chapter 9 (structured population models) and Chapter 3 (cellular automata models).

2.1.2 Example: Conceptual Model of a Lake Ecosystem

Consider a conceptual model of a whole lake. The model contains 6 state variables: Phytoplankton (PHYTO), zooplankton (ZOO), detritus (DETRITUS), ammonium (NH_4^+), fish (FISH), and bottom detritus (BotDET). The state variables are connected by 13 flows. There is one forcing function (solar radiation) and one output variable (Chlorophyll). In Fig. 2.3, the conceptual diagram is drawn.

As the model also describes ammonium, it is convenient to take nitrogen as the **model currency**, so the units of all the state variables are concentrations, expressed in mmol N m^{-3}. The law of conservation of mass applies to the balance equations of all state variables.

The components typically change seasonally, in a time window of one to a few years, so a day will be a suitable model **time unit**, and consequently, all rates will be expressed per day.

As we are not interested in spatial patterns of ecosystem changes within the lake, the **spatial scale** chosen is the entire lake.

Based on the model diagram (Fig. 2.3), we use the flows that describe exchange of matter between the state variables to generate the **conceptual model equations**. These equations simply relate the rate of change of the state variables to the flows.

For instance, the statement:

$$\frac{dNH_4^+}{dt} = \text{f11+f10} + \text{f4-f1} \tag{2.4}$$

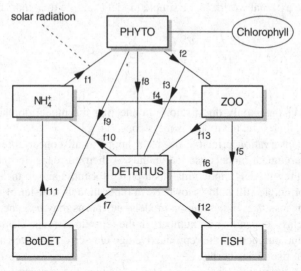

Fig. 2.3 Conceptual diagram of the lake ecosystem model. State variables are denoted by rectangular boxes, flows with arrows, forcing functions with a dashed line and the output variable with an ellipse

states that the ammonium concentration will increase as a result of mineralization of bottom detritus (flow 11), mineralization of suspended detritus (flow 10) and excretion of zooplankton (flow 4). It will decrease due to uptake by phytoplankton (flow 1).

With the state variable NH_4^+ expressed in mmol N m^{-3}, and the time unit of days, the rate of change, dNH$_4$/dt, is expressed in mmol N m^{-3} day^{-1}. Consequently, all flows are also expressed in mmol N m^{-3} day^{-1}.

The full conceptual model describes the rate of change of all state variables. The rate of change is simply the sum of all flows that enter the compartment minus all the flows that leave the compartment:

$$\frac{d\text{PHYTO}}{dt} = f1 - f2 - f8 - f9$$

$$\frac{d\text{ZOO}}{dt} = f2 - f3 - f4 - f5 - f13$$

$$\frac{d\text{DETRITUS}}{dt} = f3 + f8 + f6 + f12 + f13 - f7 - f10$$

$$\frac{d\text{FISH}}{dt} = f5 - f6 - f12$$

$$\frac{d\text{BOTTOMDETRITUS}}{dt} = f7 + f9 - f11$$

$$\frac{d\text{NH}_4^+}{dt} = f11 + f10 + f4 - f1 \tag{2.5}$$

2.1.3 *Conservation of Mass and Energy as a Consistency Check*

We now use the conservation law to check whether the model makes sense and is correctly solved (more about this later). Already at the stage of the conceptual model, we may test whether any flux leaving some state variable actually accumulates in another state variable, or leads to a known sink. The easiest way to test that is either by making mass budget calculations or by calculating the total load.

1. *Total load* can be used as a check when there is no external source or sink compartment. This is the case in the example of Section 2.1.2, where all arrows (flows) connect state variable boxes, no arrow points from or to the outside world. Then, total mass should be constant, and we can easily check for that by calculating the rate of change over time of the sum of all components:

$$
\frac{d[\text{PHYTO} + \text{ZOO} + \text{DETRITUS} + \text{FISH} + \text{BOTTOMDETRITUS} + \text{NH}_4^+]}{dt}
$$
$$
= \frac{d\text{PHYTO}}{dt} + \frac{d\text{ZOO}}{dt} + \frac{d\text{DETRITUS}}{dt} + \frac{d\text{FISH}}{dt} + \frac{d\text{BOTTOMDETRITUS}}{dt} + \frac{d\text{NH}_4^+}{dt}
$$
$$
= \text{f1} - \text{f2} - \text{f8} - \text{f9} + \text{f2} - \text{f3} - \text{f4} - \text{f5} - \text{f13} + \text{f3} + \text{f8} + \text{f6} + \text{f12} + \text{f13} - \text{f7} - \text{f10}
$$
$$
+ \text{f5} - \text{f6} - \text{f12} + \text{f7} + \text{f9} - \text{f11} + \text{f11} + \text{f10} + \text{f4} - \text{f1}
$$
$$
= 0 \tag{2.6}
$$

If mass is not conserved in this calculation, i.e. when the sum of all flows is not zero, we have an error. Every error, no matter how small, should be traced back.

2. *Mass budget* calculation is often the only available method when there is an external sink or source compartment. For instance, in pelagic ecological models that have carbon as their currency, it is not customary to include CO_2 as a state variable, as it is rarely limiting primary production. Nevertheless, CO_2 acts as a source of modelled carbon through photosynthesis, and as a sink of modelled carbon through biological respiration (that is, CO_2 is an external sink and source of C). In such models, the total amount of carbon in the model will generally not be constant. Through mass budget analysis, we inspect if the sum of all rate of changes and external sinks and sources are zero.

To illustrate the point, we have made a different model for our lake, this time expressed in carbon as the model currency, but essentially containing the same fluxes (Fig. 2.4). There are 5 flows that connect the model with the outside world (in this case CO_2, which is not modelled). They are the primary production flow (f1), a source, and four respiration flows (f4, f10, f11, f14), which remove carbon from the modelled system (sinks). The rate of change of the state variables can be easily written as:

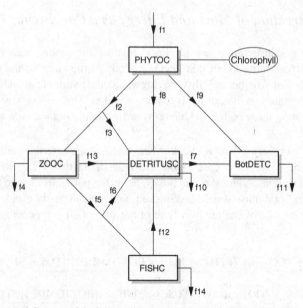

Fig. 2.4 Conceptual diagram of the lake ecosystem model, now expressed in units of carbon. Note the presence of flows from and to the outside world (f1,f4,f10,f11 and f14)

$$\frac{d\text{PHYTOC}}{dt} = f1 - f2 - f8 - f9$$

$$\frac{d\text{ZOOC}}{dt} = f2 - f3 - f4 - f5 - f13$$

$$\frac{d\text{DETRITUSC}}{dt} = f3 + f8 + f6 + f12 + f13 - f7 - f10$$

$$\frac{d\text{FISHC}}{dt} = f5 - f6 - f12 - f14$$

$$\frac{d\text{BOTTOMDETRITUSC}}{dt} = f7 + f9 - f11 \qquad (2.7)$$

while the mass budget can be written as:

$$\frac{d[\text{PHYTOC} + \text{ZOOC} + \text{DETRITUSC} + \text{FISHC} + \text{BOTTOMDETRITUSC}]}{dt}$$

$$= \frac{d\text{PHYTOC}}{dt} + \frac{d\text{ZOOC}}{dt} + \frac{d\text{DETRITUSC}}{dt} + \frac{d\text{FISHC}}{dt} + \frac{d\text{BOTTOMDETRITUSC}}{dt}$$

$$= f1 - f2 - f8 - f9 + f2 - f3 - f4 - f5 - f13 + f3 + f8 + f6 + f12 + f13 - f7 - f10+$$

$$\quad + f5 - f6 - f12 + f7 + f9 - f11$$

$$= -f4 - f10 - f11 - f14 + f1 \qquad (2.8)$$

As this is simply the sum of the external input- external outputs, we conclude that mass is conserved in the model.

2.1.4 Dimensional Homogeneity and Consistency of Units.

Not only mass conservation, but also consistency of units is an important technique to check the consistency of a model. To introduce it we must make the distinction between *quantities* and *units*, where the units are the numerical dimensions of a certain quantity. In addition, there is a distinction between *primary* (or fundamental) and *derived* quantities and units.

The fundamental units and quantities, consistent with the Système International (SI) are:

- Unit meter (m) for quantity length
- Unit kilogram (kg) for quantity mass
- Unit second (s) for quantity time
- Unit Kelvin (K) for quantity temperature
- Unit mole (mol) for quantity amount of substance

A quantity is taken as primary if it can be assigned a standard of measurement independent of that chosen for the other fundamental quantities.

Derived quantities and units are expressed as a function of the primary quantities and units. Examples:

- Unit m^2 for quantity area
- Unit $m\,s^{-1}$ for quantity velocity
- Unit $mol\,m^{-3}$ for quantity concentration
- Unit $N = kg\,m\,s^{-2}$ for quantity force

Units can be manipulated just as numbers, using the simple rules of multiplication and division.

For example, the units of mass-specific energy, Joule kg^{-1} can be written in primary units as follows: with Joule $= kg\,m^2s^{-2}$, joule $kg^{-1} = (kg\,m^2s^{-2})kg^{-1} = m^2s^{-2}$.

An equation is *dimensionally homogeneous* and has *consistent units* if the units and quantities on two sides of an equation balance.

- It is not allowed to add, for instance, length to area, or concentration to flux in one equation (dimensional homogeneity).
- Moreover, it is not allowed to add grams to kilograms, or to add $mol\,m^{-3}$ to $kg\,m^{-3}$ (consistency of units).

We can use the principle of dimensional homogeneity and consistency of units in two ways:

- To test whether an equation makes sense
- To derive the (unknown) units of a quantity.

As an example of the latter, consider the mathematical equation which expresses the extinction of light in a water column (Lambert-Beer law):

$$\frac{dI}{dz} = -\lambda \cdot I \tag{2.9}$$

where I is light intensity, expressed in μmol m^{-2} s^{-1} (or, equivalently, μEinst m^{-2} s^{-1}), z is depth (m), dI/dz is the rate of change of light with depth and λ is a first-order decay coefficient, known as the extinction coefficient.

Dimensional analysis readily shows that units of λ must be reciprocal to units of z, therefore must be equal to m^{-1}:

$$\frac{Units(I)}{Unit(z)} = Unit(\lambda) \cdot Unit(I) \tag{2.10}$$

which, after removing the common terms on the left and right hand side:

$$\frac{1}{Unit(z)} = Unit(\lambda) \tag{2.11}$$

with Unit of z = m, units of λ must equal m^{-1}.

The analytical solution of the Lambert-Beer equation, expressing light intensity as a function of depth is:

$$I_z = I_0 e^{-\lambda z} \tag{2.12}$$

Where I_0 is the light intensity at depth 0 (the air-water interface).

As exponents can only be taken of dimensionless quantities, it again follows that the units of λ and depth must balance.

2.2 Mathematical Formulations

In the previous Section (2.1), we wrote the model as a set of conceptual balance equations. These express the rate of change of a state variable as a function of sources and sinks (the flows). The remaining problem is to write a mathematical expression for each of the flows in terms of the state variables.

In this step of model building, we make use of relationships from the literature, or from theoretical considerations. Most typically, an interaction or flow is written as a function of the state variables, forcing functions and parameters. The parameters are coefficients that control model processes but do not change in the model. Some parameters are very well known (gas constants, atomic weights...), whereas others are only known approximately, e.g. the maximal grazing rate of zooplankton, the sinking rate of phytoplankton.

In the following two sections we take a closer look at how to express ecological interactions and chemical reactions. We start with chemical reactions, as these are generally simpler to formulate.

2.3 Formulation of Chemical Reactions

2.3.1 The Law of Mass Action

Consider first the example in Fig. 2.5B, where substance A reacts with substance B to give substance C, in the reaction:

$$\alpha A + \beta B \xrightarrow{k} \chi C$$

Reaction rates are generally expressed using the law of mass action. This law states that the rate of the reaction is proportional to a power of the concentrations of all substances taking part in the reaction:

$$\text{ReactionRate} = k \cdot [A]^{\alpha} \cdot [B]^{\beta} \tag{2.13}$$

where k is the proportionality constant (units of conc $^{1-(\alpha+\beta)}$ t^{-1}) and α and β are integer powers. Note that the constants α and β are also the stoichiometric constants in the reaction equation. This is generally the case, but not always.

The hypothesis behind this formulation is that this type of elementary reaction will occur only if the molecules collide. The probability for a molecule of A to collide with a molecule of B is proportional to the concentrations of A and B, so the total number of collisions will be proportional to [A]·[B]. If more than one molecule of A is participating in the reaction (e.g. $\alpha = 2$), the probability of an A-B complex to collide a second time with an A molecule is again proportional to A, so finally the number of A-A-B complexes formed per unit of time is proportional to [A]^2·[B]. The reasoning can be simply extended for other values of α and β.

Also note our use of square brackets ([]). This is standard notation for concentrations in chemistry, so, as long as we deal with chemical models only, we will stick to this formalism. However, it is very awkward for biological quantities, thus we do not use this notation for biological models.

The *order* of the reaction is the sum of the powers.

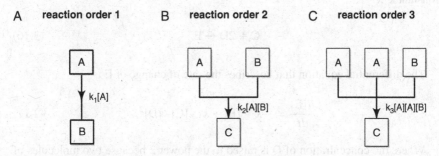

Fig. 2.5 Chemical reactions of different orders and the reaction rates

- In **first-order reactions**, the reaction rate is proportional to the concentration of one substance (Fig. 2.5.A):

$$R1 = k_1 \cdot [A] \tag{2.14}$$

- **Second-order reactions** arise either by the interaction of two substances or they are second order of the concentration of one substance (Fig. 2.5.B).

$$R2 = k_2 \cdot [A] \cdot [B]$$
$$R2 = k_2 \cdot [A]^2 \tag{2.15}$$

- Sometimes the order of the reaction formulation is lower than would be expected based on the reaction equation. This happens when one of the reactants is always present in a concentration that far exceeds that of the other reactant (water is an extreme example). In such a case, the reaction rate can be simply written as a function of the concentration of the limiting reactant only.

A special case appearing as a result of this reduction of order are **zero-order reactions**. Here the reaction rate is assumed to be constant and independent of the reactant(s). Obviously, such a formulation can only be used when the reactants are always present in sufficient quantities, and the reaction rate itself is limited by some other factor, e.g. the availability of reactive surfaces catalyzing the reaction. A model using zero-order reaction for substances in short supply can be very wrong: it will predict that the reaction will continue even when the reactant has zero concentration!

2.3.2 Example: A Simple Chemical Reaction

As stated above, the powers are determined by the stoichiometry of the reaction. Consider the following reversible chemical reaction (Fig. 2.6 A), where one mole of substance C reacts with 2 moles of substance D to form one mole of E; the forward reaction has reaction rate constant k_1; the backward reaction rate constant k_2.

$$C + 2D \overset{k_1}{\underset{k_2}{\leftrightarrow}} E \tag{2.16}$$

The differential equation that describes the rate of change of E is:

$$\frac{d[E]}{dt} = -k_2 \cdot [E] + k_1 \cdot [C] \cdot [D]^2 \tag{2.17}$$

Where the concentration of D is raised to the power 2 because two molecules of D are needed per molecule of E.

Fig. 2.6 A. Reversible reactions occur in both directions. **B.** An enzymatic reaction, where the enzyme E reacts with substance D to form an intermediate substance I. k_1, k_2, k_3 are the rate coefficients (see text)

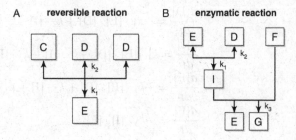

The rate of change of D is given by:

$$\frac{d[D]}{dt} = 2 \cdot (k_2 \cdot [E] - k_1 \cdot [C] \cdot [D]^2) \qquad (2.18)$$

the multiplication with 2 is because two moles of D are produced per mole of E consumed, and vice versa: two molecules of D disappear for every molecule of E formed.

2.4 Enzymatic Reactions

Enzymatic reactions are crucial to the functioning of organisms. Enzymes catalyze chemical reactions, but take no part in it, in the sense that enzymes are neither produced nor consumed in the reaction. We have depicted the scheme of an enzymatic reaction in Fig. 2.6 B:

$$E + D \overset{k_1}{\underset{k_2}{\leftrightarrow}} I$$

$$I + F \overset{k_3}{\to} E + G \qquad (2.19)$$

One of the reactants, an enzyme, E, chemically reacts with product D forming an intermediary product, I. This intermediary product then in turn enters a reaction with substance F from which the enzyme E is released, together with another product G. The first reaction is reversible with rate coefficient k_1 (forward) and k_2 (backward reaction); the second reaction is irreversible, with rate coefficient k_3.

In the model of this reaction, there are five state variables, the concentrations of the five substances. The five basic differential equations that define their rate of change are given by:

$$\frac{d[D]}{dt} = -k_1 \cdot [E] \cdot [D] + k_2 \cdot [I]$$

$$\frac{d[I]}{dt} = k_1 \cdot [E] \cdot [D] - k_2 \cdot [I] - k_3 \cdot [I] \cdot [F]$$

$$\frac{d[E]}{dt} = -k_1 \cdot [E] \cdot [D] + k_2 \cdot [I] + k_3 \cdot [I] \cdot [F]$$

$$\frac{d[F]}{dt} = -k_3 \cdot [I] \cdot [F]$$

$$\frac{d[G]}{dt} = k_3 \cdot [I] \cdot [F] \tag{2.20}$$

For instance, substance D is produced by the conversion of substrate I, which occurs at rate $k_2 \cdot [I]$; D is consumed by its reaction with the enzyme E, which occurs at a rate $k_1 \cdot [E] \cdot [D]$.

Solving this full set of reactions is awkward. Luckily, under most circumstances it can be simplified considerably. We will derive this simplification mathematically later in the book (Section 8.3.1). Here we simply bring forward the result. Essentially, the reaction can be modelled as a reaction between substances D and F, producing substance G, if the reaction rate formulation is adapted using the so-called Michaelis-Menten kinetics:

$$\frac{d[G]}{dt} = k'[F] \frac{[D]}{[D] + ks} \tag{2.21}$$

where k' and ks are parameters, recalculated from the elementary constants in the full formulation (Eq. 2.20). The Michaelis-Menten kinetics plays an important role in the formulation of biological interactions, as we will see in the next chapters.

2.5 Basic Formulation of Ecological Interactions

Ecological models usually describe the dynamics of organisms, which are significantly more complex than the simple molecules of previous section. Consequently, the formulations of ecological interactions are somewhat different from the chemical reactions. To illustrate why, we start with a simple example.

2.5.1 Example: Flows to and from Phytoplankton in the Lake Ecosystem

We take a closer look at the PHYTO compartment in the lake ecosystem example of Section 2.1.2. Four fluxes are defined in the equation for phytoplankton:

$$\frac{dPHYTO}{dt} = f1 - f2 - f8 - f9 \tag{2.22}$$

Here, f1 is the primary production flux; as this model uses nitrogen as its currency, the primary production is in fact the incorporation of ammonium. The other fluxes are loss fluxes, due to sinking and formation of bottom detritus (f9), mortality and formation of pelagic detritus (f8) and ingestion by zooplankton (f2).

2.5.1.1 A first attempt: first-order reaction?

Focusing on the primary production flux f1, we could naively represent this as a kind of chemical reaction (Fig. 2.7A):

$$NH_4^+ \xrightarrow{k} PHYTO \qquad (2.23)$$

and express the rate of this 'reaction', following the law of mass action, as a first-order reaction rate in the ammonium concentration. Now suppose that we have clear water without any phytoplankton, but with a certain concentration of ammonium, then this naïve model tells us that ammonium is going to transform spontaneously into phytoplankton. This *generatio spontanea* used to be a popular concept in biology some centuries ago, but since Pasteur we know that this is a capital mistake. No phytoplankton should be generated when there is no phytoplankton. We conclude that this simple formulation as a first-order chemical reaction is not applicable, and a basic element is missing.

That missing element is the regulation of fluxes by the activity of the organisms themselves. Thus, the formulation should be such that the flux is zero when the organism responsible for this flux has zero concentration (or numbers, biomass, depending on the currency of the model). With 'the organism responsible for this flux' we mean the organism that 'does the work', whose physiological apparatus (e.g. uptake mechanisms, enzymes) is needed for the reaction to take place. Algae actively pump ammonium from the water into their cell, involve this ammonium in a series of enzymatic reactions, to finally build N-containing organic material, such as proteins. All this would not happen if the full physiological apparatus of the algae would not be present and would not be working in a complex, ordered, way.

2.5.1.2 A Second Attempt: Second-Order Reaction?

How does the flux f1 scale with phytoplankton when it is present? Compare two situations, one with PHYTO=0.1 mmol N m^{-3}, and another with PHYTO=1 mmol N m^{-3}, everything else (light conditions, ammonium concentration etc.) being the same. In the first situation the phytoplankton in the lake will be producing at a certain rate. This production is divided over all the individual phytoplankton cells, which will each display a certain degree of activity. In the second situation, there are ten times as many phytoplankton cells present, but each cell will find itself in exactly the same external conditions as in the first case. It is, thus, natural to suppose that they will show exactly the same activity. Consequently, the total population of phytoplankton will be ten times as active in the second situation as in the first. We conclude that the rate f1 should be proportional to the biomass of the phytoplankton, although other terms could also be involved:

Fig. 2.7 Uptake of
ammonium by phytoplankton
represented as a 'first-order'
reaction (**A**) and as a
'second-order reaction'(**B**)

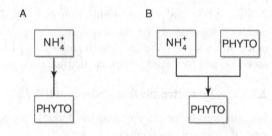

$$f_1 = a \cdot \text{PHYTO} \cdot \ldots \tag{2.24}$$

What other terms could be involved in this flux? When there is no ammonium present, nothing can be taken up and the phytoplankton will not grow (note that we do not model any other nitrogen source for phytoplankton). So a logical second attempt at modelling the flux could be to consider the process as a second-order reaction (Fig. 2.7B).

It is difficult to assign 'stoichiometric coefficients' to the 'reactants' in this reaction, but in the simplest possible approach we could put them equal to one and write for the flux f1:

$$f_1 = a \cdot \left[\text{NH}_4^+\right] \cdot \text{PHYTO} \cdot \ldots \tag{2.25}$$

With this formulation, we have found a way to limit the magnitude of the flux when there is no or little substrate.

However, problems remain. It is not a good idea to make the flux directly proportional to ammonium concentration: at a certain concentration of ammonium the phytoplankton will have plenty of nutrients, but will be limited by the intrinsic capacity of its physiological apparatus. With increasing ammonium concentration, the flux f1 will asymptotically approach this maximum capacity.

2.5.1.3 A Last Attempt: Using a Rate Limiting Term

We solve this problem by defining a *rate limiting term*, i.e. a function f(NH4) that becomes zero when NH_4^+ is zero, and tends to 1 when ammonium is sufficiently abundant. Below we will see how such a function can be formulated mathematically. The model for flux f1 now becomes:

$$f_1 = a \cdot \text{PHYTO} \cdot f\left(\text{NH}_4^+\right) \cdot \ldots \tag{2.26}$$

Further terms can be introduced in the formulation of this flux. We note that when there is no light penetrating into the water column (such as would be the case in extremely turbid waters), no algal growth would take place. Again, with plenty of light other factors would become limiting, so that we also have to define a light-limiting function g(I) (where I is light intensity in the water column) that is zero at zero light, and tends to 1 at high light levels. More details on light limitation are given in Section 2.8.2. This gives us the final formulation for the flux f1:

$$f_1 = a \cdot \text{PHYTO} \cdot f\left(\text{NH}_4^+\right) \cdot g\left(\text{I}\right) \cdot \ldots \tag{2.27}$$

Can we do the same for the flux f2, ingestion of phytoplankton by zooplankton? Again, we first note that without the zooplankton's physiological apparatus, phytoplankton will not transform spontaneously into zooplankton. Further, when zooplankton is ten times as abundant, all else being equal, ingestion will be ten times as high. Thus, flux f2 is proportional to zooplankton.

It may be limited however by phytoplankton: when phytoplankton is scarce the ingestion will be lowered, but it will not be limited by phytoplankton any more once that component is sufficiently abundant. Thus, we need a limitation function f'(PHYTO) that is zero when PHYTO is zero, and tends to one when PHYTO is sufficiently abundant. This gives us for the flux f2:

$$f_2 = b \cdot \text{ZOO} \cdot f'\left(\text{PHYTO}\right) \tag{2.28}$$

2.5.2 Maximal Interaction Strength, Rate Limitation and Inhibition

How can we generalize from these examples? In most ecological models that involve interaction between components, the first basic principle is that the component that is performing the *work* controls the *maximal strength* of the interaction. By the 'compartment performing the work', we mean the one whose physiological apparatus is needed for the flux to take place. For respiration, ingestion and mortality, the component that is performing the work is the one that is respectively respiring, ingesting, and dying.

Maximal interaction strengths are then written as proportional to the concentration of this work component. In other words: maximal interaction strength equals a maximal rate times the concentration of the work component (WORKER).

$$\text{MaxInteraction} = max\,Rate \cdot \text{WORKER} \tag{2.29}$$

The rationale behind this rule is quite simple. First, we can note that if there is no one to do the work, no work will be done. Thus, the flux has to be zero when the component responsible for the work is absent. Next, the more workers there are, the more work can be done. So the maximally attainable rate or flux will have to be proportional to the amount of workers (or the concentration of the component responsible for the work).

In many cases, the maximal interaction strengths are not attained endlessly: feedback with the environment may cause a slow-down of the interaction. Thus, the actual interaction strength includes a *rate-limiting term* that reduces the rate.

There are roughly two ways to express these rate-limiting terms:

- as a *functional response*, which explicitly includes the effect of a limiting resource (the source component of the flux).
- as a '*carrying capacity*' term, where the limitation is written as a function of the consumer or sink compartment.

Both types of formulations will be discussed in subsequent chapters.

Sometimes an interaction may be *inhibited* by the presence of some substance (the inhibitor), and this effect is also added to the ecological interaction. An example is an anaerobic process, inhibited by the presence of oxygen.

Summarising then, an ecological interaction (flow) can be written as the product of a maximal rate and the compartment performing the work, and, if appropriate, one or more rate limitation and inhibition terms:

$$\boxed{\text{Interaction} = maxRate \cdot \text{WORKER} \cdot \text{RateLimitation} \cdot \text{RateInhibition}} \qquad (2.30)$$

2.5.2.1 The Compartment Performing the Work

As the notions of the work compartments and maximal strength may be rather abstract, we illustrate the principle with three examples.

Example 1. Resource consumption processes

For consumer-resource interactions, the flux is called 'consumption' and the compartment that performs the work is the consumer. A predator for instance needs food and has to perform work to obtain it; the prey does not perform any work in order to be eaten! Thus, the consumption rate is written as:

$$\text{Consumption} = maxRate \cdot \text{CONSUMER} \cdot \text{RateLimitingTerm} \qquad (2.31)$$

where *maxRate* * CONSUMER is the maximal interaction strength.

With CONSUMER expressed as a concentration, the consumption will have units of concentration.time $^{-1}$, *maxRate* will have units of t^{-1} and the rate limiting term will be a dimensionless value between [0,1].

When the rate limiting term equals 0, the consumption rate will be 0, when equal to 1, maximal consumption rate will be achieved.

For a predator grazing on prey (Fig. 2.8 B) we can write:

$$\text{PredationRate} = maxGrazing \cdot \text{PREDATOR} \cdot \text{RateLimitingTerm} \qquad (2.32)$$

The idea behind this functionality is simple: predators or grazers will try to realise growth or reproduction, for which they need food. The higher the predator biomass, the more food is captured, hence the first-order (linear) dependence with respect to

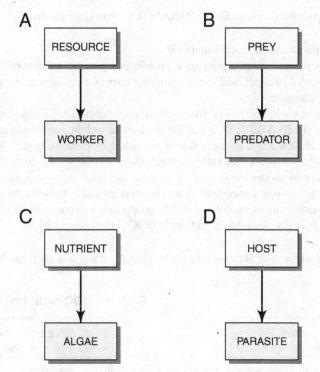

Fig. 2.8 Schematic representation (**A**) and three examples (**B–D**) of consumer-resource interactions. The compartment performing the work is the consumer (shaded in grey); the resource sets the rate-limiting term

the predator term. Because the predator must also handle the food and incorporate it into its own biomass or into eggs, there is a limit to the actual rate at which it can take in food, as expressed by the maximal rate coefficient (*maxGrazing*). Under ideal circumstances, the grazing of the predatory population will equal *maxGrazing**PREDATOR, the RateLimitingTerm will attain its maximal value (1). However, when the prey is relatively scarce, the predator will spend more of its time searching for prey. During this time it cannot handle and digest, and the total amount of prey captured and ingested per unit time and per predator will decrease. The maximal grazing rate will no longer be realised, the rate limiting term will have a value less than 1.

Similar dependency applies for algae taking up nutrients (Fig. 2.8 C): algae are performing the work (taking up the nutrients) and thus set the maximal rate; the rate limiting term will be a function of the nutrient concentration:

$$\text{NutrientUptakeRate} = max\,Rate \cdot \text{ALGAE} \cdot \text{RateLimitingTerm} \qquad (2.33)$$

or for parasites (working compartment) infesting their hosts (rate limiting compartment) (Fig. 2.8 D):

$$\text{InfectionRate} = max\,Rate \cdot \text{PARASITES} \cdot \text{RateLimitingTerm} \qquad (2.34)$$

Example 2. Biochemical transformations

Many biochemical transformations are mediated by bacteria. Their biomass controls the maximal strength, and the rate limiting term is a function of the substance that is being transformed.

For instance, the hydrolysis of dissolved organic carbon (DOC) with large molecular weight (SEMILABILE DOC) to low molecular weight dissolved organic carbon (LABILE DOC) is a process that is mediated by bacteria (Fig. 2.9 B). The bacteria excrete enzymes that catalyze the hydrolysis outside the bacterial cell. If we choose not to model the details of enzyme excretion, it is straightforward to assume that the enzyme concentration will be proportional to bacterial biomass, and thus, the maximal rate of hydrolysis will be proportional to bacterial biomass.

The flow between semi-labile and labile DOC is then described as:

$$\text{Hydrolysis} = max\,Hydrolysis \cdot \text{BACTERIA} \cdot \text{RateLimitingTerm} \qquad (2.35)$$

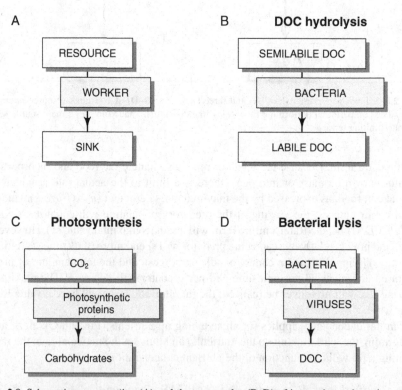

Fig. 2.9 Schematic representation (**A**) and three examples (**B–D**) of interactions where the work compartment (shaded box) is neither source nor sink. B. DOC hydrolysis, performed by bacteria. C. Photosynthesis, performed by the photosynthetic apparatus. D. Bacterial lysis performed by viruses

The difference with the previous example is that bacteria do *not* directly consume semi-labile DOC (they are not the sink compartment), but they *do* perform the work for the simple reason that they grow on the products of hydrolysis, the labile DOC.

In certain physiological models, different types of molecules are modelled. Some molecules 'perform the work' and transform other molecules. The example in Fig. 2.9 C for instance, is part of a complex physiological model describing algal growth that we will consider in Section 2.9.2. Through photosynthesis, algae transform CO_2 into low molecular weight carbohydrates, and this work is performed by the photosynthetic apparatus.

Another example of this type of interaction is bacterial lysis induced by viruses (Fig. 2.9 D).

Example 3. Loss processes

Respiration is the process that supplies an organism with the energy required for growth and maintenance. With growth we mean the incorporation of organic matter into the organism's own tissue and the production of offspring. Maintenance includes the cost of basal, cellular metabolism (to support the body functions) and the cost of regenerating lost material (moults, matter to repair the ageing of the organism,...).

The component performing the work is the animal's biomass, and maintenance respiration is modelled as proportional to this biomass. Respiration consumes oxygen, and thus a rate limiting term as a function of oxygen is added in a standard formulation. However, for organisms living in a permanently oxic world, oxygen is seldom limiting and the rate limiting term (which is always near to 1) can be dropped from the formulation. Note, however, that this is not true for animals living in environments (e.g. water columns, sediments) with a low oxygen concentration !

Thus the model formulation for maintenance respiration in general is:

$$\text{MaintenanceRespiration} = respirationRate \cdot \text{BIOMASS} \cdot \text{RateLimitingTerm}$$
$$(2.36)$$

while the model formulation in permanently oxic environments is:

$$\text{MaintenanceRespiration} = respirationRate \cdot \text{BIOMASS} \qquad (2.37)$$

Maintenance respiration is not the only form of respiration. Whenever organisms feed, they have additional energetic costs, linked to the hydrolysis of organic material in their food and re-assemblage into their own tissues. This 'growth respiration' increases with feeding rate. We will see later how to express it mathematically (Section 2.6.1).

2.5.3 One Rate-Limiting Resource, 3 Types of Functional Responses

In the examples of the previous section, the maximal interaction was modulated by a rate limiting term that we did not yet specify.

It is common in ecological models to write this rate limitation term as a function of the *resource* (or source component). Equations that include such functionalities are called *functional response* equations. For example, in the predator-prey equation (Eq. 2.32) we would write the rate limiting term as a function of the prey concentration, for the nutrient uptake equation (Eq. 2.33) as a function of the nutrients, for the hydrolysis example (Eq. 2.35) as a function of semilabile DOC. The idea is simple: if there is a surplus of prey, the predator will be able to consume all it can eat, i.e. it will not be limited by prey availability. In contrast, if prey is very scarce, the predator will be lucky to catch one, hence it will not be able to attain the maximal predation rate.

Essentially, the rate limiting term is a mathematical function that describes how the consumption rate is affected by changes in the resource. It obeys the following rules:

- the rate limiting term is dimensionless and scales between 0 and 1.
- if the resource component is abundant, the rate limiting term is large (approaches 1) and interaction is strong, near to maximum intensity,
- if the resource is scarce, the rate limiting term becomes small (approaches 0) and the effective interaction rate will be low,

Based on considerations of encounter probabilities, searching behaviour and handling time, three types of functional responses can be deduced (Holling, 1959) (Fig. 2.10).

- Functional response *type I*, is also called a linear response. It is expected when consumer and resource encounter each other at random, and therefore it is also called the 'blundering idiot search strategy'. It also assumes that handling time is negligible. Here the rate limiting term is simply proportional to the resource (R). One parameter (k) describes the functional form. The function is not allowed to exceed 1.

Functional responses

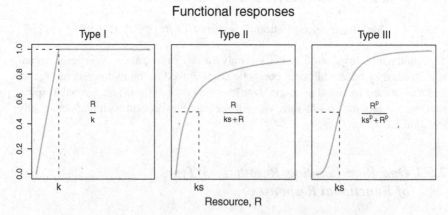

Fig. 2.10 Three types of functional responses. I. linear response. II. Monod or Michaelis-Menten response. III. Sigmoid response

$$\text{RateLimitingTerm} = MIN \left(\frac{R}{k}, 1 \right) \tag{2.38}$$

This type of response is not so often used in ecological models. The discontinuity at the concentration $R = k$ makes it difficult to handle from a mathematical
point of view.

- In a functional response *type II*, the rate increases nearly linearly at low resource
  density, and levels off at higher resource density.

  This type of equation is known in enzyme kinetics as a Michaelis-Menten
  function, and is also frequently called a Monod, or hyperbolic equation. It is the
  most common form of rate limiting term used in ecological models.

$$\text{RateLimitingTerm} = \frac{R}{R + ks} \tag{2.39}$$

When the resource (R) is scarce, the level of consumption is mainly limited
by the ability to take up or find the resource, whereas at high concentrations, the
handling time (required to assimilate the resource) becomes the limiting factor.
Alternatively, a predator may become satiated and stop feeding. We will derive
the Michaelis-Menten formulation for an enzymatic reaction in Chapter 8. To
describe the hyperbolic functionality, one parameter is required (ks), which is
called the 'half-saturation constant'. It is the resource concentration where the
rate attains 50% of its maximal value. The half-saturation coefficient determines
how fast the rate approaches the maximal rate. Low values of ks describe rapidly
rising curves, and vice versa.

- In a functional response *type III*, the rate initially increases more than linearly
  (e.g. related to a learning process, switching to more abundant prey, preferring
  high-density patches,...), after which it levels off. The simplest such function is
  a sigmoid function:

$$\text{RatelimitingTerm} = \frac{R^p}{R^p + ks^p} \tag{2.40}$$

where ks is, again, the half-saturation concentration, p is a shape factor. The
higher its value, the steeper the curve.

2.5.4 More than One Limiting Resource

Many processes are potentially limited by more than one factor. For example, phytoplankton growth requires availability of all the essential elements N,P,C,(Si), ...

Two empirical solutions are adopted to solve the problem of concomitant limiting factors: the minimum and multiplicative law.

- The *Liebig law of the minimum*, considers that the growth rate will depend only on the factor for which the corresponding growth is lowest, i.e. as a function of the element least in supply

$$\text{RateLimitingterm} = MIN\,(\text{LimitingTerm1}, \text{LimitingTerm2}) \qquad (2.41)$$

- In contrast, the *multiplicative law* considers that the different factors act simultaneously, as the product:

$$\text{RateLimitingterm} = \text{LimitingTerm1} \cdot \text{LimitingTerm2} \qquad (2.42)$$

A common example is algal growth, which is limited both by light availability and nutrient concentrations (assume N, Si). This is often expressed as:

$$\text{RateLimitingTerm} = \text{LightLimitation} \cdot MIN\left(\frac{N}{N + ks_N}, \frac{Si}{Si + ks_{Si}}\right) \qquad (2.43)$$

where the co-limitation of light and nutrients is described multiplicatively, and where concomitant limitation by inorganic nitrogen and silicate is expressed using the minimum law. Growth limitation on nitrogen and silicate is described with Monod kinetics, with ks_N and ks_{Si} the respective half-saturation constants.

Sometimes the co-limitation by light and nutrients is described using the minimum law:

$$\text{RateLimitingTerm} = MIN\left(\text{LightLimitation}, \frac{N}{N + ks_N}, \frac{Si}{Si + ks_{Si}}\right) \qquad (2.44)$$

We will see how light limitation can be expressed in Section 2.8.2.

The two alternatives presented, Liebig's law of the minimum and multiplicative multiple limitation, can be considered as two end-members of a continuum of possible expressions. In the literature many alternatives are presented that express some stronger or weaker form of interaction between different limiting processes.

For predators that have access to different food sources and that can switch from one food to the other ($FOOD_i$), it is customary to add a certain preference p_i ([0,1]) for each food item. With all $p_i=1$ there is no preference for any of the food items:

$$\text{RateLimitingTerm} = \frac{\sum_i p_i \cdot FOOD_i}{\sum_i p_i \cdot FOOD_i + ks_{FOOD}} \qquad (2.45)$$

2.5.5 Inhibition Terms

Some processes are inhibited by the presence of a certain substance. The more abundant this inhibitory substance, the lower the rates will be (Fig. 2.11A).

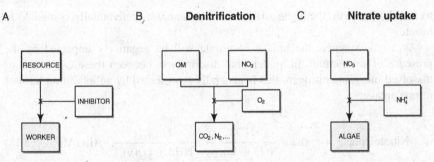

Fig. 2.11 Schematic representation (**A**) and two examples (**B,C**) of inhibition. **B**. Denitrification is inhibited in the presence of oxygen. **C**. nitrate uptake by algae is inhibited by ammonium. The compartment performing the work is denoted with a shaded box

2.5.5.1 Example: Denitrification

There are many examples of inhibition in chemical or biogeochemical models, where the occurrence of a reaction is negatively impacted by the concentration of some substance that does NOT play a direct part in the chemical reaction.

For example, the mineralization (respiration) of organic matter with nitrate, called denitrification, is a process that occurs at very low rates in the presence of oxygen: it is inhibited by oxygen (Fig. 2.11B).

A process description of denitrification should take this dependence into account. The simplest solution is to multiply, to the rate term describing nitrate consumption, a dimensionless inhibition term:

$$\text{Inhibition} = 1 - \frac{O2}{O2 + kin_{O2}} = \frac{kin_{O2}}{O2 + kin_{O2}} \tag{2.46}$$

The value of this inhibition term will decrease as the oxygen concentration increases; it has the value 1 at zero oxygen concentrations. The parameter kin_{O2} is the oxygen concentration at which the rate drops to half of its maximal value.

Assuming that mineralization is first-order with respect to organic matter availability (see above), and that denitrification is limited by nitrate and inhibited by oxygen, the denitrification rate can be expressed as:

$$\text{Denitrification} = rmax \cdot \frac{NO3}{NO3 + ks_{NO3}} \cdot \frac{kin_{O2}}{O2 + kin_{O2}} \cdot \text{ORGANICMATTER} \tag{2.47}$$

2.5.5.2 Example: Nitrate and Ammonium Uptake by Algae

Inhibition terms can also be present in biological models.

For example, algae can use ammonium as well as nitrate as a source of nitrogen for their growth (Fig. 2.11 C). However, nitrate has to be reduced before it can be assimilated in algal biomass. As this reduction presents an additional energetic cost to the algal growth, the algae will take up ammonium preferentially compared to nitrate.

As a consequence, the uptake of nitrate will be negatively impacted by the presence of ammonium. In models that discriminate between these two forms of dissolved inorganic nitrogen, this is generally represented by an inhibition term of nitrate uptake:

$$\text{NitrateUptake} = rmax \cdot \frac{\text{NO3}}{\text{NO3} + ks_{NO3}} \cdot \frac{kin_{NH3}}{\text{NH3} + kin_{NH3}} \cdot \text{ALGAE} \qquad (2.48)$$

In the example above, we used a Monod term (subtracted from 1) to account for the inhibition by ammonium (Fig. 2.12 A). It is more traditional in aquatic modelling to represent the inhibition as an exponential term (Fig. 2.12 B), declining exponentially with increasing ammonium concentration:

$$\text{NitrateUptake} = rmax \cdot \frac{\text{NO3}}{\text{NO3} + ks_{NO3}} \exp\left(-InhibitionCt \cdot \text{NH3}\right) \cdot \text{ALGAE}$$
$$(2.49)$$

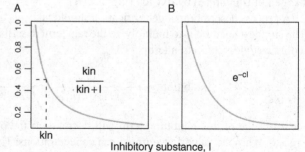

Fig. 2.12 Often-used inhibition terms in ecological models. **A**: 1-Monod function, **B**: exponential function

2.6 Coupled Model Equations

Models often contain several equations that are coupled either through the interaction between source and sink compartments (Fig. 2.13A) or via stoichiometric relationships (Fig. 2.13B). The coupling of equations is often necessary because fluxes are typically dependent on one state variable for their maximal rate, and on another one for their limitation term.

Below we give three examples of coupled model formulations. Other couplings, e.g. between physical and biological model parts, may also be realised but will not be detailed here.

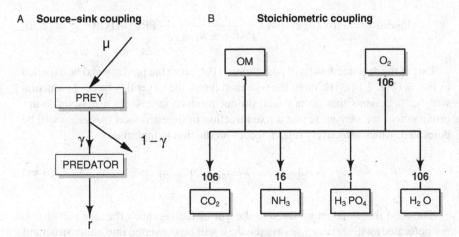

Fig. 2.13 Two examples of coupled equations. **A.** An ingestion flow links the increase of the predator to the decrease of prey and the production of detritus. **B.** The mineralization (respiration) of organic matter (OM) links the oxygen (O_2), carbon (CO_2), nitrogen (NH_3) and the phosphorus cycle (H_3PO_4), according to specific stoichiometric ratios (here: Redfield ratios)

2.6.1 Flows Modelled as Fractions of Other Flows

Not all flows are written as a function of the source and /or sink compartment. Sometimes it is more convenient to express one flow as a function of another flow.

Consider, again, the feeding of an organism (predator) on its prey. In previous sections we have seen how to describe this as first-order to predator biomass (the work compartment) and with a rate limiting term depending on the prey concentration:

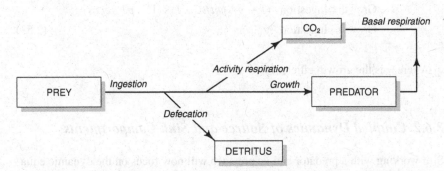

Fig. 2.14 When an organism feeds, it not only provokes a flux from the food to its own biomass, but also to detritus (Defecation) and to CO_2 + inorganic nutrients (Activity respiration). All these flows are related: an organism can not produce more faeces than the amount of food it ingested. To make the picture complete, the conceptual model also includes basal respiration

$$\text{Ingestion} = maxIngestion \cdot \frac{\text{PREY}}{\text{PREY} + ks_{PREY}} \cdot \text{PREDATOR} \qquad (2.50)$$

Part of the ingested food will pass through the gut of the predator and be expelled as faeces (Fig. 2.14). The more the organism feeds, the larger this faeces production will be; organisms that do not feed do not produce faeces. As a reasonable approximation, we may write that a fixed fraction of ingested food (*pFaeces*) will be defecated, which effectively relates faeces production to ingestion.

$$\text{Defecation} = pFaeces \cdot \text{Ingestion} \qquad (2.51)$$

The food that is taken up through the gut of the organism (i.e. the part that is not defecated) will serve organism growth: it will be converted into either structural biomass of the organism or into reproductive tissue. However, conversion of food into tissue is a costly process and the energy required will be delivered by respiration, the so-called activity or growth respiration (Fig. 2.14). The more an organism feeds, the higher this energy requirement, and the higher its respiration. It is usually assumed that a constant fraction (the *growthCost*) of food uptake through the gut is used for respiration:

$$\text{Growthrespiration} = (\text{Ingestion} - \text{Defecation}) \cdot growthCost$$
$$\text{Growthrespiration} = \text{Ingestion} \cdot growthCost \cdot (1 - pFaeces) \qquad (2.52)$$

We may then finally write the food incorporated into structural tissue as:

$$\text{Growth} = \text{Ingestion-Defecation-GrowthRespiration}$$
$$\text{Growth} = \text{Ingestion} \cdot (1 - growthCost) \cdot (1 - pFaeces)$$
$$\text{Growth} = \text{Ingestion} \cdot \gamma \qquad (2.53)$$

where γ is the growth efficiency.

2.6.2 Coupled Dynamics of Source and Sink Compartments

Still working with a predator and its prey, we will now focus on the dynamic equations for these two state variables. Obviously, the dynamics of a consumer and its resource are connected. Reflecting this, their equations are coupled.

Such coupled equations take into account the feedback between the consumer and resource and therefore make the model more realistic.

Consider for instance the coupled predator-prey dynamics:

$$\frac{d\text{PREDATOR}}{dt} = g \cdot \text{PREDATOR} \cdot \frac{\text{PREY}}{\text{PREY} + ks_{PREY}} \cdot \gamma - r \cdot \text{PREDATOR}$$

$$\frac{d\text{PREY}}{dt} = -g \cdot \text{PREDATOR} \cdot \frac{\text{PREY}}{\text{PREY} + ks_{PREY}} + \mu \cdot \text{PREY}$$

$$(2.54)$$

The grazing of the predator on the prey not only induces a proportional increase in predator biomass, but also reduces the biomass of the prey (first term). As not all of the prey grazed is converted into predator biomass, the grazing is multiplied by parameter γ, the growth efficiency ($[0, 1]$), in the predator equation.

Note that the growth efficiency parameter γ is used to represent both faeces production flux and growth respiration flux. Similarly, the parameter r in the predator equation represents mortality of the predator, as well as maintenance respiration. In contrast to biogeochemists, who are also interested in detritus, or in CO_2, and therefore separate all these fluxes, theoretical ecologists tend to lump this type of 'loss processes' as much as possible.

As the predator biomass increases, so does the predation pressure on its prey. This will slow down the growth of the prey until its biomass eventually decreases. Lower prey biomass will provoke stronger resource limitation, causing slow-down of predator growth and eventual decline of predator biomass, such that the prey biomass may recover.

2.6.3 Stoichiometry and Coupling of Element Cycles

The budgets of several main constituents on earth, e.g. C, N, P, Si are coupled through stoichiometric relations.

The *stoichiometry* of a compound is defined as the proportion of the various quantities.

For instance, in marine algae, the molar ratio of elements is remarkably constant and called the Redfield ratio. Thus, algal organic matter has the composition of

$$(CH_2O)_{106}(NH_3)_{16}(H_3PO_4) = C_{106}H_{263}O_{110}N_{16}P$$

or a stoichiometric C:H:O:N:P ratio of 106:263:110:16:1.

Most marine animals more or less follow this stoichiometric composition of algae. Marine bacteria tend to deviate in composition, with lower C:N ratios.

This interconnectedness of elements provides an additional constraint on mathematical models. Data on the nitrogen cycle can be used to cross-check data or model results on the carbon cycle, as both cycles are intimately linked.

For instance, to oxidise (respire) 106 mole of fresh algal organic matter, 106 mole of oxygen is generally consumed, whereas 16 moles of ammonium, and one mole of phosphate are released. The respiration reaction reads:

$$(CH_2O)_{106}(NH_3)_{16}(H_3PO_4)+106O_2$$
$$->106CO_2 + 16NH_3 + H_3PO_4 + 106H_2O$$

This effectively couples the cycles of C, O, N and P.

Models that describe all these elements take into account these changes, using the appropriate stoichiometric ratios.

For instance, with 'Respiration' being the amount of organic matter respired, the following terms will pop up in the rate of change of oxygen, ammonium, phosphate, carbon dioxide and organic carbon:

$$\frac{dO_2}{dt} = -\text{Respiration} \cdot OC + \ldots$$

$$\frac{dNH_3}{dt} = \text{Respiration} \cdot NC + \ldots$$

$$\frac{dPO_4}{dt} = \text{Respiration} \cdot PC + \ldots$$

$$\frac{dCO_2}{dt} = \text{Respiration} + \ldots$$

$$\frac{d\,ORGANICCARBON}{dt} = -\text{Respiration} + \ldots \tag{2.55}$$

where $OC = 1$, $NC = 16/106$, and $PC=1/106$, the stoichiometric ratios of organic matter.

2.7 Model Simplifications

As discussed in Chapter 1, a modeller tries to simplify the world as much as possible. This is not always easy, and particular tricks are sometimes needed to keep a model within reasonable bounds of complexity.

In models involving trophic relations (consumer-resource interactions) the flows are calculated as a function of both consumer and resource. This resource, in turn, may again consume its own resources, etc. – but often we do not want our model to go down all the way to the lowermost level of the food web. In that case we need an expression that approximates the dynamics of the lowermost consumer level in our model as well as possible, without the need to include its resource level explicitly in the model. Often this simplification makes use of a 'carrying capacity formulation'. (Fig. 2.15A)

Similarly, the highest consumer level in our model may be the prey of a still higher trophic level, but here again we want to stop at some particular level. We will simplify again by modelling the mortality rate of the uppermost trophic level in the model, without the need to explicitly model an even higher trophic level. We call this procedure, with a technical term, *closure* of the model. (Fig. 2.15B)

Sometimes, there exist a number of intermediate links between two compartments that we are not particularly interested in; in that case we may want to suppress the intermediate compartments and simply model the dynamics as if there were a direct interaction between the end-members. (Fig. 2.16 C)

2.7.1 Carrying Capacity Formulation

In the functional response models discussed previously, the rate limiting term was a function of the resource (the prey for predator-prey relationships, nutrients for algal uptake,...).

In population models, rate limiting feedback terms are often modelled as a function of the *consumer* population density itself. These models are called carrying capacity models. The carrying capacity is the density or concentration above which the growth rate of the consumer becomes negative.

The carrying capacity formulation is an approximation of processes not explicitly included in the model. These processes may be resource limitation, but also competition for space.

The following is a simple example of a model of population growth, with a carrying capacity formulation. It is the logistic equation, sometimes called the Verhulst model, in reference to its creator (Verhulst, 1838)

$$\frac{d\mathrm{N}}{dt} = r \cdot \mathrm{N} \cdot \left[1 - \frac{\mathrm{N}}{K} \right] \qquad (2.56)$$

where N, the state variable, is population density, dN/dt is its rate of change, r is maximal net growth rate, also referred to as the intrinsic rate of increase (t^{-1}), K is the carrying capacity, i.e. the maximum population size that can be supported; K has the same units as N. This equation has been so influential in ecological modelling, and we use it several times in this book, so that it is worthwhile to explain it in more detail.

Clearly, N is the compartment performing the work (growing), r is the maximal rate, whilst the term $\left[1 - \frac{\mathrm{N}}{K} \right]$ is the rate limiting term.

- At very low densities (N$<<K$), the limitation term $\left[1 - \frac{\mathrm{N}}{K} \right]$ will be ≈ 1 and the population grows exponentially.
- In the neighborhood of the carrying capacity K, the limitation term approaches 0, the rate of change becomes very small and density will remain quasi-constant.
- At densities much higher than the carrying capacity, the limitation term $\left[1 - \frac{\mathrm{N}}{K} \right]$ will be negative, and the population density will decrease quasi-exponentially towards K.

The carrying capacity model leads to asymptotic behaviour in time (Fig. 2.15 B). For starting values of population density much smaller than the carrying capacity, the trajectory of population density versus time is a sigmoid relationship (S-shaped curve, solid line). When initiated at very high values, population density will decline quasi-exponentially towards K (dashed line).

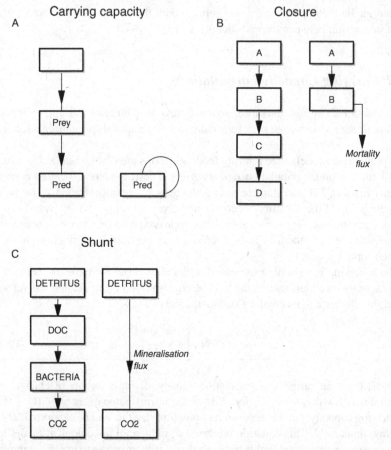

Fig. 2.15 Schematic representation of model simplifications. **A.** In simple models, the dependence of a predator on the dynamics of its prey is often not represented explicitly but mimicked by a function that only takes into account the value of the predator concentration; carrying capacity models are the most notorious examples of such simplifications. **B.** In food web models, the top-predators (here C, D) are often not represented; rather their effect on their prey (here B) is added as a mortality flux (the closure flux). **C.** If we are interested in the decay of detritus, and the eventual respiration, but not in the organisms performing the work (here bacteria), we simply model a direct flow between detritus and CO_2

In contrast to the models including functional responses, carrying capacity models do not need to include the dynamics of the limiting resource, so they are in a sense simpler. However, they are not as general as functional response models.

Gotelli (2001) clarifies how to derive the Verhulst model in a succinct way, which we find particularly informative:

We start with a very simple model that describes changes in population density (N) due to the difference between birth and death, which are both a function of density. Thus, we write the rate of change of density as:

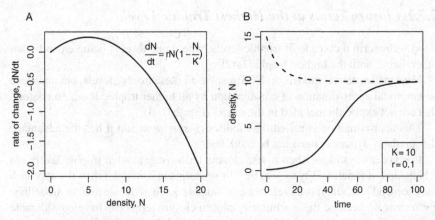

Fig. 2.16 The logistic equation. A. the rate of change as a function of density. B. Two solution curves, representing how density changes in time

$$\frac{dN}{dt} = [b(N) - d(N)] \cdot N \qquad (2.57)$$

and where b(N) and d(N) are the per capita birth and death rate (units of time^{-1}) respectively.

As population density increases, less food will be available, space and shelter may fall short, predation pressure may be higher, or stress may increase due to too frequent contacts, or to reduction of a limiting substance (e.g. oxygen in schools of fish).

These phenomena will negatively affect the birth rate, and will increase mortality. The simplest way of expressing this in a model is by making the per capita birth rate (b(N)) a linearly decreasing and the death rate (d(N)) a linearly increasing function of density. With

$$b(N) = b_0 - b_1 \cdot N \qquad (2.58)$$

and a similar description for death rate, we obtain:

$$\frac{dN}{dt} = [b_0 - b_1 \cdot N - (d_0 + d_1 \cdot N)] \cdot N \qquad (2.59)$$

After rearranging and simplifying, we obtain the logistic growth equation:

$$\frac{dN}{dt} = (b_0 - d_0) \cdot N \cdot \left[1 - \frac{b_1 + d_1}{b_0 - d_0} \cdot N\right]$$

$$\frac{dN}{dt} = r \cdot N \cdot \left[1 - \frac{N}{K}\right] \qquad (2.60)$$

2.7.2 Closure Terms at the Highest Trophic Level

Food webs form a cascade of trophic levels, where one level is being eaten by another, etc. . . until the highest trophic level.

Most mathematical models do NOT resolve all these trophic levels, but introduce an empirical approximation of consumption by all higher trophic levels (predators) that are not explicitly included in the model (Fig. 2.16 B).

This approximation is called the model's *closure term* and it has the advantage that predator dynamics need not be modelled.

It is not always clear when to use closure terms (e.g. at what trophic level) and which kind of functionality to give the closure terms in a model. Often the approach may seem ad hoc, and it is therefore a continuous source of debate among modellers. Even more so because these arbitrarily chosen closure terms can have considerable influence on model behaviour. Especially model stability characteristics are fundamentally influenced by closure terms. Therefore they cannot be used carelessly.

As the predator in our example is not modelled, the mortality of the prey, induced by this level is a function of the prey concentration rather than the predator concentration. Most often used is a linear mortality or a quadratic mortality term:

$$\text{Mortality} = k \cdot \text{PREY}$$
$$\text{Mortality} = c \cdot \text{PREY} \cdot \text{PREY} \tag{2.61}$$

In the first equation, the closure is linear, and parameter k has units of time^{-1}. Here it is effectively assumed that predation pressure (k) remains constant. In the second equation, the closure term is quadratic, units of c are conc^{-1} time^{-1} where 'conc' are the units of Prey. This equation assumes that the predatory rate (c.PREY) fluctuates proportionally with the prey biomass.

Quadratic closure tends to stabilize food web models to a considerable degree, and its parameters also influence the steady state results of the models substantially. This may or may not be a desirable property. If the model aims to simulate real observations, enhanced stability may help to find 'good' solutions. However, if the model study aims at investigating the stability properties of different interaction terms, artificial stabilization by a quadratic closure may clutter the argument.

2.7.3 Simplification by Deletion of Intermediate Levels

In biogeochemical models, the processes associated with the mineralization (respiration) of organic matter are often of interest. Whereas mineralization is performed by bacteria and higher organisms, their dynamics is usually not relevant.

By assuming that organic matter decay proceeds at a first-order rate, the necessity to model explicitly the organisms that are performing the work, is avoided.

$$\text{OrganicMatterRespiration} = k \cdot \text{ORGANICMATTER} \tag{2.62}$$

This effectively assumes that, whatever organism is metabolizing, the rate of decomposition is controlled by the amount and reactivity (palatability) of the organic matter, and not by organism biomass. It is a closure term because it closes off the model formulation for the flux to CO_2, without considering all factors, other than organic matter concentration, influencing this flux. (Fig. 2.16 C).

2.8 Impact of Physical Conditions

Many physical quantities, such as temperature, light, water flow, and wind have an effect on ecosystems. These physical factors are either computed by physical models that are coupled to the biological descriptions or, more likely, they are imposed as forcing functions.

2.8.1 Temperature

Many rates are modulated by temperature. This applies as much to physical processes (e.g. molecular diffusion), as to chemical (reaction) rates, and physiological rates (growth, respiration, feeding, excretion,..). In aqueous environments, temperature also affects the solubility of many substances and therefore its exchange across the air-water interface.

The response of individual *organisms* to temperature is one where rates gradually increase towards a clearly defined temperature optimum, above which rates decline. This response of individual organisms should not be confused with the *ecosystem* response. Throughout the different seasons, ecosystems are characterised by a succession of species that are acclimated to different thermal conditions. As a result rates scale more or less exponentially to temperature, without the presence of a clear optimum, when viewed at the ecosystem level (Fig. 2.17). Typically, biological processes will more or less double in rate for a temperature increase of 10°C, as long as temperature is within acceptable ranges. Below and above, rates will drop strongly and organisms may die because essential enzymes denaturalise.

Thus, for *multispecies* assemblages, an exponential increase with temperature is most often used.

The formulation for multispecies assemblages comes in many different shapes, the most well-known is the *Q10* formulation:

$$\text{TempFactor} = \exp\left(\frac{T - Tref}{10}\log_e(q10)\right) \tag{2.63}$$

where $Tref$ = reference temperature at which TempFactor = 1. This is the temperature at which the rate parameters are expressed. Parameter $q10$ is the factor of increase for every increase of the temperature with 10°C. A typical value for $q10$ is 2.

Fig. 2.17 Response to
temperature for individual
species (*thin lines*) and for
the ecosystem
(*thick line*)

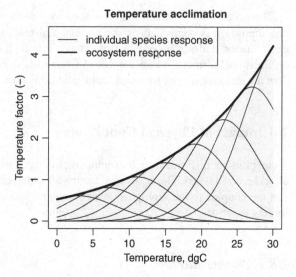

For instance, with R_{20} the rate at 20°C and a $q10$ of 2, the value of the rate at 10°C will be:

$$R_{10} = R_{20} \cdot \exp\left(\frac{10 - 20}{10} \log_e(2)\right) = R_{20} \cdot 0.5 \qquad (2.64)$$

or a halving of the rate.

Another description of the same exponential dependence is given by:

$$\text{TempFactor} = \exp\left(Tcoeff \cdot \text{T}\right) \qquad (2.65)$$

where the rate parameters must now be defined at 0°C. A value for $Tcoeff = \log_e(2)/10 = 0.069$ corresponds to a $q10$ of 2.

In most models, temperature is imposed as a forcing function, although, in aquatic environments it can be dynamically described by physical models that are driven by atmospheric conditions (wind, air temperature, air humidity, solar radiation).

2.8.2 Light

Being the energy source of photosynthesis, the availability of light in the photosynthetically active range (PAR=photosynthetically active radiation) drives most ecological systems. In addition to its functioning as an energy source, solar radiation heats up the water and exposed sediments.

Within water masses, the decline of solar radiation with depth (Fig. 2.18B) effectively divides the water column into an autotrophic and a heterotrophic part. This is essential for understanding the vertical distribution of the biotic (algae, zooplankton) and abiotic components (nutrients) in the water column. In summer, when oceanic water columns are stratified, the upper water layer is typically depleted in nutrients, whereas the deeper layers are rich in nutrients but too dark for algal growth. As algae need both light and nutrients for growth, a chlorophyll maximum often develops at intermediate depth, where some light is available and moreover some nutrients are mixed in from below (Fig. 2.18D).

Similarly the light intensity declines as light penetrates in a dense plant canopy.

The decline of light (I) with depth (z) can be described as a simple differential equation, expressing the rate of change of light with depth (dI/dz). Light intensity is lost in constant proportion to the available light, and the proportionality coefficient is the extinction coefficient, k (units of m^{-1}):

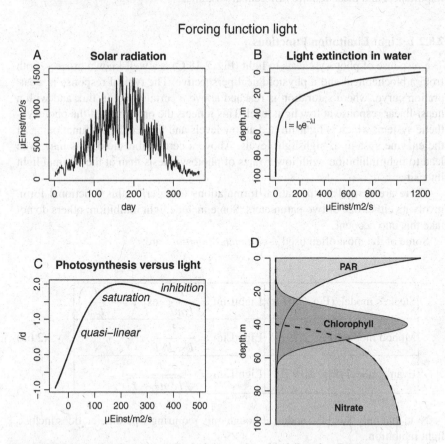

Fig. 2.18 Light forcing function. **A.** Typical time series with daily solar radiation, representative for temperate areas. **B.** Light intensity declines as a function of water depth. **C.** Typical photosynthetic response to light. **D.** The decrease of light with water depth divides the water column into distinct functional zones. See text for explanation

$$\frac{dI}{dz} = -k \cdot I \quad ; \quad I|_{z=0} = I_0 \tag{2.66}$$

With I_0 the light intensity at the upper (e.g. air-water) interface ($z=0$), it is possible to calculate the light intensity at any depth. The solution to this differential equation (see Chapter 5) is the Lambert-Beer relationship:

$$I|_z = I_0 \cdot \exp(-kz) \tag{2.67}$$

Light is always imposed as a forcing function, either using actual measurements (from the meteorological office) or using special algorithms that approximate light intensity as a function of latitude and season. Depending on the type and scale of the model, seasonal, diurnal and within-day variations in light intensity may be of importance, and these require different algorithms.

2.8.2.1 Light Limitation Functions

The response of photosynthesis to light (Fig. 2.18 C) is a well-known process, both from a biochemical and a physiological perspective. The typical response is a saturation curve, where saturation is reached above a certain photon flux and with a quasi-linear response at low light level. This reflects the properties of the photosynthetic system, which is light-limited at low levels and limited by the functioning of the enzyme system at high light levels. Above a certain threshold, this may even lead to light inhibition, with lower rates of photosynthesis than at the optimal light intensity.

There are several mathematical formulations to describe this functional form, involving either one or two parameters. Some include light inhibition; others do not take this into account.

Some of the most often used *1-parameter functions* are:

Steele's model, (Fig. 2.19 A)	$LightLim = \dfrac{I}{Iopt} \cdot \exp\left(1 - \dfrac{I}{I_{opt}}\right)$	
Monod model, (Fig. 2.19 B)	$LightLim = \dfrac{I}{I + ksI}$	(2.68)
Evan's model (Fig. 2.19 B)	$LightLim = \dfrac{I}{\sqrt{Iopt^2 + I^2}}$	

Note that only Steele's model, although only requiring 1 parameter, does include light inhibition.

For marine algae, typical values for Iopt are 50-300 μEinst m^{-2} s^{-1} ; ks is typically in the range of 50-150 μEinst m^{-2} s^{-1}.

2-parameter functions (Fig. 2.19 C) are significantly more complex:

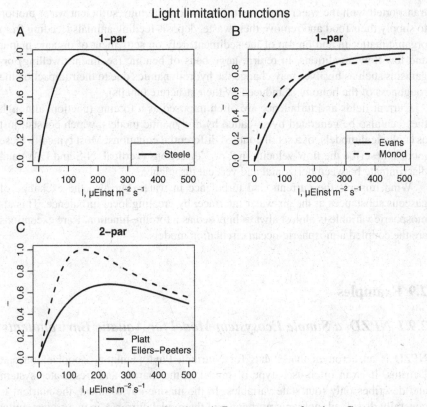

Fig. 2.19 Examples of light-limitation functions. **A–B**: one-parameter functions, **C**: two-parameter functions. See text for more details

The Platt model:	$\text{LightLim} = \left(1 - \exp\left(\dfrac{-\alpha \cdot I}{pMax}\right)\right) \cdot \exp\left(\dfrac{-\beta \cdot I}{pMax}\right)$
Eilers-Peeters model	$\text{LightLim} = \dfrac{2 \cdot (1 + \beta) \cdot I/Iopt}{(I/Iopt)^2 + 2 \cdot \beta \cdot I/Iopt + 1}$

With pMax the maximal photosynthesis rate (d^{-1}). Typical values for α and β are 0.005-0.01 (μEinst m^{-2} s^{-1})$^{-1}$ d^{-1}. (Note: in the Platt model, α and β do not occur independently from $pMax$, so there are effectively only 2 independent parameters).

2.8.3 Other Physical Impacts

Currents and turbulence obviously have a large impact on pelagic animals with limited swimming capacity. However, they also affect sediment-inhabiting (benthic) animals in various ways. Many benthic animals have pelagic larvae that are transported with the water movement; filter feeders require sufficient water motion to supply their food and remove their waste, deposit-feeding animals (feeding from organic matter in and on top of the sediment) rely on settlement of organic matter and low-flow conditions. In return, large beds of benthic (sediment-dwelling) organisms such as bivalves may change the hydrodynamics due to their impact on the roughness of the bottom, as induced by their structure (shells).

Current fields and turbulence are often imposed as a forcing function, although they can also be generated by so-called hydrodynamic models, which consist, just as ecological models, of a set of coupled differential equations. Most typically these models describe the flow velocity in 2 or 3 directions (vertical, N-S and E-W) and discriminate between horizontal and vertical turbulence.

Wind impacts the currents and turbulence in water, but also the exchange of gaseous substances at the air-water interface, by creating local turbulence. This atmospheric variable is almost always imposed as a forcing function. Rare exceptions are the coupled atmospheric-ocean circulation models.

2.9 Examples

2.9.1 NPZD, a Simple Ecosystem Model for Aquatic Environments

NPZD is an acronym that stands for Nutrient, Phytoplankton, Zooplankton and Detritus. It is an often-used type of model in the marine and freshwater system and describes only four state variables. In the marine environment, the nutrient is generally dissolved inorganic nitrogen and the model currency is in nitrogen units; freshwater models generally deal with phosphorus.

2.9.1.1 Step 1. Flowchart

To create an NPZD model, we start by drawing the flowchart, comprising the state variables (square boxes), the flows (arrows connecting the state variables), the forcing functions and the ordinary variables. The flowchart of an NPZD model is given in Fig. 2.20, where:

- f1 = net nitrogen uptake of algae
- f2 = zooplankton grazing
- f3 = zooplankton faeces production
- f4 = zooplankton excretion
- f5 = zooplankton mortality
- f6 = detritus mineralization

All four state variables are expressed in mmolN.m^{-3}, solar radiation is a forcing function and Chlorophyll is an output variable.

Fig. 2.20 Flowchart of the NPZD (nutrient, phytoplankton, zooplankton, detritus) model

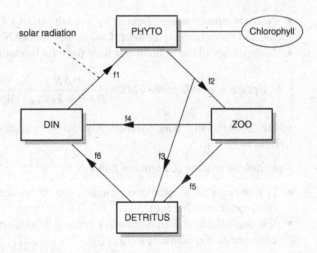

2.9.1.2 Step 2. Conceptual Model

Next, we write the conceptual model, which specifies the rate of change of the four state variables as the sum of all inflows (arrows directed into the boxes) minus all outflows,

For the NPZD model the conceptual model is:

$$\frac{d\,\text{PHYTO}}{dt} = f1 - f2$$

$$\frac{d\,\text{ZOO}}{dt} = f2 - f3 - f4 - f5$$

$$\frac{d\,\text{DETRITUS}}{dt} = f3 + f5 - f6$$

$$\frac{d\,\text{DIN}}{dt} = f4 + f6 - f1 \tag{2.69}$$

2.9.1.3 Step 3. Mathematical Formulations

Thirdly, we create a mathematical expression for all the flows. We distinguish between:

1. Flows that express ecological interactions between two components. These flows are modelled as: maxRate*WORKER*RateLimitingTerms.

   Net nitrogen uptake (f1, units of mmol N m^{-3} d^{-1}):

   - The phytoplankton (PHYTO) performs the work and sets the maximal rate (parameter *maxUptake*).
   - Nitrogen uptake is limited by light availability, expressed as a Monod function (parameter ks_{PAR}, units μEinst m^{-2} s^{-1})

- Nutrient uptake is also limited by availability of DIN, also expressed by a Monod formulation (parameter ks_{DIN}, units mmol N m^{-3}).
- Liebig's law of the minimum is used for total limitation:

$$N\_Uptake = maxUptake \cdot \text{MIN}(\frac{PAR}{PAR + ks_{PAR}}, \frac{DIN}{DIN + ks_{DIN}}) \cdot PHYTO$$

$$(2.70)$$

where PAR is the photosynthetically active radiation, a forcing function (μEinst m^{-2}s^{-1}).

Zooplankton grazing (f2, units of mmol N m^{-3} d^{-1}):

- Is first-order to zooplankton biomass, the component performing the work (parameter *maxGrazing*)
- We use a Monod formulation for grazing limitation by the phytoplankton concentration (parameter ks_{PHYTO})

$$Grazing = maxGrazing \cdot \frac{PHYTO}{PHYTO + ks_{PHYTO}} \cdot ZOO \qquad (2.71)$$

2. Flows that are expressed as a function of another flow.

Zooplankton faeces production, f3:

- This is expressed as a constant fraction of total grazing (parameter *pFaeces* (<1)):

$$FaecesProduction = Grazing \cdot pFaeces \qquad (2.72)$$

3. Flows which are simply first-order to the source compartment.

Zooplankton excretion, f4; this flow is first-order to zooplankton biomass:

$$Excretion = excretionRate \cdot ZOO \qquad (2.73)$$

4. Closure terms, which take into account the action of components that are not explicitly modelled. Here there are two closures:

- *Zooplankton mortality*, f5, caused by predators that are not explicitly modelled. We can describe that as second-order to zooplankton biomass:

$$Mortality = mortalityRate \cdot ZOO^2 \qquad (2.74)$$

- *Detritus mineralization*, f6, performed by bacteria that are not part of the model; this is described as first-order to detritus concentration:

$$Mineralization = mineralizationRate \cdot DETRITUS \qquad (2.75)$$

Finally, we convert from phytoplankton biomass (in mmol N m^{-3}) to chlorophyll (mg Chl m^{-3}) by assuming a constant proportionality factor, *Chl Nratio* (units of mgChl (mmol N)$^{-1}$):

$$Chlorophyll = Chl\_Nratio \cdot PHYTO \qquad (2.76)$$

The following table lists typical parameters values:

Parameter	Value	Units
maxUptake	1.0	Day^{-1}
ks$_{PAR}$	120	μEinst m^{-2} s^{-1}
ks$_{DIN}$	0.5	mmol m^{-3}
maxGrazing	1.0	Day^{-1}
ks$_{PHYTO}$	1	mmol N m^{-3}
pFaeces	0.3	–
mortalityRate	0.4	(mmolN m^{-3})$^{-1}$ day^{-1}
excretionRate	0.1	Day^{-1}
mineralizationRate	0.1	Day^{-1}
Chl_N_ratio	1	mg chl (mmolN)$^{-1}$

2.9.1.4 Step 4. Testing the Correctness of the Equations

We first test the consistency of units. We restrict ourselves to one equation, the grazing of the zooplankton:

$$\text{Grazing} = maxGrazing \cdot \frac{\text{PHYTO}}{\text{PHYTO} + ks_{PHYTO}} \cdot \text{ZOO} \qquad (2.77)$$

With ZOO in mmolN m^{-3}, and the model time unit in days, the grazing is expressed in mmol N m^{-3} d^{-1}.

The dimensions relate as follows:

$$mmolN \cdot m^{-3} \cdot d^{-1} = d^{-1} \cdot \frac{mmolN \cdot m^{-3}}{mmolN \cdot m^{-3} + mmolN \cdot m^{-3}} \cdot mmolN \cdot m^{-3} \quad (2.78)$$

which is dimensionally consistent.

We then test for mass conservation. As there are no external sinks nor sources, total nitrogen concentration should be constant. In other words: the rate of change of total nitrogen should be 0. Total nitrogen is the sum of phytoplankton, zooplankton, DIN and detritus, and the rate of change of total nitrogen is just the sum of the rates of changes of these state variables:

dtotalN/dt = dPHYTO/dt + dZOO/dt + dDETRITUS/dt + dDIN/dt

dtotalN/dt = f1 − f2 + f2 − f3 − f4 − f5 + f3 + f5 − f6 + f4 + f6 − f1 = 0

2.9.1.5 Step 5. Solving the Model

We will learn how to solve such models in next chapters. With a light intensity that varies as a sine-wave with the season, the output in Fig. 2.21 is obtained:

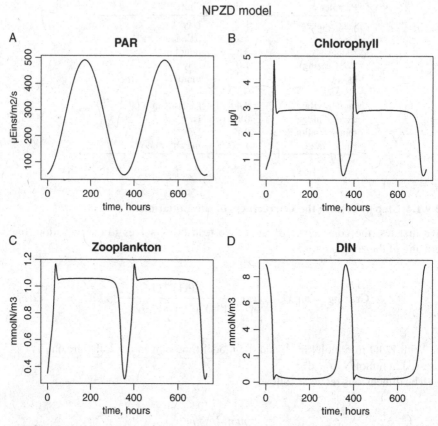

Fig. 2.21 Solution of the NPZD model, obtained after 3 years of simulation. Only the last two years are depicted. **A**. The photosynthetically active radiation (forcing function). **B**. Chlorophyll, **C**. Zooplankton and **D**. Dissolved Inorganic Nitrogen concentration

2.9.2 AQUAPHY, a Physiological Model of Unbalanced Algal Growth (**)

AQUAPHY (Lancelot et al., 1991) is a complex, physiological model that describes the growth of algae in response to nutrients and light.

Algae need carbon (CO_2), light energy and elements such as nitrogen and phosphorus for growth. Carbon is assimilated through the process of photosynthesis where CO_2 is incorporated into organic form (carbohydrates of low molecular weight), using light as an energy source. Nitrogen is taken up in inorganic form, usually as ammonium or nitrate, and built into complex molecules such as proteins, that form the biosynthetic and photosynthetic apparatus of the algae. Carbohydrates are also essential for synthesising these proteins, as they act both as a carbon source and as an energy source (through respiration). The surplus of low molecular weight carbohydrates that is not channelled into protein production is converted into storage molecules (such as starch), that serve as a source of carbon and energy during times when photosynthesis stops (night).

Thus carbon assimilation is limited by light intensity (CO_2 is almost never limiting in the water column), whereas nitrogen assimilation is limited both by the availability of nutrients and of carbohydrates.

The conditions for carbon and nitrogen assimilation may not be simultaneously fulfilled. For instance, during night, photosynthesis will stop (due to lack of light energy), but nitrogen assimilation may continue as long as both nutrients and carbohydrates are available.

Due to this uncoupling of carbon and nitrogen assimilation, algae have, to a certain degree, a variable stoichiometry: their carbon to nitrogen (C:N) ratio is higher under high light or low nutrient conditions (high light will favour photosynthesis, low nutrients will reduce nitrogen assimilation). However, algae also try to keep their C:N ratio within certain physiologically tolerable limits. Thus, if there is excess carbon, the acquisition of new carbon through photosynthesis will be downgraded, whilst under carbon shortage, photosynthesis will be performed at maximal rates, as set by the prevailing light conditions.

Models that describe this uncoupling between carbon and nitrogen assimilation are called *unbalanced growth* models. They differ from the more often used *balanced growth* models, where nitrogen and carbon assimilation occurs simultaneously and where algae have fixed stoichiometry.

In AQUAPHY, algal biomass is described via 3 different state variables: low molecular weight carbohydrates (**LMW**) that are the product of photosynthesis, storage molecules (**RESERVE**) and the biosynthetic and photosynthetic apparatus (**PROTEINS**). All state variables are expressed in mmol C m^{-3}.

Only proteins contain nitrogen and chlorophyll, at a fixed amount (i.e. using a fixed stoichiometric ratio). As the relative amount of proteins changes in the algae, so does the N:C and the Chl:C ratio.

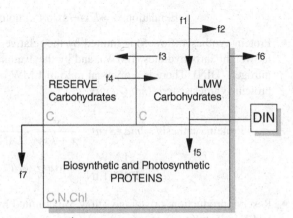

Fig. 2.22 Flowchart of the AQUAPHY model

An additional state variable, dissolved inorganic nitrogen (**DIN**) has units of mmol N m^{-3}.

The model formulation proceeds, in three steps.

Step 1. The *flowchart* is depicted in Fig. 2.22. Fluxes are:

Where: f1=Photosynthesis, f2=Exudation, f3=Storage, f4=Catabolism, f5=Protein Synthesis, f6=Respiration, f7=Loss due to mortality

Step 2. The *conceptual model equations* are:

$$\frac{d\text{LMW}}{dt} = \text{Photosynthesis+Catabolism}$$

$$-\text{Exudation-Storage-Respiration-ProteinSynthesis-LMWMortality}$$

$$\frac{d\text{RESERVE}}{dt} = \text{Storage-Catabolism-ReserveMortality}$$

$$\frac{d\text{PROTEIN}}{dt} = \text{ProteinSynthesis-ProteinMortality}$$

$$\frac{d\text{DIN}}{dt} = -\text{ProteinSynthesis} \cdot NCratio\_Protein \qquad (2.79)$$

Step 3. To write the actual *mathematical equations*, we specify each of the constitutive processes, as a function of the state variables, and the forcing function (light). Proteins are performing most of the work, thus many rates are first-order with respect to the protein concentration.

The following functional dependencies apply:

- Photosynthesis (flow 1) is described using the Platt equation; there is no light inhibition.

$$\text{Photosynthesis} = max PhotoSynt \cdot (1 - \exp^{\frac{\alpha \cdot IPAR}{maxPhotoSynt}}) \cdot \text{PROTEIN} \qquad (2.80)$$

- Exudation (flow 2) is a fixed fraction of photosynthesis

$$\text{Exudation} = pExudation \cdot \text{Photosynthesis} \qquad (2.81)$$

- Protein synthesis (flow 5) is limited by the relative availability of low molecular weight carbohydrates (LMW), and by the availability of dissolved inorganic nitrogen (DIN). There is a minimal ratio of LMW over PROTEIN at which the protein synthesis stops (*minQuota*).

$$\text{ProteinSynthesis} = maxSynt \cdot \frac{\gamma}{\gamma + ks\gamma} \cdot \frac{\text{DIN}}{\text{DIN} + ks_{DIN}} \cdot \text{PROTEIN}$$

$$\gamma = \frac{\text{LMW}}{\text{PROTEIN}} - minQuota \qquad (2.82)$$

- Reserve production, or storage, (flow 3) is limited by the relative availability of LMW molecules (γ), where (γ) is defined as above:

$$\text{Storage} = maxStorage \cdot \frac{\gamma}{\gamma + ks_\gamma} \cdot \text{PROTEIN} \qquad (2.83)$$

- Respiration (flow 6) includes basal respiration (1st term) and growth respiration (2nd term), the latter a fraction of protein synthesis.

$$\text{Respiration} = respRate \cdot \text{LMW} + p \cdot \text{ProteinSynthesis} \qquad (2.84)$$

- Catabolism (flow 4) follows first-order kinetics with respect to reserve molecules

$$\text{Catabolism} = catabolismRate \cdot \text{RESERVE} \qquad (2.85)$$

- Finally, mortality impacts all algal state variables similarly:

$$\text{LMWMortality} = mortRate \cdot \text{LMW}$$
$$\text{ReserveMortality} = mortRate \cdot \text{RESERVE}$$
$$\text{ProteinMortality} = mortRate \cdot \text{PROTEIN} \qquad (2.86)$$

Step 4. We will check the *consistency of units* of the photosynthesis equation only (and leave it as an exercise to the reader for the other equations):

$$\text{Photosynthesis} = maxPhotoSynt \cdot (1 - e^{\frac{\alpha \cdot \text{IPAR}}{maxPhotoSynt}}) \cdot \text{PROTEIN}$$
$$mmolC \cdot m^{-3} \cdot d^{-1} = d^{-1} \cdot (-) \cdot mmolC \cdot m^{-3} \qquad (2.87)$$

Moreover, as the exponent $\alpha \cdot \text{IPAR}/maxPhotoSynt$ should yield a dimensionless quantity, and with Ipar in μ Einst m^{-2} s^{-1}, $maxPhotoSynt$ in d^{-1}, it follows that the units of α should be $d^{-1}(\mu\text{Einst } m^{-2} s^{-1})^{-1}$

We also check the *mass balance*.

There are a number of external carbon flows (f1, f2, f6 and f7) such that the total load will not be constant:

$$\frac{d\,\text{TotalC}}{dt} = \frac{d\,\text{LMW}}{dt} + \frac{d\,\text{PROTEIN}}{dt} + \frac{d\text{RESERVE}}{dt} = f1 - f2 - f6 - f7 \quad (2.88)$$

We also check consistency for nitrogen. Within the algal cells, only proteins contain nitrogen:

$$\frac{d\,\text{TotalN}}{dt} = \frac{d\,\text{PROTEIN}}{dt} \cdot NCratio\_protein + \frac{d\,\text{DIN}}{dt} = -\text{ProteinMortality} \qquad (2.89)$$

Thus, the rate of change of total nitrogen = net export of total nitrogen, and the rate of change of total carbon = net carbon production, and mass is conserved.

Step 5. To *solve* the model, rather complicated techniques are necessary, and the model is implemented as a computer programme (see later chapters).

AQUAPHY

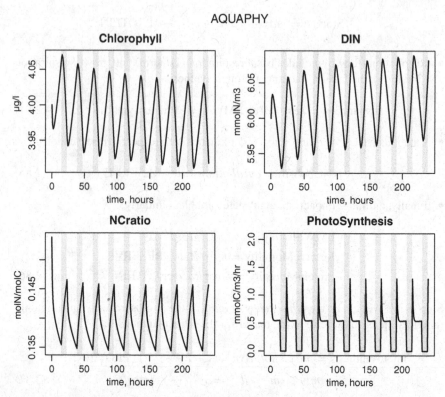

Fig. 2.23 Solution of the AQUAPHY model. The *light* and *dark* periods are denoted by *blank* and *shadowed* areas respectively. Note the daily variation in all depicted variables. See text for more explanation

In Fig. 2.23, the model has been run for 10 days under a 15 h light, 9 h darkness illumination regime; there is constant inflow of DIN and outflow of culture water (i.e. it is a dilution culture, see Chapter 3). The model output demonstrates the diurnal and long-term change of the algal chlorophyll concentration and DIN concentration. Also, the effect of the light-dark cycle on the algal stoichiometric N:C ratio is clear: the Nitrogen to Carbon ratio of the algae decreases during the day when carbon is assimilated, and increases during the night when the algae have stopped photosynthesising but continue to assimilate nitrogen. Finally, the photosynthesis process is triggered as soon as the light is switched on. Soon after that, the accumulation of low molecular weight carbohydrates inhibits photosynthesis and the rate declines. Photosynthesis stops during night.

This model has been used in a number of environmental model applications, e.g. to model the phytoplankton growth in the Weddell sea, Mediterranean, and North Sea.

Fig. 2.24 Rate of change as a function of density for the carrying-capacity formulation. R-code, see text

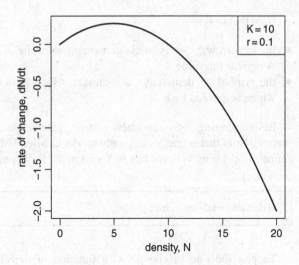

2.10 Case Studies in R

Even during the initial phase of the modelling process, before we even start to solve the models, we can make good use of the R software: R is extremely well suited for making graphs, and thus visualising the mathematical formulations.

2.10.1 Making Sense Out of Mathematical Formulations

The easiest way of understanding mathematical formulations is to make graphs. We will first do so for the carrying-capacity formulation.

$$\frac{dN}{dt} = r \cdot N \cdot \left[1 - \frac{N}{K} \right] \tag{2.90}$$

We make a graph, depicting the rate of change $\frac{dN}{dt}$ as a function of density (N, ind m^{-2}), and using the following parameter values: $r = 0.1 \, \mathrm{d}^{-1}$, $K = 10 \, \mathrm{ind} \, \mathrm{m}^{-2}$.

R-code

We first assign a value to the parameters r and K;

```
r  <- 0.1   # intrinsic rate of increase /day
K  <- 10    # carrying capacity
```

Note that in R

- the left arrow `'<-'` is an assignment operator; i.e. if we write `'A <- 3'` then A obtains the value 3.
- the symbol '#' demarcates a comment, which lasts till the end of the line and which is ignored by R.

Before plotting, we open a new window, (the height and width are expressed in inches). Note that `windows()` only works in Microsoft windows ®. Instructions for other operating systems can be found in the R documentation.

```
windows(width=5,height=5)
```

To plot the rate of change as a function of density (x), we use R-function `'curve'` which is particularly handy to plot an expression (`expr`), i.e. a function of x. Parameters `'from'` and `'to'` set the x-axis limits, `lwd=2` sets the line width to double the default, `xlab` and `ylab` specify labels of the x- and y axes. A `legend` adds the parameter values to the graph. The result is depicted in Fig. 2.24:

```
curve (expr= r*x*(1-x/K),from=0,to=20,lwd=2,
        xlab="density, N",ylab="rate of change, dN/dt")
legend ("topright",c("K=10","r=0.1"))
```

2.10.2 One Formula, Several Parameter Values

To really demystify a mathematical formulation, it should be plotted with different parameter values.

Here is how to do this in R. We plot a type-II (Monod or Michaelis-Menten) rate limiting term for describing grazing of a consumer on its food. We plot the function for several values of the half-saturation constant ks.

We start by defining the values of *food* and the parameter values for which we want to depict the relationship. R's function `seq(0,30,by=0.1)` creates a sequence, from 0 to 30 and with interval 0.1. Then we calculate the Monod function for every combination of the food and ks values. R's function `outer` does that; the results are stored in a matrix called `foodmat`. Next, all columns of `foodmat` are plotted using `matplot`. Note that the title (`main`) has been written as an expression. Also, we overrule the default colors (`col`). By default, `matplot` uses a different color for each column plotted; here we use black for all lines. We also set the line width to double the default (`lwd`). Finally a `legend` clarifies the plot.

```
food <- seq(0,30,by=0.1)
ks   <- seq (1,10, by=2)

foodmat  <- outer(food,ks,function(x,y) x /(y +x))

matplot(x=food,foodmat,type="l",col="black",
        xlab="food",ylab="-",lwd=2,
        main= expression (frac(food ,food+ks)))

legend ("bottomright", as.character(ks), title="ks=",
        lty=1:5,lwd=2)
```

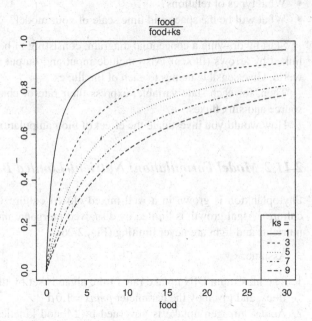

Fig. 2.25 Output generated by the R-code from example 2.10.2

2.11 Projects

2.11.1 Conceptual Model: Lake Eutrophication

Phosphorus is the nutrient that is generally limiting primary production in lakes. Increasing the input of phosphorus increases the concentration of phytoplankton, which may have a radical effect on water quality. Cladoceran grazers (zooplankton) are the main consumers of lake algae and may reduce algal biomass.

To overcome the negative impacts of eutrophication, the concept of biomanipulation was introduced in the seventies, which consisted of reducing the predation pressure on these large cladocerans. When successful, this treatment resulted in a reduced phytoplankton biomass and a higher zooplankton biomass dominated by large organisms. However, several cases were reported where this manipulation failed to give the desired results. Close examination revealed that failure was most likely in lakes that received a phosphorus input above a certain critical level.

· *Tasks:*

Make a conceptual model that serves to investigate the effect of biomanipulation on a lake ecosystem.

- Review the components that should be included in the model.
- What state variables will be in your model?
- What forcing functions?
- What types of relations?
- What will be the space and time scale of your model?

Start by drawing a conceptual diagram, consisting of boxes (the state variables), linked by arrows (fluxes). Also include input and output from and to the external world. Give sensible names to each of the fluxes.

Then, for each state variable, express their rate of change as a function of the source and sink fluxes.

How would you investigate the effect of biomanipulation?

2.11.2 Model Formulation: Nutrient-Limited Batch Culture

Phytoplankton is grown in a well-mixed closed culture vessel (a so-called batch culture). Algal growth is limited by dissolved inorganic nitrogen (DIN) only, other nutrients and light are never limiting (Fig. 2.26).

Assume:

1. The maximum DIN uptake rate of the algae is set by the prevailing light conditions, and given by the parameter $pmax = 1.0 \, d^{-1}$.
2. Actual nitrogen uptake is governed by Monod kinetics with parameter $ks = 1.0 \, \mu mol \, N.dm^{-3}$.

Task:

1. Write the mathematical formula that describes uptake of DIN by the algae. Give the units of the parameters and state variables
2. Make a model that describes the interaction of algae and DIN.
   Start with a conceptual model.

   - What are the state variables in this model?
   - Which are the flows? Draw the flow chart.
   - Write the rate of change of the state variables as the sum of influxes – effluxes.
   - Write the mathematical equations for the fluxes.
   - Check the mass balance of the model. Is mass conserved by the equations?

Fig. 2.26 Components of the batch culture model

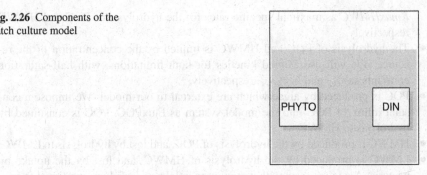

2.11.3 Model Formulation: Detritus Degradation

Fig. 2.27 State variables in the detritus degradation model

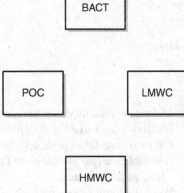

Detritus is degraded by the action of heterotrophic bacteria. This is not a one-step process: bacteria cannot 'eat' detritus!

A model that is closer to the reality of the process considers that particulate detritus (POC) is first degraded by the action of bacterial exoenzymes to high-molecular-weight dissolved organic carbon (HMWC). This in turn is attacked by enzymes to yield low-molecular-weight dissolved organic carbon (LMWC), which can then be taken up by the bacteria (BACT) which grow on it.

Assumptions:

- We will not model the exoenzyme concentration explicitly in the model. Instead, we assume that the maximal rate of hydrolysis (degradation) of POC and of HMWC is proportional to bacterial biomass [note: it are the bacteria that perform the work, they set the maximal rate]. We will use the parameters *KmaxPOC* and

KmaxHMWC as maximal specific rates for the hydrolysis of POC and HMWC respectively.

- The hydrolysis of POC and HMWC is limited by the concentration of the resource. We will use Monod kinetics for both limitations, with half-saturation constants ks_{POC} and ks_{HMWC} respectively.
- POC is produced by algae which are external to our model. We impose a constant influx of POC into the model system as FluxPOC. POC is consumed by hydrolysis to HMWC.
- HMWC is produced by the hydrolysis of POC, and lost by hydrolysis to LMWC.
- LMWC is produced by the hydrolysis of HMWC, and lost by the uptake by bacteria. Again we assume that maximum uptake is directly proportional to bacterial biomass, with rate parameter *upMax*, and limited by substrate availability: Monod kinetics with parameter ks_{UP}.
- Bacteria grow by uptake of LMWC, but loose carbon by basal respiration (*rBas*) and by activity respiration: they respire a fraction *pLoss* of the uptake. Moreover, bacteria are subject to predation, and this is modelled as a quadratic closure term, with parameter *rClos*.

Tasks :

Make a coupled model of this process.

Use four state variables as depicted in Fig. 2.27.

For each state variable sketch the influxes and effluxes in a flow chart.

Assemble the rate of change of the state variables as the sums of these positive and negative fluxes.

Write the formulations for each of these fluxes. Remember the two basic rules:

1.   The compartment that is doing the work sets the maximal flux: maxflux =maxrate*WORKER
2.   The effect of the resource that is limiting is expressed as a functional response. Use Monod kinetics.

Check the dimensionality of your model.

What are the units of *rClos*, *rBas* and *pLoss*?

Fig. 2.28 Schematic
representation of an
autocatalytic reaction

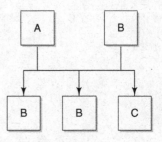

2.11.4 Model Formulation: An Autocatalytic Reaction

An autocatalytic reaction is one where the reaction product is itself the catalyst for
the reaction. For instance:

A+B → 2B +C

Write the rate of change of the concentrations [A], [B] and [C] (Fig. 2.28).

Chapter 3
Spatial Components and Transport

Much of ecological theory focuses on the interactions between organisms in a homogeneously mixed, non-spatial environment. This makes the mathematics and the set-up of experiments easier, but its scene is a world without landscapes (or seascapes).

The outcome of interactions among individuals, populations, or chemical substances, is often fundamentally different if one takes into account how they are distributed over space, compared to a model in which this spatial distribution is not represented. For instance, models that describe competition or predation where individuals only interact when they are close enough in space, may predict stable co-existence whereas non-spatial models would lead to rapid exclusion of one species. In a partially-mixed water column where light comes from the surface and nutrients have to be mixed into the illuminated (euphotic) layer from below, strong gradients in algal concentration and primary production will develop, and the total rates of production and consumption of substances will strongly differ from those modelled in a well-mixed water column.

When developing spatially-explicit models, one has to make a number of fundamental decisions on how to represent objects within this spatial context. Space, as we observe it daily, is a three-dimensional world. In this world, a co-ordinate system can be defined and the position of an object is characterized by (x, y, z) coordinates. Space is also continuous: an object can occupy an infinite number of positions. Finally, objects can, in principle, move in space. All these characteristics, when represented in a model, may lead to complex and computationally cumbersome formulations. Thus modellers are keen to simplify the representation of space in a model in different ways.

When a three-dimensional model domain has certain spatial symmetries, it can be represented in a model with fewer than three dimensions. Consider a river as an example: although concentrations of nutrients in the river will differ along the length, the depth and the width of the river, often the concentration differences along the length axis will be much more important than the concentration gradients with depth or width. We may then choose to represent the river as a 1-dimensional domain, assuming that concentrations are *only* a function of length (i.e. do not vary along the other two dimensions, which therefore need not be represented in the model). In this chapter, we will mainly deal with 1-dimensional models, as they are

K. Soetaert, P.M.J. Herman, *A Practical Guide to Ecological Modelling*,
© Springer Science+Business Media B.V. 2009

much simpler to explain than full 3-dimensional ones. However, basic reasoning is the same (but not mathematical clutter in the formulas!).

Some models also treat space as discontinuous, i.e. objects are only allowed to occupy certain positions, which can be identified with integer numbers. We will discuss some examples of such models, although most of our effort will be devoted to models that represent space in a continuous way.

An important aspect of spatial models is that object can move, or be transported, from one location to another. The formulation of transport processes is an essential characteristic of spatial models.

This chapter will first consider some fundamental options in representing space in models, and then focus on how to formulate transport processes and link them to the rest of the interactions in a model.

3.1 Microscopic and Macroscopic Models

When dealing with space, it often becomes necessary to decide whether to focus on individuals or particles (microscopic or individual-based models) or to describe average variables of the population such as concentration, biomass or density (macroscopic models).

In microscopic models, the individual interaction between entities (e.g. organisms or molecules) or between one entity and the environment is of interest. These models incorporate descriptions of the behaviour of individual organisms to an external stimulus. In order to describe meaningfully what happens to the population as a whole, large numbers of individual entities are modelled, and usually summary statistics are produced at the end (e.g. concentration fields, total numbers etc...).

Macroscopic models describe those 'average' variables such as concentration, density or biomass directly. The model formulations specify how these quantities change by processes, which are formulated as a function of the concentrations or biomasses only, i.e. not resulting from the interactions between individual entities. These models have arisen by the necessity to make models more analytically tractable and simple.

In order to grasp the difference between the two approaches, consider the example of diffusion of gas molecules (Fig. 3.1). Imagine that all gas molecules have initially been introduced in a thin rod-shaped volume in the middle of a cylindrical container. Then the molecules start to move in random directions, a process called Brownian motion. Each molecule will make a random walk, which we observe as a leap in a certain direction for a while, followed by a turn and moving in a different direction. In a microscopic model this leaping behaviour is described for a large number of molecules. Per molecule, and per time step, it is recorded how its position in the container changes. After a sufficiently long time, the density field of the molecules in the container is estimated by (statistically) averaging over space. As long as the molecules have not reached the cylinder wall, this density field is expected to look like a normal distribution (see Fig. 3.1). A macroscopic model, however, will express how the density field itself changes. It will not take

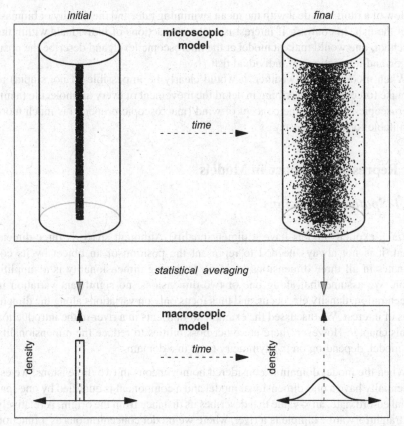

Fig. 3.1 Microscopic models (*above*) predict future distributions of particles by describing interactions or movement, while macroscopic models (*below*) consider average quantities such as density, and predict future density from actual density. When properly done, the macroscopic predictions can be retrieved by statistical averaging of the microscopic model predictions

into account the characteristics and movement of individual molecules. Rather, it describes how the density field will change in time in a way that depends *only* on the density field itself. We will give an example of both a micro- and macroscopic model of diffusion at the end of this chapter. This example will also make clear that, although the macroscopic and the microscopic models take a fundamentally different approach, they are also related: microscopic models, when averaged over sufficiently long temporal (and spatial) scales will produce the same output as the macroscopic model.

Here are some other examples of micro- and macroscale models.

- The flow of blood in our veins could be described either as the movement of the individual blood cells (microscopic) or by considering the cell density (macroscopic).
- The movement of a school of fish can also be modelled both ways. At the macroscopic level, one might treat the behaviour of a school as one would model the

flow of a fluid, and deal with the mean swimming rate, and the density or biomass of the fish. In contrast, if interest is in the interactions of individuals within the school, one would make a model at the microscopic level, and describe the position and interactions of individual fish.

- When modelling the weather, it would clearly be impossible (if not stupid) to make forecasts by describing in detail the movement of every air molecule (a microscopic property); the concept of wind (macroscopic property) is much more suitable for that purpose.

3.2 Representing Space in Models

3.2.1 Spatial Dimensions

Spatially-explicit models have a dimensionality. Although space is three-dimensional, it is not always needed to represent the position of an object by its coordinates in all three dimensions (x, y, z). Reducing dimensionality is a simplification. We assume that along one or two dimensions, no significant variation in concentration, density etc. occur, and thus focus only on variations along the dimensions of interest. We discussed the example of nutrients in a river in the introduction of this chapter. However, there are other possibilities to reduce the dimensionality of a model, depending on the symmetry rules in a domain.

- When the model domain is considered homogeneous in two dimensions, we essentially have a one-dimensional model and a component is qualified by one spatial co-ordinate, an x-value that describes its distance from the origin. A relatively straightforward example is a river, where we model concentrations as a function of position along the length axis only. As you will see later (Section 3.4.5), one can be very creative in the assumptions of homogeneity and use 1-dimensional models to represent processes in a variety of domains. For instance, if a domain is spherical, and concentrations vary only (or mainly) as a function of distance from the centre, it suffices to represent a point's position with a single number (distance from the centre) to predict the concentration at that point.
- In a two-dimensional model, the playground is a plane. Positions are characterized by co-ordinates in two directions, i.e. two horizontal coordinates, or length and depth.
- In three-dimensional models, three coordinates position objects in space.
- When no spatial dimension is described, the model is zero-dimensional.

3.2.2 Discrete Spatial Models

In discrete spatial models, space is considered as discontinuous, i.e. objects can only occupy some positions in space. These positions can be characterized by integer numbers $(0, 1, \dots n)$ in one dimension, or pairs of integer numbers in two dimensions.

Landscape models describe the evolution of populations or individuals in land-scapes (Fig. 3.2A). They subdivide the landscape into a number of distinct equally sized *cells*. The properties of these cells are typically determined by a Geographic Information System (GIS).These models are applied amongst other things to model the spatial dynamics of large organisms. They require a substantial database to char-acterise all cells.

Patch models describe dynamics in a number of *patches* (Fig. 3.2.B). Patch mod-els can be tightly linked to reality, in which case they are similar to landscape models and have similar data requirements, but patches need not be regular. However, there

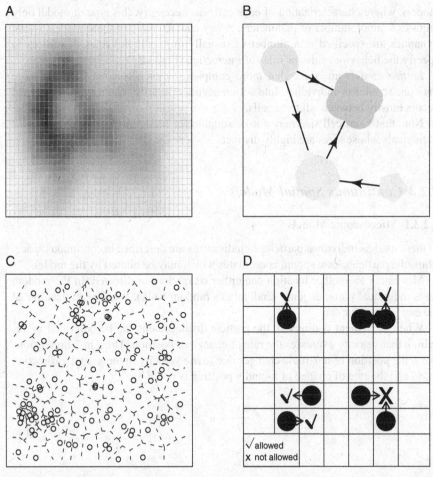

Fig. 3.2 Examples of spatial configurations in models. **A**. In this landscape model, space is divided in discrete cells that have distinct properties. **B**. Patch model with 3 discrete patches. **C**. In some models, so-called Delaunay triangulation is used to discretise space, for instance to model the terri-tory of birds. **D**. transition rules in a cellular automaton model. (●) occupied cells (v) = transition allowed; (x) = not allowed

are also highly idealised patch models, where all patches have equal characteristics and are internally well mixed. The spatial relationships between patches are described e.g. by their distance, the probability of an organism to move from one patch to another, etc... Many meta-population models fall into this class. Usually the exchanges between patches are crucial for the model behaviour.

Some patch models do not focus on the internal dynamics within the patches, but rather on the form, size and position of the patches themselves. This is the case for some models of bird territories (e.g. Fig. 3.2.C).

In *cellular automaton models*, the environment is divided in a large (huge) number of equally sized *cells* that may either be occupied (1) or not (0). Interaction is more likely between neighbouring cells (Fig. 3.2.D). In contrast to landscape models, where characterization of each cell was necessary, this type of model only requires a small number of parameters. A key characteristic of these models is that dynamics are specified in a number of (usually quite simple) rules. It suffices to specify the behaviour rules or rules of interaction (Fig. 3.2.D).

Lattice models are somewhat more complex, as each cell may contain more than one species or individual. Interaction occurs primarily within cells, movement occurs mostly between adjacent cells.

Note that fixed cell size may not be suitable for modelling spatial dynamics of individuals whose sizes are highly distinct.

3.2.3 Continuous Spatial Models

3.2.3.1 Microscopic Models

In these models individual particles or individuals are described in continuous space. Thus, the particles have spatial coordinates which may be altered by the model.

Movement to another location can either be random or directed. In the random case, individual particles jump randomly (a random walk). We give an example at the end of this chapter.

When movement is directed, the particle displacement can be described by recalling that velocity expresses the rate of change of position. Thus, in 2-dimensions (x,y), the position of a particle changes due to the velocity in the x- and y-direction (u, v) and the rate of change of x- and y position is given by:

$$\frac{dx}{dt} = u$$
$$\frac{dy}{dt} = v$$

(3.1)

A special type of continuous space models are *Neighbourhood* and *zone-of-influence* models which describe the explicit position of individuals that do not move. A circular zone around the individual determines the area where it interacts with others and where it affects the environment. The larger the degree of overlap of

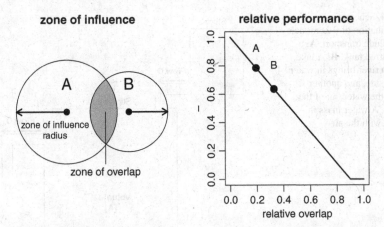

Fig. 3.3 The zone of influence for two individuals, A and B, and the overlap area (*left*). The relative degree of overlap is larger for B, which is the smaller individual and therefore the performance of individual B will be lower (*right*)

these zones, the more the relative performance of the individuals is affected. These models typically consider growth, and recruitment. This is for instance applied in forest modelling (see Fig. 3.3).

3.2.3.2   Macroscopic Models

Many continuous-space models are *macroscopic* models.

The change of macroscopic variables as a function of space is generally represented by the *advection-dispersion equation*. This equation can be derived mathematically, by appropriate averaging over space and time of the microscopic models, but the derivation is complex. Therefore, in the next chapter, we present a more qualitative explanation.

A special application of the advection-dispersion equations is in models for *taxis*, i.e. the more or less directed movement of organisms in response to certain stimuli. Several formulations are available in the literature, describing movement driven both by random factors (which make the movement diffusive-like) and by stimuli, which bias the pure random walk underlying diffusion and are described macroscopically by parameters modifying the diffusive process.

3.3   Transport in a Zero-Dimensional Model

A classic example that cannot be omitted in any modelling book is the continuously-stirred or well-mixed tank, to which fluid is continuously added on one side and continuously removed on the other side (Fig. 3.4). With the fluid comes a conservative substance A. (note: a substance is called conservative if it does not react).

Fig. 3.4 Schematic representation of 0-D models that include transport. **A**. a well-stirred tank. **B**. A lake, where a river brings in water on one side, and another carries the water out of the lake. **C**. A water mass in contact with the air

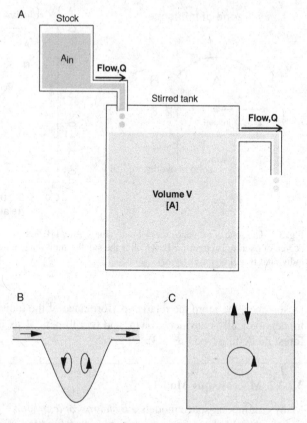

The fluid inside the tank is well-stirred and hence the concentration of A within the tank (which we denote by [A], units of mol m^{-3}) is spatially uniform. In this model, it is not necessary to describe the concentration explicitly as a function of space. There is just one number for the entire tank (the concentration [A]), thus the model is zero-dimensional. In a zero-dimensional model, transport only concerns exchange with the external world, while the model domain itself is implicitly assumed to be homogeneously and instantaneously mixed.

The example is classic because it is a faithful representation of certain laboratory set-ups, in particular the chemostat (Fig. 3.4A). However, it can also be considered to be a basic representation of a lake, where a river brings in water at one side, and another river exports the overflow at another point (Fig. 3.4B). As long as the lake is well-mixed, the volume remains constant and it does not develop too strong spatial concentration gradients, its dynamics can be captured in a zero-dimensional model.

Continuing the stirred-tank example, we assume that the tank has a constant volume V (m^3), and a constant inlet flow rate Q (m^3 s^{-1}). As the volume is constant, the flow rate at the outlet must equal the flow rate at the inlet.

Assume that the concentration of substance A in the inflowing water is A$_{in}$ (mol m^{-3}). The total mass flow (units of mol s^{-1}) into the container then equals $Q \cdot A_{in}$, whilst the total mass flowing out of the container is given by $Q \cdot [A]$.

As substance A does not react, it is simple to write the balance equation of the *total mass* of A in the tank as:

$$\frac{d\,M}{dt} = Flow_{in} - Flow_{out} = Q \cdot A_{in} - Q \cdot [A] \tag{3.2}$$

or:

$$\frac{dV \cdot [A]}{dt} = Q \cdot \left(A_{in} - [A]\right) \tag{3.3}$$

where the concentration ([A]) times the volume (V) is total mass (M).

In general, this equation is rewritten in terms of the *concentration* of A, ([A]) rather than total mass:

$$\frac{d[A]}{dt} = \frac{Q}{V} \cdot \left(A_{in} - [A]\right) = d_r \cdot \left(A_{in} - [A]\right) = \frac{\left(A_{in} - [A]\right)}{t_r} \tag{3.4}$$

where d_r (time^{-1}) is called the *dilution rate*; it is the flow rate divided by the volume.

Note that the volume V of the tank must remain constant (or $dV/dt = 0$) for this replacement to be valid. This is an important condition for the use of this type of chemostat model.

The inverse of the dilution rate is the *residence time* ($t_r = V/Q$), the time it takes for a volume V of fluid to flow into (or out of) the tank.

3.4 Transport in a One-Dimensional Model

We now use the continuum mechanics approach *sensu* Newton to derive several transport equations in one dimension. We do this in four steps.

- First we introduce the flux-divergence equation, which is the basic equation we will need for the further development of our transport equations. The concept may not be so easy to grasp, but it is absolutely fundamental to the rest of the section.
- Then we find macroscopic expressions for the fluxes, where we make a distinction between fluxes linked to random movement (dispersion) and to directed movement (advection).
- Combining the flux-divergence equation with the diffusive and advective formulations and adding a reaction term we obtain the 'general one-dimensional advection-dispersion-reaction equation'.
- Finally we apply this formulation to a diversity of different model domain types, making use of symmetry rules that allow us to represent the domains in one-dimensional formulations.

3.4.1 Flux Divergence

Suppose we have a box with a certain volume (ΔV). In the box, a substance has concentration C (units of Mass Volume^{-1}). This substance is conservative, i.e. its concentration changes only due to transport. We want to derive how the concentration of the substance will change over time due to fluxes of material into, and out of the box. If we call these fluxes I and O, respectively, the mass balance of the substance will be given by:

$$\frac{\partial \, \text{Mass}}{\partial t} = I - O \tag{3.5}$$

As before, this mass balance equation states that what enters a box (flux in, I) must either leave the box (flux out, O) or stay in the box ($\partial \, \text{Mass}/\partial t$, the rate of change, or accumulation).

To simplify the problem, we consider one-directional transport, in the x-direction only. Figure 3.5 gives a schematic image of the box we are considering. Note that it is characterized by a length (Δx), a starting position (x), a surface area A_x at position x and a (potentially different) surface area $A_{x+\Delta x}$ at position $x + \Delta x$.

The substance enters the box at point x, with a flux equal to J_x.

Similarly, a flux equal to $J_{x+\Delta x}$ leaves the box at position $x + \Delta x$.

A *flux* expresses the amount of material (*Mass*) that passes a unit surface per unit of time. It thus has dimension Mass Surface^{-1} Time^{-1}. Note that fluxes have a sign: they are positive when going in the direction of the x-axis, negative in the opposite direction.

The *total mass* that flows into the box at position x per unit time can now be estimated as the flux times the surface over which the flux occurs, $J_x \cdot A_x$ and a similar formula applies for the total mass outflow, $J_{x+\Delta x} \cdot A_{x+\Delta x}$.

The total mass influx (Mass Time^{-1}) minus the total mass outflux must equal the rate of change of total mass in the control volume.

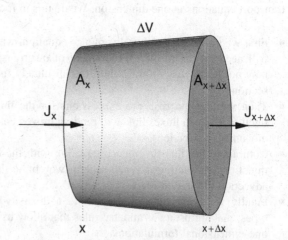

Fig. 3.5 Deriving one-directional transport in a small box. x, x+Δx: position along the X-axis, A: surface, ΔV: volume of the box, J: flux. See text for details

$$\frac{\partial \text{Mass}}{\partial t} = \text{Mass influx} - \text{Mass outflux}$$

$$\frac{\partial \text{Mass}}{\partial t} = J_x \cdot A_x - J_{x+\Delta x} \cdot A_{x+\Delta x} \tag{3.6}$$

In general, models do not describe the temporal evolution of total mass, but rather of concentration (Mass Volume^{-1}), so the volume of the box (ΔV) should be taken into account. We assume that the control volume is sufficiently small such that any material added is immediately homogenized. We further assume that the volume of the box remains constant in time. The concentration in the box changes as the change in the total mass of the constituent in the box, divided by the box volume ΔV.

$$\frac{\partial C}{\partial t} = \frac{\text{Mass influx} - \text{Mass outflux}}{\text{volume}}$$

$$\frac{\partial C}{\partial t} = \frac{J_x \cdot A_x - J_{x+\Delta x} \cdot A_{x+\Delta x}}{\Delta V} \tag{3.7}$$

As volume equals surface * length, we can also write:

$$\Delta V = \frac{A_x + A_{x+\Delta x}}{2} \cdot \Delta x = A_{x+\Delta x/2} \cdot \Delta x \tag{3.8}$$

where we used the surface halfway between x and x+Δx as an approximation of the average of the surfaces at both ends (i.e. using linear interpolation), and we obtain:

$$\frac{\partial C}{\partial t} = \frac{J_x \cdot A_x - J_{x+\Delta x} \cdot A_{x+\Delta x}}{A_{x+\Delta x/2} \cdot \Delta x} \tag{3.9}$$

Note that in the nominator, the fluxes are defined on the box interfaces, whereas in the denominator, we use surface area in the centre of the box.

This formulation finalizes the first step in our reasoning. For a (small, homogeneous) box, we have derived how the concentration changes over time as a function of influx and outflux.

However, as we are dealing with models that are continuous in space, we do not want to subdivide our model domain in a number of discrete, finite-sized boxes. Doing so would give us an estimate of the rate of change of concentration at a number of discrete locations, but not in between them. Thus, we shrink the length of the boxes, Δx, to infinitesimally small size. At any point x in our model domain, we can imagine such an infinitesimally narrow box. Technically, we can do that by taking the limit of the above equation for Δx tending to zero. We then obtain the partial differential equation that expresses the rate of change of concentration C as a function of fluxes J and surface area A at any point x in our model domain:

$$\frac{\partial C}{\partial t} = -\frac{1}{A} \cdot \frac{\partial(A \cdot J)}{\partial x} \tag{3.10}$$

Note the negative sign. If the flux increases with x (positive derivative), the outflux will be larger than the influx and the concentration will decrease, i.e. the rate of change will be negative.

The first term ($\partial C/\partial t$) is the *storage term* (the rate of change of C in time), the second term is known as the *flux divergence*, i.e. the change of the flux with space x.

You may wonder at this point why this equation is useful. It derives the rate of change of concentration at any point x, as a function of unknown fluxes J (also variable over x) and (known) surfaces A. But as long as we do not have a formulation for the fluxes, we cannot use this formula in practice! However, as we will see in the next sections, it is not too difficult to derive (macroscopic) formulations for the fluxes based on some assumptions about transport processes. Equipped with these formulations, the fundamental flux divergence equation will allow us to complete the one-dimensional transport equations.

3.4.2 Macroscopic Formulation of Fluxes: Advection and Dispersion

To estimate fluxes in *microscopic* transport models, it suffices to record the number of particles passing through the surface areas per time. However, in *macroscopic* models the fluxes are expressed as a function of the density, concentration or biomass. To make the transport equation complete, we need to find a macroscopic formulation for the fluxes (J).

Fluxes are either linked to directed movement (advection-like fluxes) or to random movement (diffusion-like or dispersive fluxes). The distinction is fundamental, but not always easy to make for a novice modeller.

3.4.2.1 Advection

Microscopically, advective fluxes are related to a directed movement that can, in every point in space, be characterized by a velocity. Particles move in space with this velocity, and the velocity is sufficient to describe their movement (Fig. 3.6A). Advective movement is an idealization of such processes as flow in a river, sinking of particles in the ocean, directed deliberate movement of animals etc..: any movement that can be characterized by its velocity. Note that this velocity should neither be constant in time, nor constant in space. However, it should apply uniformly to all (microscopic) individuals that can be found at point x at time t.

Macroscopically, the fluxes can be expressed as the product of the velocity (u, units of Length Time^{-1}) times the concentration.

Thus, the flux (J) due to advection can be written as:

$$J\big|_{advection} = u \cdot C \tag{3.11}$$

and the flux divergence due to advection is given by:

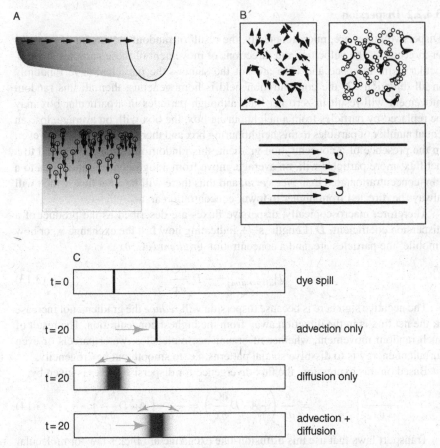

Fig. 3.6 A. Two types of advection: flow in a river or estuary(*above*) and sinking of particles out of a water column (*below*). **B.** Three types of dispersion: molecular diffusion induced by random motion of particles (*top left*), eddy diffusion caused by turbulent mixing of particles (*top right*) and mechanical dispersion, induced by variations in flow velocities. **C.** Effect of advection and diffusion on a dye spill in a river

$$\frac{\partial (A \cdot J)}{\partial x}\bigg|_{advection} = \frac{\partial (u \cdot A \cdot C)}{\partial x} \tag{3.12}$$

Consider for instance the flow of a river, where gravity directs water with its associated concentrations in the downstream direction. A peak in concentration at some position along the river, e.g. a cloud of added dye, will flow downstream with the water, and the water velocity will determine how fast it flows (Fig. 3.6 C). As another example, particles sinking through a water column will fall with a certain velocity (Fig. 3.6 A). Thus, particles in a particular box in the water column will be replaced by particles from above, while the box itself will lose particles to lower boxes.

3.4.2.2 Dispersion

Dispersive fluxes are, microscopically, the result of random movement of particles or molecules. The velocities and directions of movement of these particles at a particular point in space and time are not the same – the particles move randomly in all directions. If the concentration field is homogeneous, then all this random movement will result in zero net flux: although particles in a particular box may be replaced by particles from a neighbouring box, the box will, on average, lose an equal number of particles to this neighbouring box and there is no net flux. However, in the presence of a concentration gradient, this random movement *does* lead to a net flux: more particles will, on average, move from a high-concentration area to a low-concentration area than *vice versa*, and thus there will be a net flux, which will always be directed from higher to lower concentration areas.

Therefore, macroscopically, dispersive fluxes are described as the product of a dispersion coefficient, D (Length^2.s^{-1}), indicating how fast the exchange is, or how 'mobile' the particles are, and a concentration *gradient* ($\partial C/\partial x$).

$$J|_{dispersion} = -D\,\frac{\partial C}{\partial x} \tag{3.13}$$

The negative sign here is because dispersion will *reduce* the gradient, not increase it: the net flux is in the direction away from the highest concentration. The result of such random movement, whether of organisms, molecules, water parcels or even turbulent energy is to dissolve spatial patterns, i.e. to smooth out heterogeneity.

Based on this expression, the flux divergence for dispersive fluxes is given by:

$$\frac{\partial A \cdot J}{\partial x}\bigg|_{diffusion} = \frac{\partial}{\partial x}\left(-A \cdot D\frac{\partial C}{\partial x}\right) = -\frac{\partial}{\partial x}\left(A \cdot D\frac{\partial C}{\partial x}\right) \tag{3.14}$$

Transport laws that use this diffusion-like exchange are Fick's law for molecular diffusion, Newton's viscosity law for transport of momentum, Fourier's law for heat transport and Darcy's law for fluid flow in a porous medium. The corresponding fluxes are called mass flux, Reynold stress, heat flux, and water flux respectively.

The process where molecules or other constituents move randomly from a high concentration to a lower concentration is usually referred to as diffusion. The mixing which is due to flow processes, i.e. to changes in velocity, is generally called dispersion, whilst turbulent mixing is called eddy diffusion (Fig. 3.6 B). Their effect on concentrations is largely the same. We will use the general term 'dispersion' to refer to all transport processes of this type. Due to dispersion, the added dye along the river or estuary will be smoothened in both directions, but the centre of dye will stay at the same location (Fig. 3.6 C).

3.4.3 The General 1-D Advection-Dispersion-Reaction Equation

We can now combine the flux divergence formulation with the macroscopic formulation for advective and dispersive flux.

When we also add a reaction term, we obtain the general 1-dimensional *advection-dispersion-reaction* equation:

$$\frac{\partial C}{\partial t} = -\frac{1}{A} \cdot \frac{\partial (u \cdot A \cdot C)}{\partial x} + \frac{1}{A} \cdot \frac{\partial}{\partial x}\left(A \cdot D \frac{\partial C}{\partial x}\right) + reaction \qquad (3.15)$$

This equation forms the basis for 1-D reaction transport modelling in various disciplines (geosciences, atmospheric or ocean sciences, etc...). From this equation, we can derive several other formulations, assuming different shapes. In the next sections we will illustrate some of these adaptations.

3.4.4 The 1-D Advection-Dispersion-Reaction Equation in Estuaries, Rivers and Lakes

3.4.4.1 Estuaries, Rivers

For water bodies like estuaries or rivers, the 1-D advection-dispersion-reaction equation can be used to describe transport and reaction along the length axis, provided that vertical and lateral gradients are negligible (Fig. 3.7).

As the cross-sectional surface (A) along the length axis usually is very variable, it is, in general, not possible to simplify the equation. However, it is common to rewrite the equation using flow rates Q, rather than velocities:

$$\frac{\partial C}{\partial t} = -\frac{1}{A} \cdot \frac{\partial (Q \cdot C)}{\partial x} + \frac{1}{A} \cdot \frac{\partial}{\partial x}\left(A \cdot E \frac{\partial C}{\partial x}\right) + reaction \qquad (3.16)$$

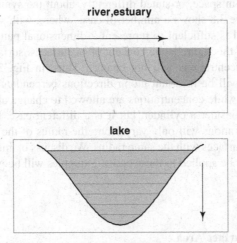

Fig. 3.7 The transport in rivers, estuaries and lakes can often be represented by the 1-D advection-diffusion equation. For rivers and estuaries, the 1-D axis is the length axis, while for lakes it is the depth axis

where Q is the volumetric flow rate (L^3T^{-1}), defined as the flow velocity (u) times the cross-sectional surface (A), E is the (horizontal) dispersion coefficient (L^2T^{-1}), and A is the cross-sectional surface area, x is the length axis of the estuary or river. This formulation will be used in Sections 6.6.5 and 7.8.5 when we implement the transport, growth and decay of zooplankton in an estuary.

3.4.4.2 Lakes

For lakes that can be considered horizontally homogeneous, it is more common to model processes as a function of water depth (denoted by z) (Fig. 3.7). Here A is the horizontal cross-sectional surface area at a certain water depth, K_z is the vertical dispersion or eddy diffusion coefficient, and u is the sinking rate (for particles) or the upwelling rate (for fluids).

$$\frac{\partial C}{\partial t} = -\frac{1}{A} \cdot \frac{\partial (u \cdot A \cdot C)}{\partial z} + \frac{1}{A} \cdot \frac{\partial}{\partial z}\left(A \cdot K_z \frac{\partial C}{\partial z}\right) + reaction \qquad (3.17)$$

3.4.5 The 1-D Advection-Dispersion-Reaction Equation in Shapes with Different Symmetries

As stated before, 1-dimensional models can be used to model processes in diverse (3-dimensional) domains, depending on our assumptions about how concentrations will change in space, or stated differently, about the symmetry applicable to the domain. Figure 3.8 shows a number of cases where, with some creativity, a 1-dimensional model is sufficient to represent 3-dimensional patterns. Crucial for the understanding of this figure is the concept of isosurfaces: surfaces along which we assume that concentrations will not change. Thus, in Fig. 3.8A we assume that concentrations will be invariant along directions perpendicular to the length axis of the cylinder, while concentrations are allowed to change along this axis. In Fig. 3.8B we also consider a cylinder, but it is a different case, because here we assume that concentrations will only vary along the radius of the cylinder, but remain constant over surfaces with the same radius. We discuss in this section how the general equation can be applied to these cases. Examples will be given later in this book.

3.4.5.1 Constant Surface Area

In a first simplified case, we model transport in a one-dimensional column, assuming that the surface area of this column remains constant. (Fig. 3.8A).

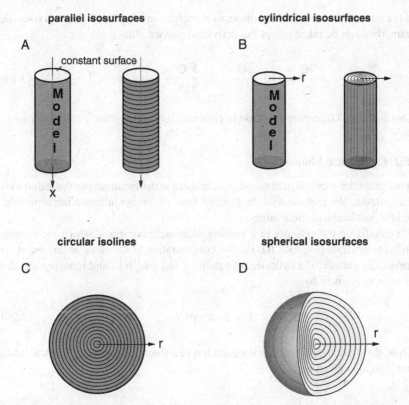

Fig. 3.8 Schematic representation of 'one-dimensional' spatial volumes as used in models. *Grey lines* denote isosurfaces. **A**. One-dimensional shape with constant surface area. **B**. Cylindrical shape, with non-zero cylinder length. **C**. cylindrical shape with zero length of the cylinder. **D**. Spherical shape

Examples of such models are

- Vertical water column models, where the components are typically described in a water column of 1 m² surface.
- Vertical sediment models where typically a sediment column of 1–100 cm² surface is considered. We give several examples of sediment models, e.g. in Section 3.6.5.
- Flow of food through the gut of an animal.
- Diffusion of substances in and out of a flat organism. We will apply the formulation in a model that describes the oxygen consumption in a 'sandwich-shaped' small organism, in Section 5.4.3.1.

With the surface area (A) constant, we can simplify Eq. (3.15) and write:

$$\frac{\partial C}{\partial t} = -\frac{\partial (u \cdot C)}{\partial x} + \frac{\partial}{\partial x}\left(D \cdot \frac{\partial C}{\partial x}\right) + reaction \qquad (3.18)$$

If the advection rate u and the dispersion coefficient D are also constant over the domain, they can be taken out of the derivative and we obtain:

$$\frac{\partial C}{\partial t} = -u \cdot \frac{\partial C}{\partial x} + D \cdot \frac{\partial^2 C}{\partial x^2} + reaction \tag{3.19}$$

This is the 1-D transport equation in cartesian (or 'rectangular') coordinates.

3.4.5.2 Cylindrical Shapes

Now we consider a cylindrical shape, where there is diffusion across the radial axis of the cylinder. We assume that the top and base of the cylinder are impermeable: there is no exchange at these edges.

We can divide the cylinder in a number of telescoping tubes, which we assume to be homogeneous (Fig. 3.8 B), i.e. the concentration is the same at any point on the tube. The surface of a cylinder with radius r and length L, and ignoring top and base surface, is given by:

$$A = 2 \cdot \pi \cdot r \cdot L \tag{3.20}$$

Thus, the 1-D dispersion-reaction equation in a (homogeneous) cylindrical coordinate system is given by:

$$\frac{\partial C}{\partial t} = \frac{1}{2 \cdot \pi \cdot r \cdot L} \cdot \frac{\partial}{\partial r} \left(2 \cdot \pi \cdot r \cdot L \cdot D \frac{\partial C}{\partial r} \right) + reaction$$

$$\frac{\partial C}{\partial t} = \frac{1}{r} \cdot \frac{\partial}{\partial r} \left(r \cdot D \frac{\partial C}{\partial r} \right) + reaction \tag{3.21}$$

and where r is the radius. Note that we can eliminate the term $2\pi L$ because it is a constant.

This formulation will be applied when we model the oxygen budget in a cylindrical organism (Section 5.4.3).

A special case of a cylindrical model is one where the length of the cylinder, L, equals 0. In this case, the model describes transport in a plane, along concentric circles (Fig. 3.8C). In Section 5.4.2, we will model organisms, growing on a flat surface in concentric circles

3.4.5.3 Spherical Shapes

Finally, we consider a spherical shape, which we divide in a number of concentric shapes ('onion rings'). By noting that the surface of a sphere with radius r equals $4\pi r^2$ we derive the 1-D dispersion-reaction equation for (homogeneous) spherical coordinates:

$$\frac{\partial C}{\partial t} = \frac{1}{4 \cdot \pi \cdot r^2} \cdot \frac{\partial}{\partial r} \left(4 \cdot \pi \cdot r^2 \frac{\partial C}{\partial r} \right) + reaction$$

$$\frac{\partial C}{\partial t} = \frac{1}{r^2} \cdot \frac{\partial}{\partial r} \left(r^2 \cdot D \frac{\partial C}{\partial r} \right) + reaction \qquad (3.22)$$

Where the constant 4π has been eliminated in the latter equation.

This formulation will be applied when we model the oxygen budget in a spherical organism (Section 5.4.3).

3.4.6 One-dimensional Diffusion in Porous Media (Sediments) (**)

In porous media, gases, liquids and particles are mixed together in comparable quantities. Examples of such 'porous media' are sediments and soils, but also certain organism tissues (bones, lungs), sponges, aerenchym of macrophytes,...

In water-saturated sediments, we have two phases: solids ('sediment grains'), permeated by a network of pores, which are filled with liquids (interstitial water). This combination of phases plays a very important role in sediment models, especially if we describe a mix of substances that are either dissolved in the water or belong to the solid phase, or even both (e.g. when dissolved substances are partially adsorbed).

Typically, we are not interested in the behaviour of the system at the pore scale, but rather at the scale of the average over many pores.

The blend of particles and liquids is called the '*bulk sediment*', and the volumetric proportion of liquids relative to bulk sediment is called the *porosity* (ϕ). Porosity is an important quantity describing the properties of the porous medium.

$$\phi = \frac{cm^3 \ liquid}{cm^3 \ (liquid + solid)} = \frac{cm^3 \ liquid}{cm^3 \ bulk} \qquad (3.23)$$

Clearly, $0 <= \phi \quad <= 1$.

The relative (volumetric) proportion of solids to bulk sediment is given by: $(1 - \phi)$.

In general, porosity is not constant but, due to *compaction*, it decreases quasi-exponentially with increasing depth into the sediment. The porosity of the overlying water is ≈ 1, whilst porosity in the sediment is smaller than 1.

Because of these different phases (liquid and solid), concentrations of substances (or number of organisms) in the sediment can be expressed in several ways:

- Solid substances (e.g. organic matter, chlorophyll, particulate silicate, ...) are generally expressed as concentration per volume of *solid* sediment (e.g. mol m^{-3} solid), or as a concentration or weight per weight of solid sediment (e.g. mol g^{-1} solid, g g^{-1} solid, weight %).
- Dissolved substances (e.g. oxygen, nitrate, dissolved silicate,...) are generally expressed as a concentration per volume of *liquid* (e.g. mol m^{-3} liquid).

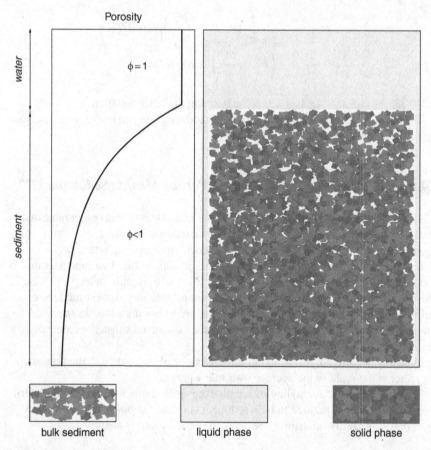

Fig. 3.9 Schematic representation of sediments and overlying water with liquid and solid phase and bulk sediment. Porosity (φ) is the volumetric proportion of liquid over bulk sediment. Sediment models are generally more complex than water column models, because the transport and reaction equations have to take into account the conversion between these phases

- Dissolved and solid substances can also be expressed as concentrations per volume *bulk* sediment (e.g. mol m^{-3} bulk).

The units can be easily converted from either the liquid or solid to the bulk phase or between liquid and solid phases. For example:

- If C is a concentration expressed in mol cm^{-3} liquid, then ϕ C is the concentration in mol cm^{-3} bulk (where ϕ is porosity).
- If S is the concentration in mol cm^{-3} solid, then $(1-\phi)$S gives the concentration in mol cm^{-3} bulk.
- If R is the carbon mineralization rate, expressed in mol C cm^{-3} solid day^{-1}, then $\gamma \cdot R(1-\phi)/\phi$, (where γ is the conversion factor expressing mol O$_2$ consumed

per mol C mineralized), will be the corresponding oxygen consumption rate, expressed in mol O_2 cm^{-3} liquid day^{-1}.

The predominance of two phases, solids and liquids, and the conversions between concentrations expressed per liquid and per solid make multi-phase models somewhat more complex compared to other models.

We now derive the transport equation in sediments. To keep it relatively simple, we do not consider sediment or porewater advection. Advection is due to the deposition of particles on the sediment surface, and, for most sediments it is insignificant compared to sediment mixing processes.

The confounding factor when deriving the 1-D sediment transport equation is that we have to consider the *mass balance* for the total, *bulk* sediment fraction, while *mixing* operates on the *liquid* or *solid* phase[1].

Thus, for a dissolved substance C (concentration, units mol cm^{-3} liquid) we specify the rate of change for the bulk concentration $\partial \phi C / \partial t$, and we have to formulate the fluxes crossing the bulk surface area.

However, molecular diffusion only takes place in the liquid phase and thus depends on the concentration gradient expressed per volume unit of liquid. To make both compatible, we convert the diffusive fluxes per area liquid to total fluxes per area bulk sediment, using the porosity:

$$Flux|_{liquid} = -D_s \cdot \frac{\partial C}{\partial x}$$

$$Flux|_{bulk} = -\phi_x \cdot D_s \cdot \frac{\partial C}{\partial x} \tag{3.24}$$

where D_s is the 'sediment diffusion coefficient'; this is the molecular diffusion coefficient, corrected for the hindrance by particles. Note that we used the subscript 'x' to denote that the porosity changes with sediment depth x only.

The diffusion equation in sediments, for constant surface area A, and for dissolved substances C is then written as:

$$\frac{\partial \phi_x C}{\partial t} = -\frac{\partial\, Flux|_{bulk}}{\partial x} = \frac{\partial}{\partial x}\left(\phi_x \cdot D_s \cdot \frac{\partial C}{\partial x}\right) \tag{3.25}$$

Assuming that porosity remains constant in time, it can be taken out from the derivative of the left hand side ($\partial \phi_x C / \partial t = \phi_x \partial C / \partial t$). Note however that we cannot play the same trick with the spatial partial derivative, because we assume that porosity *does* change with x. Thus we obtain:

$$\frac{\partial C}{\partial t} = \frac{1}{\phi_x} \frac{\partial}{\partial x}\left(\phi_x \cdot D_s \cdot \frac{\partial C}{\partial x}\right) \tag{3.26}$$

---

[1] Here we assume for simplicity that mixing operates on the solid phase only – there are other equations that assume that mixing of particles affects both solid and liquid phases. The interested reader is referred to expert literature, given in the last chapter.

This is the more general form in which sediment equations are written. Here $\partial C/\partial t$ is expressed as a concentration per volume liquid per time.

For solid substances, S, (concentration per volume solid), we derive a similar equation:

$$Flux|_{bulk} = -(1 - \phi_x) \cdot D_b \cdot \frac{\partial S}{\partial x} \tag{3.27}$$

$$\frac{\partial S}{\partial t} = \frac{1}{1 - \phi_x} \frac{\partial}{\partial x} \left((1 - \phi_x) \cdot D_b \cdot \frac{\partial S}{\partial x} \right) \tag{3.28}$$

where D_b is the bioturbation (or sediment mixing) coefficient.

3.4.7 The 3-D Advection-Dispersion-Reaction Equation (*)

Deriving the transport equation in 3 dimensions would take too far so we will just scratch the surface.

Basically, to extend the one-dimensional transport equations into three dimensions, we need to consider how we will represent a point in three dimensions.

- Most often, we use cartesian (or 'rectangular') coordinates which, in 3-D, use three numbers representing distances in the X-, Y- and Z- direction. With a constant grid size, the 3-D transport equation is an extension of the constant-surface 1-D equation.
- In the cylindrical coordinate system, a point is represented by a distance from the origin, a height and an angle. The full equation is based on the one-dimensional equation in a cylindrical shape, but taking into account potential gradients along the angle θ and the height (length-axis z).
- In the spherical coordinate system, a point is represented by a distance from the origin and two angles. The 3-D transport equation in spherical coordinates extends the 1-D spherical case with the gradients in both angles.

3.5 Boundary Conditions in Spatially Explicit Models

Before a spatial model is entirely specified, the following questions have to be resolved: (1) is the spatial domain bounded, or does it extend to infinity. (2) If the modelled space is finite, what happens to substances and individuals at the boundaries? Do individuals fall off or bounce back; are substances not allowed to cross, . . .

Answering this amounts to specifying the spatial boundary conditions, i.e. we determine what happens at the borders that separate the model domain from the outside world.

Spatially resolved models have one or more spatial boundaries.

- In a vertical 1-D water column model for instance, the upper boundary is often the air-water interface, the lower boundary the sediment (Fig. 3.10A).
- In a 1-D sediment model, the upper boundary is the sediment-water interface, the lower boundary is generally taken at infinite depth (or, more specifically, deep enough such that nothing of interest happens below this depth).
- In a 1-D estuarine model, the upstream boundary is generally the river, the downstream boundary the sea.
- For cylindrical and spherical volumes, the internal 'boundary' is the central axis of the cylinder and the central point of the sphere, the external boundary is the external surface of the volume (Fig. 3.10 B-D).

Transport models can only be solved if it is specified what happens at these physical boundaries. This requirement is similar to the need of an initial condition for fully specifying temporal dynamics. We will explain that in later chapters.

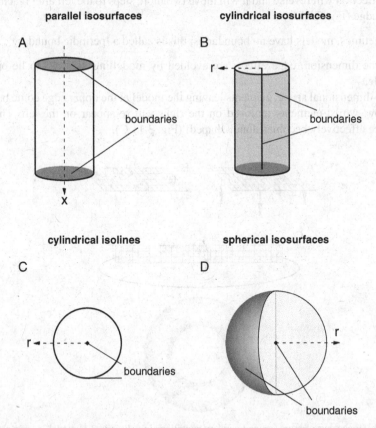

Fig. 3.10 Boundaries in one-dimensional models of various shapes. **A**. One-dimensional shape with constant surface area. **B**. Cylindrical shape, with non-zero cylinder length. **C**. Cylindrical shape with zero length of cylinder. **D**. Spherical shape

Boundary conditions are required both for microscopic and macroscopic models, for discrete and continuous models.

3.5.1 Boundary Conditions in Discrete Models

When movement of organisms in a discrete number of cells is described, we need to specify what happens when these organisms have reached the edges. There are different possibilities:

• Absorbing boundary: animals reaching the edge 'drop off' and are lost. Note that this leads to a decrease in the total number of individuals in the model, which can be undesirable (Fig. 3.11 A).
• Reflecting boundary: the total flux of individuals over the boundary is set to zero. Every individual reaching the edge 'bounces back' into the model domain. When it wants to move 4 units to the right, but reaches the edge after two units, its direction will reverse and it will move two more steps to the left after reaching the edge (Fig. 3.11 B).

Sometimes, models have no boundaries; this is called a 'periodic boundary':

• In one dimension, edge effects are avoided by modelling the cells to lie on a circle.
• In 2-dimensional space, elements leaving the model at the upper edge come back below, whilst elements removed on the right will re-appear on the left. Thus, space effectively becomes donut-shaped! (Fig. 3.11 C).

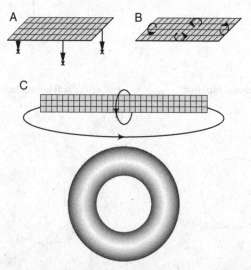

Fig. 3.11 Three ways of representing boundary conditions in a discrete 2-D model. **A.** organisms moving outside the model domain are removed. **B.** Organisms reaching the end bounce back. **C.** Organisms are displaced at the other side. This is equivalent to folding the surface such that the edges are removed, and a donut-shape is obtained

3.5.2 *Boundary Conditions in Continuous Models*

Continuous models are described by differential equations. As a rule of thumb, the number of spatial boundary conditions needed is simply equal to the highest spatial derivative in the equation, the order of the differential equation (see Appendix A.3).

For instance, in the advection-dispersion-reaction equation:

$$\frac{\partial C}{\partial t} = -u \cdot \frac{\partial C}{\partial x} + \frac{\partial}{\partial x}\left(D \cdot \frac{\partial C}{\partial x}\right) - k \cdot C \tag{3.29}$$

the highest derivative is second-order (the 2nd, dispersive term), thus this is a second-order differential equation, which requires two boundary conditions to be fully specified.

In the absence of dispersion the order of the differential equation:

$$\frac{\partial C}{\partial t} = -u \cdot \frac{\partial C}{\partial x} - k \cdot C \tag{3.30}$$

equals one, and only one boundary condition is needed.

The mathematical foundation for this simple rule is explained in the chapter on analytical model solution (Section 5.2). However, one can also develop an intuitive feeling for it. We explain it for concentration boundaries, but it can be extended for other types of boundary conditions.

Advective fluxes are estimated as the product of a velocity (assumed known throughout the model) and a concentration just upstream of a (very thin) box. Downstream concentrations need not be known to calculate the fluxes. Thus at the downstream boundary of the model, the flux can be modelled without knowing the outside concentration. However, in order to know the influx at the upstream boundary, one needs to know the concentration just upstream of the model domain. Thus, advection models require one boundary condition, specified only at the upstream boundary.

Diffusive fluxes are estimated as the product of a diffusion coefficient (again, assumed known everywhere) and a concentration gradient. To estimate this gradient, we need two concentrations, one on each side of the interface. Consequently, at both boundaries one needs to know the outside concentration, so that the value of the concentration gradient there can be established. Two boundary concentrations are needed (for this second-order derivative).

In continuous models, the boundary conditions may be represented as.

- a *concentration* (α) in the outside world ($x = b$).

$$C|_{x=b} = \alpha \tag{3.31}$$

This is usually referred to as a Dirichlet condition or boundary condition of the first kind.

- an imposed *gradient* (Y), also called a Neumann condition (or boundary condition of the second kind). If the gradient is set to 0, then there is no diffusive flux across the boundary.

$$\left.\frac{\partial C}{\partial x}\right|_{x=b} = \gamma \tag{3.32}$$

- Finally there also exist so-called boundary conditions of the third kind, or Robin's conditions, in which a linear combination of a Dirichlet (concentration) and a Neumann (concentration gradient) is specified. It is used to impose *fluxes* across the boundaries. The general form is:

$$u \cdot C|_{x=b} - D \cdot \left.\frac{\partial C}{\partial x}\right|_{x=b} = \beta \tag{3.33}$$

A special case of the Robin's conditions are the so-called 'surface evaporation conditions', which prescribe the flux as a function of the difference between actual concentration at the surface, and some reference concentration C_s, e.g. the saturated concentration of a gas in water:

$$\beta = p \cdot (C_s - C|_{x=b}) \tag{3.34}$$

Here p is called the 'piston velocity', units of Length Time-1

3.5.2.1 Example 1: 1-D Model of Organic Matter Dynamics in Sediments

Consider a 1-dimensional (vertical) model that describes the dynamics of organic matter in sediments (Fig. 3.12 A).

Organic matter is mineralised (respired), which is modelled using first-order decay (see Section 2.7.3). The deposition of sediment particles on top of the sediment surface causes net movement of the organic matter towards deeper layers; this is sediment advection. The activity of organisms randomly displaces the organic matter, similar as a diffusive process. This is called sediment bioturbation.

The equation that describes the change of organic matter as a function of time and space is the advective-diffusive-decay equation with constant surface area:

$$\frac{\partial C}{\partial t} = -u \cdot \frac{\partial C}{\partial x} + D_b \cdot \frac{\partial^2 C}{\partial x^2} - k \cdot C \tag{3.35}$$

where x is the distance below the sediment-water interface (cm), u is the sediment advection rate (cm yr^{-1}), D_b is the constant bioturbation rate (cm^2 yr^{-1}) and k is the first-order decay rate (yr^{-1}).

As we deal with a vertical sediment column, there are two boundaries, one at the sediment-water interface (the upper boundary, at x=0), one at large distance below the sediment-water interface (the lower boundary at infinite depth).

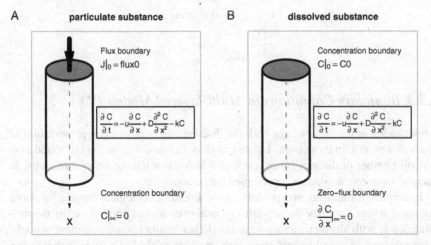

Fig. 3.12 Model description for 1-D sediment biogeochemical models, with typical boundary conditions. **A**. For a particulate substance such as organic matter, an upper flux boundary condition is often prescribed. **B**. For a dissolved substance, such as oxygen, the upper boundary is more often prescribed as a concentration. J denotes the flux, C the concentration. Boundary conditions are in bold, model equations are enclosed in a box. It is assumed that porosity is constant, thus it can be removed from the equation

If there is no *in situ* organic matter production, then the entire system is driven by deposition of organic matter at the sediment-water interface. This deposition flux forms one boundary condition, at x=0. The flux enters the sediment column through sediment advection and bioturbation:

$$Flux_{x=0} = u \cdot C|_{x=0} - D_b \cdot \frac{\partial C}{\partial x}\bigg|_{x=0} \qquad (3.36)$$

It is customary to set the lower boundary of the sediment model at a sufficiently large distance; more specifically at infinite distance (∞). This ensures that all organic matter has been consumed there. We can impose this condition by specifying:

$$C|_{x=\infty} = 0 \qquad (3.37)$$

3.5.2.2 Example 2: 1-D Model of Oxygen Dynamics in Sediments

When the modelled substance is a dissolved substance rather than a solid, the upper boundary is generally imposed as a concentration boundary (the bottom-water concentration); the lower boundary is specified as a zero-gradient boundary (Fig. 3.12 B). The latter ensures that the consumption of the dissolved substance at the lower boundary is zero, but not necessarily its concentration. Thus we have, for the upper and lower boundary of oxygen respectively:

$$O2|_{x=0} = BW\_conc \tag{3.38}$$

$$\left.\frac{\partial O2}{\partial x}\right|_{x=\infty} = 0 \tag{3.39}$$

3.5.3 Boundary Conditions in Multi-layered Models (**)

Sometimes it is useful to merge both the discrete and continuous representation of space in a model. In these multi-layered models, the model domain is subdivided in a small number of distinct zones, each of which has a unique set of equations. In each of these zones, space is represented continuously.

Examples of multi-layered models can be found in ecological models for water columns, where one may distinguish layers above, around and below the thermocline, each with distinct mixing regimes. Other examples are sediment models, where distinct biogeochemical zones can be distinguished, e.g. oxic, suboxic and anoxic zones which have distinct metabolic pathways (Fig. 3.13).

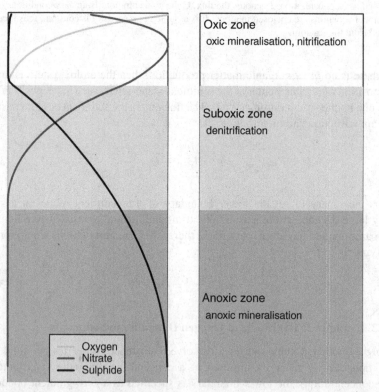

Fig. 3.13 Vertical profile of oxygen, nitrate and sulphide in the sediment. The coloured layers represent three zones where different processes take place. Multilayered models consider distinct zones, but within a zone space is represented continuously

In these multi-layered models, we need to specify the boundary conditions at the interface between the layers (the internal boundary), in addition to the boundaries with the external world.

There are different requirements for the internal compared to external boundary conditions: we expect the separate models for the different layers to be mutually consistent where they meet, so as to give a coherent picture of what we are modelling. We would not accept for instance that concentrations would suddenly jump at the transition between layers.

Usually the internal boundary conditions are of the type:

- Continuity of concentration. This means, that the concentrations at the interface, when calculated from the equations from either of the adjacent layers must be the same.
- Continuity of flux. This means that whatever is leaving one layer (and is calculated from its constitutive equation) enters the next layer.

For each layer, the number of boundary conditions needed equals the order of the differential equation.

We give two examples related to bioturbation. The first is an extension of the organic matter dynamics in previous section, the second an example of rather unusual boundary conditions. Other examples can be found in the chapter that deals with analytical solutions (Chapter 5).

3.5.3.1 Example 1: To Bioturbate or Not to Bioturbate

Burrows and holes dug by organisms are typically found in the upper section of the sediment (say in the upper 10 cm), while they are absent deeper down. The animal movement and feeding activity mixes the sediment, a process called bioturbation, making it more porous, in contrast to the more compact deeper layers. Thus, as transport processes differ, it makes sense to distinguish between an upper bioturbated (mixed) sediment layer, overlying a layer that is not bioturbated. We now make a 1-dimensional model of organic matter dynamics in sediments that includes an upper bioturbated layer of thickness L, and a lower non-bioturbated layer (Fig. 3.14 A).

We thus have two layers (I, II), with different model equations:

$$\left.\frac{\partial C}{\partial t}\right|^{I} = -u \cdot \left.\frac{\partial C}{\partial x}\right|^{I} + D_b \cdot \left.\frac{\partial^2 C}{\partial x^2}\right|^{I} - k \cdot C^I \quad x <= L \qquad (3.40)$$

$$\left.\frac{\partial C}{\partial t}\right|^{II} = -u \cdot \left.\frac{\partial C}{\partial x}\right|^{II} - k \cdot C^{II} \qquad\qquad x >= L \qquad (3.41)$$

where x is the distance below the sediment-water interface (cm), L is the thickness of the bioturbated layer, u is the sediment advection velocity (cm yr^{-1}), D_b is the upper-layer bioturbation coefficient (cm^2 yr^{-1}) and k is the first-order decay rate (yr^{-1}).

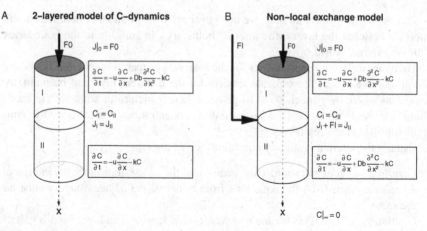

Fig. 3.14 Schematic representation of two multilayered models; two layers are described (I,II). **A.** model of C-dynamics. **B.** Non-local exchange model. Boundary conditions are in *bold*; model equations are in a *box*

The first layer (I) considers both mixing and advection; as the order of the differential equation is 2 there is need for 2 boundary conditions, one at the sediment-water interface, one at the interface with layer II. There is no mixing in layer II and the order of the differential equation is 1, so there is no need for a lower boundary, but the upper boundary at the interface with layer I needs to be prescribed.

The boundary condition at the sediment-water interface is the same as before, except that it now applies to the first layer.

$$Flux_{x=0} = u \cdot C|_0^I - D_b \cdot \left.\frac{\partial C}{\partial x}\right|_0^I \tag{3.42}$$

The internal boundary conditions are continuity of concentration and continuity of flux at depth L

$$C|_{x=L}^I = C|_{x=L}^{II} \tag{3.43}$$

$$-D_b \left.\frac{\partial C}{\partial x}\right|_{x=L}^I + u \cdot C|_{x=L}^I = u \cdot C|_{x=L}^{II} \tag{3.44}$$

The solution of this model is left as an exercise in Section 5.5.3, dealing with analytical solutions.

3.5.3.2 Example 2: A Non-Local Exchange Model (***)

In the previous example, animal activity was modelled analogous to diffusion, where the diffusion coefficient was called a 'bioturbation coefficient'. However, some animals ingest organic matter from the sediment-water interface and rapidly inject it at certain depth into the sediment; this phenomenon is known as 'non-local exchange'.

Here we discuss a simple model that takes into account both diffusive mixing and non-local injection of organic matter (Soetaert et al., 1996a) (Fig. 3.14 B).

The model is atypical in the sense that two layers of sediment are considered, but both have the *same* dynamics; the non-local exchange is included in the internal boundary condition.

The constitutive equations in both layers (i=1, 2), layer 1 extending from the sediment surface till depth L, layer 2 from depth L till infinite depth are:

$$\frac{\partial C}{\partial t} = -u \cdot \frac{\partial C}{\partial x}\bigg|^i + D_b \cdot \frac{\partial^2 C}{\partial x^2}\bigg|^i - k \cdot C|^i \tag{3.45}$$

With two second-order differential equations, we need two times two boundary conditions for the model to be fully specified.

The upper external boundary condition is imposed as a deposition flux, whereas the lower external boundary condition imposes that all reactive organic matter is absent

$$Flux|_{x=0} = u \cdot C|^{i=1}_{x=0} - D_b \cdot \frac{\partial C}{\partial x}\bigg|^{i=1}_{x=0} \tag{3.46}$$

$$C|^{i=2}_{x=\infty} = 0 \tag{3.47}$$

At the interface between both layers (depth L), the continuity of concentration is imposed.

$$C|^{i=1}_{x=L} = C|^{i=2}_{x=L} \tag{3.48}$$

The continuity of flux is more complex, and considers that part of the flux from layer 1 to 2 is directly injected at the interface, i.e. it does not result from diffusive mixing and advection.

$$u \cdot C|^{i=1}_{x=L} - D_b \cdot \frac{\partial C}{\partial x}\bigg|^{i=1}_{x=L} + injectFlux = u \cdot C|^{i=2}_{x=L} - D_b \cdot \frac{\partial C}{\partial x}\bigg|^{i=2}_{x=L} \tag{3.49}$$

This model will be solved and implemented in R, in Section 5.4.4, when we deal with analytical models.

3.6 Case Studies in R

3.6.1 An Autocatalytic Reaction in a Flow-Through Stirred Tank

We start with a zero-dimensional model that includes transport across the model boundaries.

Two chemicals A and B are fed into a stirred tank by continuous inflow on one side and equal outflow at the other side of the tank, such that the tank's volume remains constant. In the tank, an autocatalytic reaction occurs between A and B (Fig. 3.15).

An autocatalytic reaction is one where the reaction product is itself the catalyst for the reaction, for instance:

$$A + B \rightarrow 2B + C \tag{3.50}$$

Based on the law of mass action (Section 2.3.1), the rate of the autocatalytic reaction can be written as:

$$rate = k \cdot [A] \cdot [B] \tag{3.51}$$

where k is the reaction rate constant. Chemical A is consumed, and chemical C is produced by the autocatalysis reaction, while one mole of B is consumed for two moles of B produced, so the net effect is one mole B produced.

Adding the in- and outflow of the chemicals, the rate of change of the concentrations [A], [B] and [C] can be written as:

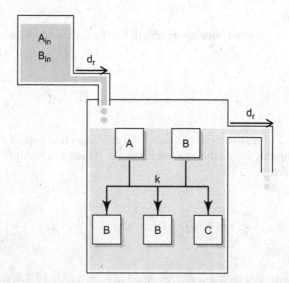

Fig. 3.15 Schematic representation of an autocatalytic reaction in a flow-through, stirred tank

$$\frac{d[A]}{dt} = d_r \cdot (A_{in} - [A]) - k \cdot [A] \cdot [B]$$

$$\frac{d[B]}{dt} = d_r \cdot (B_{in} - [B]) + k \cdot [A] \cdot [B]$$

$$\frac{d[C]}{dt} = -d_r \cdot [C] + k \cdot [A] \cdot [B] \tag{3.52}$$

where d_r is the dilution rate; B_{in} and A_{in} are the concentrations in the inflowing water.

In R we define a function, called autocatalysis, and which has as input the time (t), the values of the state variables (state) and the values of the parameters (pars). The function calculates the rate of change of the state variables (dA, dB and dC) and returns those as a list. The R-statement with (as.list (c(state,pars)),... ensures that the state variables and parameters can be addressed by their names.

```
autocatalysis <- function(t,state,pars) {

with (as.list(c(state,pars)),
{
dA <- dr*(Ain-A)-k*A*B
dB <- dr*(Bin-B)+k*A*B
dC <- -dr*C +k*A*B

return (list(c(dA,dB,dC)))
})

}
```

We will see later how to solve such models; therefore, the solution of this model is left as an exercise in Section 6.7.1 (numerical solutions).

3.6.2 A 1-D Microscopic and Macroscopic Model of Diffusion

We now model the process of diffusion in continuous space, both by taking a microscopic and a macroscopic approach.

Microscopic models of diffusion follow the fate of individual particles or individuals as they jump randomly through space. Here we follow 100 individuals that move randomly in one dimension. Each individual takes 200 random steps with step length between -0.5 and 0.5 (i.e. in both directions).

Implementing this in R takes only 5 lines of code!

Matrix pos contains the positions of all individuals, and at all time steps, including the initial position. All individuals are initially positioned at x=0 (pos[1,] <-0). (Note that pos[1,] refers to all elements of the first row – this is at the initial time step ; pos[,1] would refer to all elements on the first column).

The random walk is performed for 200 steps by adding, to the original position of each individual at time step i (pos[i,]) a random step inbetween −0.5 and 0.5. Statement runif(nind) efficiently generates nind uniformly distributed random numbers, between 0 and 1; subtracting 0.5 causes the steps to be confined between [−0.5,0.5]. Finally the individual random walks are plotted, using R's function matplot which plots all columns separately (Fig. 3.16 left upper). By default matplot uses different line colors and line types for each data series; here this is overruled (col="black", lty=1).

```
nind      <- 100
nsteps    <- 200

pos       <- matrix (nrow=nsteps+1,ncol=nind)
pos[1,] <- 0
for (i in 1:nsteps) pos[i+1,] <- pos[i,]+runif(nind)- 0.5

matplot(pos,type="l",col="black",lty=1,xlab="step",ylab
="Position")
```

To analyze the behaviour of this model, we need statistical treatment of its output. As R was originally a statistical package, it is very flexible in this respect.

Here we let R estimate, for each time step, the probability density function of the individual positions. A probability density function can be considered a smooth, continuous version of a histogram (see Appendix C.5 for a more formal definition).

The statement density(pos[i,],from=-10,to=10,n=100) generates the probability density distribution of the individual positions from the i-th step and over a regular grid, extending from −10 to +10, composed of 100 grid points. By density(...)$y the probability density values are extracted only. The density functions are stored in a matrix called 'posmicro' and then plotted using function matplot (Fig. 3.16 left below).

```
posmicro <- matrix(ncol=nsteps+1,nrow=100)

for (i in 1:(nsteps+1))
  posmicro[,i] <- density(pos[i,],from=-10,to=10,n=100)$y

matplot(posmicro[,seq(1,100,by=4)],type="l",lty=1,

col=" black", axes=FALSE,frame.plot=TRUE,ylab="Density")
```

The result of the random movement is to smoothen out spatial heterogeneity, and this is represented by the *macroscopic* diffusion equation:

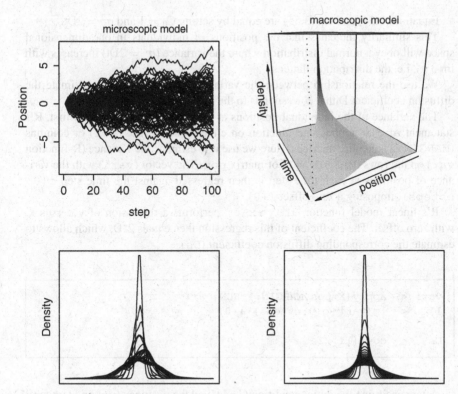

Fig. 3.16 Model output generated with the R-code of the 1-D microscopic (*left*) and macroscopic (*right*) model of diffusion

$$\frac{\partial C}{\partial t} = D \cdot \frac{\partial^2 C}{\partial x^2} \tag{3.53}$$

As we will see later, the solution of some models are simple algebraic functions (so-called analytical solutions), which can be found in books. This is the case for the diffusion model, whose solution is given in Appendix A.3:

$$C(x, t) = \frac{C_0}{2\sqrt{\pi D t}} \cdot \exp^{-\frac{x^2}{4Dt}} \tag{3.54}$$

Where C_0 is the initial concentration at position x=0 and $C(x,t)$ is the concentration at position x and time t.

The resemblance of this solution with the probability density function of the normal distribution with mean 0 and standard deviation σ is striking:

$$f(x; 0, \sigma) = \frac{1}{\sqrt{2\pi\sigma^2}} \cdot \exp^{-\frac{x^2}{2\sigma^2}} \tag{3.55}$$

Equation (3.54) and eq. (3.55) are equal by setting $C_0 = 1$ and $\sigma^2 = 2Dt$.

This similarity indicates that the positions of individuals in one-dimensional space will obey a normal distribution where the variance ($\sigma^2 = 2Dt$) increases with time (t), i.e. the distribution flattens.

We use the relationship between the variance σ^2 and D and t to estimate the diffusion coefficient D that corresponds to the microscopic model.

The variance of the individual positions at each time step is estimated first. R's statement `apply` applies one function on either rows (MARGIN=1) or columns (MARGIN=2) of entire matrices. Here we use it to estimate the variance (R-function `var`) on the rows (MARGIN =1) of matrix `pos`. The vector (`vari`) with the variance of positions at each time step is then regressed against the time step `c(0:nsteps)`, imposing a zero offset (`+0`).

R's linear model function $\text{lm}(y \sim x+0)$ performs a regression of y versus x, with zero offset. The coefficient of this regression then equals 2*D, which allows to estimate the corresponding diffusion coefficient (`Ds`).

```
vari  <- apply(X=pos,MARGIN=1,FUN=var)
ll    <- lm(vari~c(0:nsteps) + 0)

Ds    <- coef(ll)/2
```

After specifying the initial condition (`ini`), and the positions (`xx`) and temporal step (`tt`) at which output is wanted, eq. 3.54 is estimated for all spatio-temporal combinations. The simplest method is to use R's function `outer` to estimate these function values. Results are stored in a matrix called `posmacro`, which is then plotted in perspective using R's function `persp`. Single density versus distance lines are also plotted (`matplot`) for comparison with the microscopic model.

```
ini   <- 1 xx    <-seq(-10,10,length=100) tt
<-seq(1,100,by=1)

posmacro <- outer(xx,tt,
FUN =function (xx,tt) ini/(2*sqrt(pi*Ds*tt))*exp(-
xx^2/(4*Ds*tt)))

persp(xx,tt,z=posmacro,theta=150,box=TRUE,axes=TRUE,
      xlab="position",ylab="time",zlab="density")
mtext("macroscopic model",side=3,line=2,cex=1.2)

matplot(posmacro[ ,seq(1,100,by=4)],type="l",lty=1,
col="black",axes=FALSE,frame.plot=TRUE,ylab="Density")
```

3.6.3 Cellular Automaton Model of Diffusion (**)

In the previous example, space was considered continuous. Here we implement diffusion in a cellular automaton. The model is a simplified version of a model from Wilson (2000).

The cellular automaton grid is one dimensional, consisting of 200 cells, some of which are occupied ($\bullet$), while others are empty (Fig. 3.17). At each time step, all particles try to move, either one grid cell to the right or to the left; movement is only allowed to a new cell when this cell is unoccupied and when no other particle is moving to the same cell (v=transition allowed; x=not allowed). Objects that want to move to a forbidden position have to stay in their current position. The model has periodic boundary conditions: particles moving out of the domain at one side come back in from the other side.

For the implementation of the model one could take two different points of view. One is to follow all 200 cells (or 'positions') in the model domain, document whether they are occupied or not and adjust this list at the next time step. The other is to follow all particles in the model, document where they are located, and adjust these locations at the next time step. Here we have chosen the first approach, for reasons of speed of execution. In the second approach, for every particle wanting to move at every time step, one has to loop over all other particles to see what is permitted and what not. Such looping can be avoided in the first approach.

The implementation in R starts with a function (called `diffuse`) that takes one diffusion step. It takes as input the vector (`Particles`) with length `ncell`, documenting which cells are occupied; cells that are occupied take on a value of 1, empty cells are 0. The function `diffuse` outputs the updated values of the cells.

The function first finds the cells that are occupied and stores the positions of these cells (in fact their serial number or 'index' in `Particles`) in vector x. Then, for each occupied cell, a uniformly distributed random number is generated (`runif`). If the random number is smaller than 0.5, the particle tries to move one cell to the left, and its index will decrease with 1. If the random number is larger or equal to 0.5, it tries to move to the right, and the index will increase with 1. Vector `move` contains the new positions to which the particles *desire* to move. To avoid that the particles disappear when crossing the edges (<1, > ncell), they are wrapped around if necessary.

The new positions that are not free ('notfree') and those that are targeted by two particles ('duplo') are determined next, and removed. This is rather complicated and will be explained below. Vector `free` then contains the movement positions that are allowed; vector `source` contains the corresponding source positions. Finally, the source positions are zeroed (emptied), and the new positions are set to 1.

In order to determine the target positions that are not free, we use the R statement 'which'. It returns the index of those elements in a vector for which the conditional statement is TRUE. Therefore, in this statement we obtain a vector (`notfree`) containing all the positions to which a particle wants to move, but

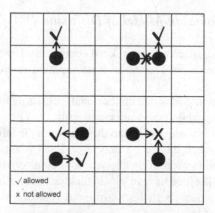

Fig. 3.17 Schematic representation (*above*) and model output (*below*) of the cellular automaton diffusion model

where a particle is already present. `Particle [move]` is a vector of zero's and ones, stating for each of the target cells (positions of the target cells are in vector move) whether it is occupied or not. If it is occupied, it is larger than zero, it is picked up by the 'which' statement and its position is stored in 'notfree'.

In order to determine duplicate target cells (i.e. target cells of more than one particle), we proceed in a number of steps. First, we determine which target cells have already been targeted before, using R's statement 'duplicated'. Thus, `duplicated(move)` returns a vector with values TRUE if the corresponding element in move has already been encountered before (when going from the first to the last element of move), FALSE otherwise. We call the result `already_targeted`.

`Move[already_targeted]` contains those elements of move (i.e. target positions) that have been targeted before. Note that this is not yet a complete list of duplicate targets: a duplicate is only picked up when it occurs the second time in move, not the first time! We call this vector `bad_targets`.Our next step is then to make a list of all elements in move that occur in `bad_targets`. We use the R statement `A %in% B`, which returns a vector with value TRUE if the element of A is present in B. Thus, `move%in%bad_targets` yields a vector with value TRUE when the corresponding element in move occurs in `bad_targets`. `Which(move%in%bad_targets)` gives all index numbers (in move) that are to be removed because they contain a bad (duplicate) target. These are stored in `duplo`.

In order to determine which positions are free, we make use of the possibility to index a vector with negative subscripts. In R, when indexing a vector with negative subscripts, the elements are removed: thus `move[-c(duplo,notfree)]` removes the elements whose index numbers are given in `duplo` and `notfree`.

```
Diffuse <- function (Particles)

{
  x                  <- which(Particles>0)

  rnd                <- runif(length(x))
  move               <- c(x[rnd<0.5]-1,x[rnd>=0.5]+1)

  move[move<1]       <- move[move<1]     +ncell
  move[move>ncell]   <- move[move>ncell]-ncell

  notfree <- which (Particles[move]>0)

  already_targeted<- duplicated(move)
  bad_targets      <- move[already_targeted]
  duplo  <- which(move%in%bad_targets)
```

```
   free          <- move[-c(duplo,notfree)]
   source        <- c(x[rnd<0.5],x[rnd>=0.5])[-
c(duplo, notfree)]

   Particles[source] <- 0
   Particles[free]   <- 1
   return(Particles)

}
```

The simulation starts off with a pattern of 5 bands, each 2 cells wide. The model
is run for 1000 diffusion steps; at each step, the particle positions stored in matrix
Grid.

```
ncell       <- 200 Particles <- rep (0,ncell)
Particles [c(10:30,50:70,90:110,130:150,170:190)] <- 1
nsteps      <- 1000

Grid <- matrix(ncol=nsteps, nrow=ncell)

for (j in 1:nsteps) {
  Particles <- Diffuse(Particles)
  Grid[,j]  <- Particles
           }
```

Finally, R's function image is used to create the output (see Fig. 3.17).

```
windows(width=4,height=8)
par(oma = c(0,0,3,0))
image(y=1:nsteps,z=Grid, col = c(0,1), axes=FALSE,
       ylim=c(nsteps,1),xlab="",ylab="")
box()
mtext(outer=TRUE,side=3,"DIFFUSION in cellular
automaton", cex=1.5)
```

3.6.4 Competition in a Lattice Grid

For plants that root in the soil, space is of vital importance, and competition for
this resource can be vast. Different plant species have developed different strategies

to ensure victory: some produce huge amounts of seeds, such as to rapidly cover up empty spaces and prevent other species from invading these patches. This strategy may be fruitful if empty spaces are available, but is rarely successful in closed stands. Other plants simply invade their neighbourhood by expansion (lateral clonal growth) or by overgrowth, thereby driving their neighbouring species to extinction. The grasses *Agrostis*, *Poa*, *Holcus*, *Cynosurus* and *Lolium* cover a rich spectrum of these strategies.

Silvertown et al. (1992) developed a model in order to investigate whether and how spatial configuration of the different plant communities affected the ultimate outcome of competition, which species finally dominates, and which species become extinct.

The model is a cellular automaton model, treating space as a regular lattice of cells. Each cell may either be vacant or be occupied by a group of individuals of one species. The five species of grasses occupy a 40*40 lattice, where each grid cell contains only one of the species belonging to the genera *Lolium*, *Agrostis*, *Holcus*, *Poa* or *Cynosurus*. Within one time step, there is a certain probability of an individual to be replaced by any of its four immediate neighbors (left, right, above, below).

The input data consist of the names of the species and the probabilities that a species in a grid cell (column) will be replaced by a neighboring species (row) if all neighboring cells are occupied by the species:

```
species <-
c("Lolium","Agrostis","Holcus","Poa","Cynosurus")

replacement <- matrix(ncol=5,byrow=TRUE,data=c(
1.0,0.02,0.06,0.05,0.03,
0.23,1.0,0.09,0.32,0.37,
0.06,0.08,1.0,0.16,0.09,
0.44,0.06,0.06,1.0,0.11,
0.03,0.02,0.03,0.05,1.0 ) )
```

For instance, if an *Agrostis* individual (species 2) has one neighboring cell inhabited by *Lolium* (species 1), two cells consisting of *Holcus* (species 3) and one of *Agrostis*, then the probability that *Agrostis* will be replaced by *Lolium* is given by replacement[1,2]/4 = 0.02/4=0.005, whilst the probability that it is replaced by *Holcus* will be given by replacement[3, 2]/4*2 = 0.08/2 = 0.04. The cell will remain as *Agrostis* otherwise (i.e. with probability 1−0.005−0.04 =0.955). Whether or not such event will occur depends on chance, as determined by a uniformly generated random number (see below).

The competition algorithm works as follows: for each step (ss), we start by determining the neighbors of all cells: they are composed by shifting the cells downward (dn), upward (up), to the left (le) and to the right (ri), ensuring that there is continuity at the edges (i.e. we consider a donut-shape). For instance, the statement rbind(cells[ncell,],cells[1:(ncell-1,)]) puts the last row of cells (cells[ncell,]) in front of the remaining rows (cells[1:(ncell-1),]).

Then for each grid cell, a uniform random number is generated and stored in a matrix (rnd); it is most efficient if all random numbers are generated at once. R's function runif(ncell*ncell) generates these random numbers, uniformly distributed in the interval [0,1]. Next we loop over all rows (i) and columns (j) of the grid, locating the neighbors of each cell (neigb) and the probability of replacement of the plant by its neighbors (p). The chance effect of replacement is exerted by comparing the random number (rnd[i,j]) by the cumulative probabilities of replacement (cump); the interval in which the random number is located gives the replacement species. R's function cumsum creates cumulative sums.

At each time step (ss), the total density of all five species is estimated. The statement "cells = = i" returns true (1) or false (0), so the sum gives the total density of species i, which is stored in matrix dens.

```
competition <- function(cells,nstep=100)

{
 nind  <- nrow(replacement)
 ncell <- nrow(cells)
 dens  <- matrix(nrow=nstep,ncol=nind)

 for (ss in 1:nstep)
   {
     dn <- rbind(cells[ncell,]  ,cells[1:(ncell-1),])
     up <- rbind(cells[2:ncell,],cells[1,]   )
     le <- cbind(cells[,2:ncell],cells[,1]   )
     ri <- cbind(cells[,ncell]  ,cells[,1:(ncell-1)])

     rnd <- matrix(nr=ncell,nc=ncell,runif(ncell*ncell))

     for (i in 1:ncell)
     { for (j in 1:ncell)
       {  ii     <- cells[i,j]
          neigb <- c(up[i,j],ri[i,j],dn[i,j],le[i,j])
          p     <- replacement[neigb,ii]
          cump  <- c(cumsum(p/4),1)
          rep   <- min(which(cump>=rnd[i,j] ))
          if (rep<5) cells[i,j] <- neigb[rep]
       }
     }

       for (i in 1:nind) dens[ss,i]<-sum(cells == i)
   }
 return(list(cells=cells,density=dens))

}
```

The model is run for a 40*40 grid, where the species are initially aligned in horizontal bands (8 cells wide) in the following succession: Poa, Cynosurus, Lolium, Holcus and Agrostis. The model is then run in two steps, for 100 steps each.

```
ini    <-c(4,5,1,3,2)
cells <- matrix(40,40,data=0)

cells[,1:8]  <-ini[1] # species 4 : Poa
cells[,9:16] <-ini[2] # species 5 : Cynosurus
cells[,17:24]<-ini[3] # species 1 : Lolium
cells[,25:32]<-ini[4] # species 3 : Holcus
cells[,33:40]<-ini[5] # species 2 : Agrostis

A100   <- competition(cells)
A200   <- competition(A100$cells)
```

The graphs are aligned in two rows, two columns (mfrow), and the margins of each figure are reduced (mar). R's function 'image' generates a filled contour plot, the colors of which have been specified before (col); note the use of zlim=c(1,5), which enforces that the same color scheme is used, even if certain species are absent. Function matplot plots several columns of a matrix; it is used here to plot the time evolution of all species densities. None of the graphs requires drawing the axes (axes=FALSE) (Fig. 3.18).

```
par(mfrow=c(2,2),mar=c(2,2,2,2))
col    <-c("grey","lightblue","blue","darkblue","black")

image(cells,col=col,zlim=c(1,5),axes=FALSE,main="initial")

text(x=rep(0.1,5),y=seq(0.1,0.9,length.out=5),
      labels=species[ini],col="white",adj=0,font=2)

image(A100$cells,col=col, zlim=c(1,5),axes=FALSE,
      main="100 steps")
image(A200$cells,col=col, zlim=c(1,5),axes=FALSE,
      main="200 steps")

matplot(rbind(A100$density,A200$density),type="l",lwd=2
,lty=1,

col=col,xlab="time",ylab="",axes=FALSE,frame.plot=TRUE)
```

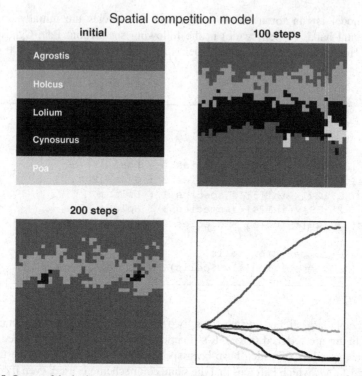

Fig. 3.18 Output of the lattice grid competition model

One final note: the function 'competition' includes 3 nested loops (over steps, rows, and columns). As R is not very efficient when it comes to looping, this model takes a while to execute.

In these cases, it is worthwhile to implement model dynamics in a lower-level language, and use R for post-processing. We show how to do that in Appendix A.3.

3.6.5 Transport and Reaction in Porous Media: Silicate Diagenesis

In aquatic systems, diatoms form an important part of the food chain, and play a major role in the sequestration of carbon dioxide from the atmosphere, transferring it into deeper waters. Diatoms incorporate silicon both in their cell walls and in their cell content. As they sink through the water column, part of the biogenic silicon (BSi) is dissolved, but some fraction is deposited onto the sediment where it will further dissolve. As a result, the concentration of dissolved silicon (DSi) in the porewater of the sediment increases, while the concentration of particulate silicate decreases with increasing depth below the sediment-water interface. Especially in sediments where the deposition of biogenic silicon is high, not all

BSi is dissolved, and a variable fraction may be preserved in the sedimentary column.

We will make a simple model that describes these so-called 'diagenetic' reactions. The model (Schink et al., 1975) was one of the first dealing with the early diagenesis of silica. The model will be implemented and solved in Section 7.8.4.

Biogenic silicate (BSi), expressed in μmol l^{-1} solid, is affected by transport (1st term) and dissolution (2nd term); the latter is first-order with BSi concentration (dissolution rate λ) and decreases linearly with increasing dissolved silicate concentration, until an equilibrium concentration ($eqSi$) is reached at which dissolution stops. At the sediment-water interface, an amount, $Flux_0$, of biogenic silicate is deposited per time (i.e. we define a flux boundary condition); the deep boundary condition (at x=∞) is a zero-gradient condition.

$$\frac{\partial BSi}{\partial t} = \frac{1}{1 - \phi_x} \frac{\partial}{\partial x}[(1 - \phi_x) \cdot D_b \frac{\partial BSi}{\partial x}] - \lambda \cdot BSi \cdot (1 - \frac{DSi}{eqSi}) \qquad (3.56)$$

$$Flux_0 = -(1 - \phi_0) \cdot D_b \left. \frac{\partial BSi}{\partial x} \right|_0$$

$$\left. \frac{\partial BSi}{\partial x} \right|_\infty = 0 \qquad (3.57)$$

Dissolved silicate (DSi) is expressed in μmol l^{-1} liquid, and impacted by molecular diffusion (1st term), and produced by dissolution.

At the upper boundary, a bottom water concentration is prescribed; at large depths, a zero-gradient boundary is imposed.

$$\frac{\partial DSi}{\partial t} = \frac{1}{\phi_x} \frac{\partial}{\partial x}[\phi_x \cdot D_s \frac{\partial DSi}{\partial x}] + \lambda \cdot BSi \cdot (1 - \frac{DSi}{eqSi}) \cdot \frac{1 - \phi_x}{\phi_x} \qquad (3.58)$$

$$DSi_0 = BW$$

$$\left. \frac{\partial DSi}{\partial x} \right|_\infty = 0 \qquad (3.59)$$

Note the factor $(1 - \phi)/\phi$ to convert the dissolution rate from l^{-1} solid to l^{-1} liquid. The magnitude of this factor increases with decreasing porosity, i.e. the impact of biogenic silica dissolution on the dissolved silica will be relatively larger deeper into the sediment, where porosity is generally lower (and the relative importance of solids compared to liquids higher) than near the sediment-water interface.

Chapter 4
Parameterization

One of the consequences of writing a quantitative (mathematical) model is that we introduce parameters and constants that must be given a specific value. How modellers deal with that is the subject of this chapter.

Model parameterization can proceed in several ways:

- By measurements, where specific experiments are performed to establish the value of a parameter.
- By literature searches.
- By calibration, where parameters are estimated by fitting the model to data.

4.1 In Situ Measurement

This is the most-straightforward way of obtaining model parameters. One just measures what one needs.

For instance,

- Photosynthesis-light response curves (also known as P-I curves) can be used to parameterise the response of phytoplankton photosynthesis to light (see Section 4.4.1; Fig. 4.1).
- Sediment accumulation rates can be measured and used in a model that describes the biogeochemical cycles of carbon and nitrogen in the sediment.

Note, however, that in both examples the parameters of interest to the model are *derived* from the observations, not directly observed. A statistical analysis of the observations is needed, and care should be taken to fit a function to the data that is consistent with the formulation in the model where the parameters are to be used.

Also note that the result of such measurements is not an absolutely valid, fixed value for the parameter. Even if the statistical fit of the model to the data is excellent (Fig. 4.1), there always remains some variation of the observations around the fitted curve. The statistical fitting procedure yields information about the magnitude of this variation, e.g. in the form of standard errors of the fitted parameters (see Section 4.4.1) but also in the form of the residual variation. This information is

K. Soetaert, P.M.J. Herman, *A Practical Guide to Ecological Modelling*,
© Springer Science+Business Media B.V. 2009

Fig. 4.1 The relationship between photosynthesis and light intensity, fitted with the Eilers-Peeters model. Data courtesy of Jacco Kromkamp, NIOO

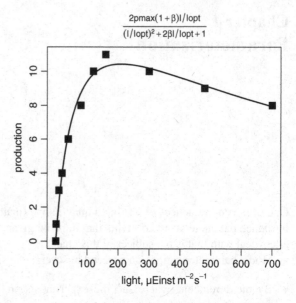

$$\frac{2pmax(1+\beta)I/Iopt}{(I/Iopt)^2 + 2\beta I/Iopt + 1}$$

production

light, µEinst m^{-2}s^{-1}

highly relevant for the uncertainty and sensitivity analysis of the model (Chapter 11) and should be kept wherever possible.

4.2 Literature-Derived Parameters

The most frequent source of model parameters is the scientific literature, where many rate parameters and mathematical equations can be found, either from other modelling papers or from field and laboratory measurements.

Sometimes, a specific parameter value cannot be found, in which case modellers may use *relationships with other parameters* as a last resort. In Fig. 4.2 we give two examples:

- Allometric relationships, where rates are expressed as a function of the size of the organisms. Several biological rates, on a log scale are highly significantly related to the log of the mass of the organisms, and this relationship is very near to linear, with a slope close to −0.25. Examples include: growth rates (Fig. 4.2 A), respiration rates, grazing intensities, rates of increase, Production versus Biomass ratios...
- Biogeochemical characteristics of the sediments in relation to water depth. Certain quantities such as sediment bioturbation rates, sediment community oxygen consumption rates (SCOC, Fig. 4.2 B), etc..., when transformed on a log scale correlate very well with the log of water depth.

Remark that, notwithstanding the often-high r^2 reported, these relationships do NOT allow accurate estimation of the parameters, as for any given X, the total range of possible Y's may span almost one, sometimes more than one order of

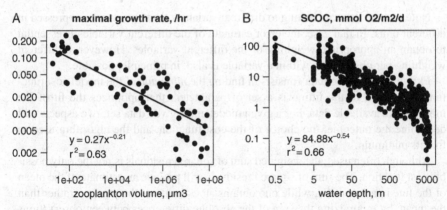

Fig. 4.2 (**A**) log-log relationships describing maximal growth rate of zooplankton as a function of its biovolume (data as in Hansen et al. 1997) and (**B**) Depth variation of sediment community oxygen consumption (SCOC; data compiled by Andersson et al. 2004). Both relationships are much better fitted with a power compared to a simple linear function. See the original articles for more elaborate fit models

magnitude. This is due to the double log transformation such that minor scatter along the regression line in fact represents considerable error in the relationship. This range of uncertainty might greatly affect the outcome of models, which is why these relationships should be used only as a last resort.

Fortunately, one generally finds a stronger allometric relationship within taxonomic groups and similarly, the relationship with depth of various biogeochemical sediment properties is much clearer in a more restricted geographical area. Such interpolations based on a more focused data set may reduce the uncertainty introduced in the model.

4.3 Calibration

In model calibration, the output of the model (after it has been solved, see next chapters) is compared to observations and the mismatch between model and data is reduced by fine-tuning the least-known parameters.

The mismatch of model to data is termed the 'model cost' and is usually evaluated as a weighted sum of squared residuals:

$$ModelCost = \sum_i \frac{(Modelvalue_i - Observedvalue_i)^2}{error_i} \quad (4.1)$$

where i is a data point, and error is a measure of the accuracy of the data (i.e. the data variance). Data that are more precisely known must be fitted more precisely and therefore get more weight (they have a smaller error).

Note that the data can belong to different variables, and can even be expressed in different units. In that case, the error estimate of the different variables is essential to obtain an appropriate weighing of the different variables. However, a different weighting per observation within a variable is also, in principle, possible.

Fitting the model to data consists in finding the *minimum* of the model cost function. The result of the fitting is a set of parameters that optimizes the fit of the model to the available data. For a given model and a given data set, two aspects will determine the outcome: the choice of the cost function, and the algorithm used to find its minimum.

Although often used, the weighted sum of squared residuals is not the only possible cost function. The sum of squared residuals will provide an estimate of the mean of the fitted parameter(s), while one obtains an estimate of the median, rather than the mean, by minimizing the sum of the absolute differences between observations and model output. What cost function to use, depends on the (assumed) statistical distribution of the data points. In Bayesian statistics, it also depends on the prior information one has about the parameters and the model functions. We refer to advanced statistical handbooks for more information on these problems.

Finding the minimum of the cost function, using some appropriate algorithm, sounds much simpler than it is! Most ecologists are used to fitting linear models for which there exist clear-cut equations, so-called linear least-squares fits. However, many ecological models are non-linear and as a consequence there may not be a simple relation that expresses the dependence of the model fit to a model parameter.

4.3.1 Linear Regression

A linear model is one where all equations are of the type:

$$y = ax + b \tag{4.2}$$

where a is the slope and b is the intercept.

Such equations are solved for the unknown parameters a and b by linear regression techniques.

Several nonlinear relationships can be transformed to linear form, and such transformations were (and still are) often used to estimated the parameters. For instance, power functions are linearized by a double log-transformation (Fig. 4.3 A,B):

$$y = a \cdot x^b$$
$$\log(y) = \log(a) + \log(x^b) = a' + b \cdot \log(x) \tag{4.3}$$

By regressing $\log(y)$ against $\log(x)$, b can be estimated as the slope of the regression and then a is calculated from the antilog of the intercept ($a = \exp(a')$).

Similar transformations have also been used to estimate the parameters of inverted functions, e.g. the function

$$y = \frac{ax}{b+x} \tag{4.4}$$

can be transformed to:

$$\frac{1}{y} = \frac{b}{a} \cdot \frac{1}{x} + \frac{1}{a} \tag{4.5}$$

and a linear regression of $1/y$ versus $1/x$ gives the required parameters (Fig. 4.3 C,D).

There are however several reasons why linearization of non-linear models may not be the best solution. Classical linear regression makes a number of strong assumptions on the data. If these assumptions are not met, the results are invalidated. The most important assumption is that for any particular value of x, the observations y are normally distributed around the estimated value (on the regression line), with a constant variance and independence of the distributions from the value of x. This assumption does not remain untouched by data transformation. For instance, by inverting the observations as $1/y$, very small deviations for very small values of y

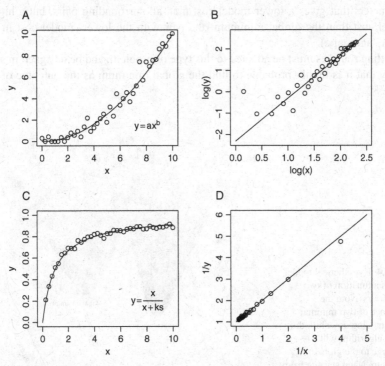

Fig. 4.3 Two examples where a non-linear model is transformed to linear form. **A,B**. power model. **C,D**. Monod equation. Pseudo-observations (*dots*) were generated by adding normally-distributed (Gaussian) noise with constant variance, around the non-linear model (A,C). Note that, after linear transformation, the distribution of the observations around the regression line has changed (B,D)

are 'blown up', such that the variance of the observations around the regression line, after transformation, is usually far from homogeneous (Fig. 4.3 B, D).

The methodology of General(ized) Linear Models (GLM) can solve a large number of these problems. Routines for GLM estimation are available in most statistical software packages, and these are the preferred method to estimate parameters of non-linear, but linearizable relations.

4.3.2 Nonlinear Fitting

A nonlinear model contains higher order terms or exponents, logarithms etc... Examples are:

$C=ax^2+b$, $C=\log(x)$, $C=x/(x+b)$

Finding the parameter values that minimise the model cost in non-linear models is not easy. Figure 4.4 illustrates a (fictitious) 'model cost landscape', expressing the cost as a function of the values of two parameters in a model. When starting from an arbitrary point and going 'downhill' (i.e. changing the parameter values in the direction of lower model cost) one can easily be trapped in a local minimum (e.g. when starting from point 2). A local minimum is a combination of the two parameters that gives a lower model cost than all surrounding pairs, but a higher model cost than the global minimum (the pair with the lowest model cost in the whole landscape).

Fitting routines must be adapted to this type of problem, and be designed in such a way that it is highly probable to find the global minimum as the outcome of the process.

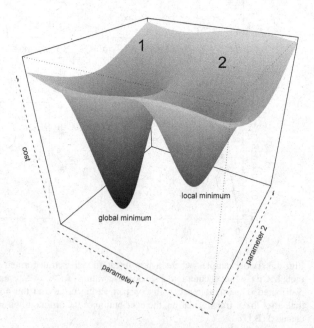

Fig. 4.4 Hypothetical model cost as a function of two parameters. Note the existence of two minima; when moving downhill, the fitting algorithm will converge to the global minimum when starting from parameter combination (1); when initiated at combination (2) the downhill path will end up in the local minimum

Nonlinear fitting routines work iteratively. This means they perform a large number of model calculations, where model fit successively improves, until a 'best' fit is found.

Unfortunately, there is usually no way of knowing whether the 'best fit' actually represents a truly global minimum or a local minimum, because in practice too many model runs would be required to generate the complete 'model cost landscape'.

Iterations may be guided either by:

1. *Mathematical* considerations: steps are taken in the direction of a better fit (a decrease of the cost function). As these methods often get trapped in local minima, they are not necessarily the best for ecological applications.
2. *Random-based* techniques. Here the parameter domain is explored with a certain degree of randomness. In Section 4.4.3 an example of such a procedure is shown. Two other, often-used techniques are:

   - *Simulated annealing*, in which the best model fit is sought in a similar way as in nature a perfect crystal is formed: liquid material at a very high temperature is cooled slowly, ensuring thermal equilibrium at each temperature, giving the system time to create a perfect crystal. The simulated annealing algorithm adapts the above process to minimise a model cost function: starting with a randomly chosen parameter set, each new parameter configuration with a lower cost is always accepted. However, there is also a non-vanishing probability for a configuration at higher cost to be accepted. At the start, parameters are allowed to change drastically and the probability of accepting a worse fit is high. Both the change rate of parameters and the probability of accepting decrease as the method proceeds.
   - *Genetic algorithms* mimic evolutionary processes. The 'population' consists of a set of parameters; their values are the 'genome'. Each of these values relates to certain 'fitness' (the negative of the model's cost function). From this population, a child population is created by allowing 'breeding' between members, cross-over and random mutation of the 'genome'. When the number of children equals the number of parents, the former replace the latter and a new generation is 'born'. As the probability of a parent to be selected for reproduction depends on its fitness (the fitter, the larger the probability of propagation), the average fitness of each generation increases as the calibration proceeds.

4.4   Case Studies in R

4.4.1   Nonlinear Parameter Estimation: P-I Curve

We first implement the fitting of a photosynthesis-light (P-I) curve.

Assume that primary production was measured with added ^{14}C in incubated phytoplankton samples, at different light intensities. The resulting production estimates are fitted with the 3-parameter Eilers-Peeters equation (see Section 2.8.2) using a non-linear estimation routine.

The Eilers-Peeters light-limitation function ([0-1]) is:

$$\text{LightLim} = \frac{2 \cdot (1 + \beta) \cdot I / Iopt}{(I / Iopt)^2 + 2 \cdot \beta \cdot I / Iopt + 1} \tag{4.6}$$

where I is light and β and *Iopt* are two parameters. Photosynthesis (pp) can be calculated as:

$$pp = pmax \cdot \text{LightLim} \tag{4.7}$$

where *pmax*, the maximal photosynthesis rate, is another parameter to be fitted.

We start by inputting the data and plotting primary production (pp) versus light (11). We use filled square symbols (pch=15), enlarged 50% (cex=1.5). The x-axis label is entered as an expression, which allows using superscripts.

```
ll <- c(0.,1,10,20,40,80,120,160,300,480,700)
pp <- c(0.,1,3,4,6,8,10,11,10,9,8)

plot(ll,pp,xlab= expression("light, μEinst"~ m^{-2}~s^{-1}),
     ylab="production",pch=15,cex=1.5)
```

Non-linear fitting of simple functions in R can be done using function nls, which requires as input the formula (y~function(x,parameters)) and starting values of the parameters.

Good starting value of the parameter *pmax* and *iopt* are the maximal value of measured primary production rates (pmax=max(pp)), and the light intensity (11) at which this maximal rate was achieved (iopt = 11[which.max(pp)]).

We end the fitting by giving a summary of the fitting parameters and store the values of the coefficients, as a list, in pars. In R, a 'list' is easier to work with, as it allows us to refer to the parameters by their names. Statement (with(pars, curve(...))) does this; it adds the best-fit curve to the graph (add=TRUE); the line width is twice the default (lwd=2). Finally, a title is added as an expression.

```
fit<-nls(pp ~ pmax*2*(1+b)*(ll/iopt)/
                      ((ll/iopt)^2+2*b*ll/iopt+1),
        start=c(pmax=max(pp),b=0.005,iopt=ll[which.max(pp)]))

summary(fit)
pars <- as.list(coef(fit))

with(pars,
     curve(pmax*2*(1+b)*(x/iopt)/((x/iopt)^2+2*b*x/iopt+1),
     add=TRUE,lwd=2)
  )
title(expression (frac(2*pmax*(1+beta)*I/Iopt,
              (I/Iopt)^2+2*beta*I/Iopt+1)),cex.main=0.8)
```

The resulting figure can be found as Fig. 4.1.

The outputted summary table below shows the estimates of the parameters and their standard errors. Standard errors of the parameters are very useful for sensitivity analysis of the model (see Chapter 11), as they indicate the uncertainty associated to the parameter values used.

```
Estimate          Std. Error t value    Pr(> |t|)

pmax   10.4351    0.3171     32.909     7.93e-10   ***
b       1.5998    0.4353      3.676     0.00626    **
iopt  209.6325   15.8052     13.264     9.96e-07   ***
```

Note though that the standard deviations just represent the uncertainty due to fitting this one set of primary production values, derived from one sample of algae. In practice, one has to consider all important sources of variability, including temporal and spatial effects.

4.4.2 Linear Versus Non-Linear Parameter Estimation: Sediment Bioturbation

By their movement and feeding activity, animals inhabiting aqueous sediments redistribute particles in sediments. In many cases, this animal activity can be modelled analogous to dispersion, and where the dispersion coefficient is called a bioturbation coefficient (see also previous chapters).

Radio-active tracers, such as Th^{234} or Pb^{210}, are also subjected to mixing and decay and the decay rate is well known and fixed: it is 0.029 d^{-1} and 0.031 yr^{-1} respectively.

By fitting activity versus sediment depth profiles of these tracers, the bioturbation coefficient can be estimated.

The underlying model is the 1-dimensional steady-state dispersion-reaction equation with constant surface area, and with an imposed concentration boundary at the upper interface (sediment-water interface), a zero-gradient boundary at infinite depth (advection is considered negligible):

$$0 = D_b \frac{\partial^2 C}{\partial x^2} - \lambda \cdot C \tag{4.8}$$

with boundary conditions:

$$C|_{x=0} = C_0$$

$$0 = \left. \frac{\partial C}{\partial x} \right|_{x=\infty} \tag{4.9}$$

and where D_b is the unknown mixing coefficient, λ is the (known) decay rate.

Fig. 4.5 Hypothetical measurements of Pb²¹⁰ activity versus sediment depth (symbols) and the best-fit line, obtained by nonlinear fitting (*dashed line*) and with a linear fit, after log transformation (*solid line*). The inset shows the same data and fits, but on a log scale for the x-axis

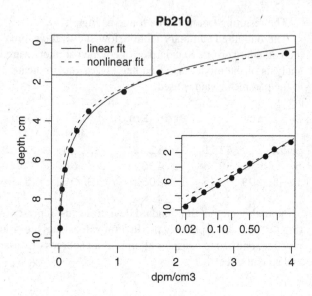

The solution to this equation is given by (see Chapter 5):

$$C(x) = C_0 \cdot \exp^{-\sqrt{\frac{\lambda}{D_b}} \cdot x} \qquad (4.10)$$

Where C(x) is the activity of the tracer at depth x.
After log transformation, we obtain:

$$\log_e(C(x)) = \log_e(C_0) + \sqrt{\frac{\lambda}{D_b}} \cdot x \qquad (4.11)$$

In the following R-script, we fit the mixing model to a hypothetical data set consisting of Pb²¹⁰ versus sediment depth measurements; Pb²¹⁰ data are generally expressed in dpm/cm³ sediment (and where dpm=disintegrations per minute). The unknown parameters are the Pb²¹⁰ value at the interface (C_0) and the bioturbation rate (D_b). We fit the data both using linear regression on the log-transformed data and using non-linear regression. The results are depicted on a linear scale (main plot in Fig. 4.5) and on a log scale (inset).

After inputting and plotting the data, we first fit the log transformed data using R's linear fit model function `lm`; `coef(LL)` retrieves the linear fitting parameters which are back transformed to retrieve the desired parameters C_0 and Db. A line with the obtained fit is added to the plot (`lines`).

```
# the data:
x     <- 0.5:9.5
y     <- c(3.9,1.7,1.1,0.5,0.3,0.2,0.1,0.05,0.03,0.02)

plot(y,x,pch=16,ylab="depth,cm", xlab="dpm/cm3",
     main="Pb210", ylim=c(10,0))

# the decay rate
lam   <- 0.031

# linear fit
LL    <- lm(log(y)~ x)
C0    <- exp (coef(LL)[1])
Db    <- lam/(coef(LL)[2])^2

xx    <-seq(0,10,0.1)
lines(C0*exp(-sqrt(lam/Db)*xx),xx)
```

Performing the nonlinear fit is only slightly more complex and uses R's non-linear fitting routine nls. This routine also requires reasonable initial guesses (start) of the parameters. The non-linear fit is added to the graph, using a dashed line type (lty). A legend is also added.

```
fit<-nls(y ~C0*exp(-sqrt(lam/Db)*x),
         start=c(C0=5,Db=0.1))

C02   <- coef(fit)[1]
Db2   <- coef(fit)[2]

lines(C02*exp(-sqrt(lam/Db2)*xx),xx,lty=2)

legend("topleft",lty=c(1,2),
       c("linear fit", "nonlinear fit"))
```

Finally, the results are also depicted on a log x-scale (log="x"). We put these in an inset, which is put in the right–lower corner of the graph (fig).

```
par(new=TRUE,fig=c(0.5,1,0.1,0.6))

plot(y,x,pch=16,ylab="",xlab="",ylim=c(10,0),log="x")
lines(C0*exp(-sqrt(lam/Db)*xx),xx)
lines(C02*exp(-sqrt(lam/Db2)*xx),xx,lty=2)
```

Note that the nonlinear fitting algorithm reproduces more faithfully the (large) activity in the upper part of the sediment, whilst the linear fitting of log-transformed data stays closer to the data at the lower activities (mainly visible in the inset) (Fig. 4.5).

4.4.3 Pseudo-Random Search, a Random-Based Minimization Routine

Both random-based techniques that were discussed in Section 4.3.2, the method of simulated annealing and the genetic algorithm are included in the R software. The former is standard in R's optimization function 'optim', the latter is included in a specific R-package.

Notwithstanding the broad success of these two methods, the pseudo-random search algorithm of Price (1977) is our favourite random-based minimization routine. Although it is only seldom used, it is appallingly simple, yet very effective. We implement it to illustrate the principles of a random-based optimization algorithm. In addition, this example serves to demonstrate how to create a function in R.

The algorithm of the pseudo-random search method is as follows:

- Start with a 'population' of parameter vectors (with parameter vector, we mean a set of parameter values, one value for each parameter to be fitted). For each parameter vector the model cost is estimated. This step is similar in a genetic algorithm.
- During each calibration step, three parameter vectors are randomly selected from the population, the centroïd (mean) calculated, and mirrored over a fourth, also randomly drawn parameter vector (Fig. 4.6).
- The model cost is calculated with this new parameter vector.
- In case the obtained model cost value is lower than the worst cost value in the population, the new parameter vector replaces the worst parameter vector in the population, else it is discarded.
- The whole procedure is repeated until a requested number of runs have been performed. As better parameter values replace worse values, the average cost function of a population will decrease in time.

The function below that implements the algorithm of Price in R is called 'pricefit'; it takes as input at least 8 arguments. Some of these arguments are given a default value (minpar, maxpar, npop, numiter, centroid, varleft). When calling the function, these arguments need not be specified, if we agree with their default value. Other function arguments are not specified (par, func). These *must* be given a value in the call to this function.

Note that we also pass a function as an argument (func). This function should calculate the model cost value based on the model parameters (and possible other arguments). As we cannot foresee which extra information will be passed to this function, we use a '...' argument, which allows passing all arguments that are specific to func.

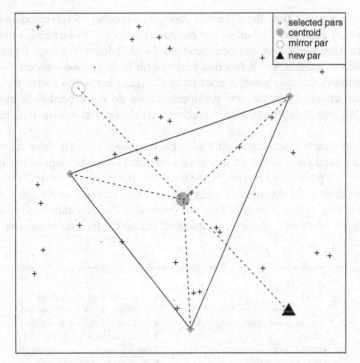

Fig. 4.6 Strategy for generating a new parameter set (vector) in the population during the pseudo-random search algorithm (Price, 1977). Here a parameter set consists of a parameter pair (x-y). Three parameter sets are randomly selected (*grey diamonds*), their centroid estimated (*grey circle*) and mirrored over a fourth randomly selected parameter (*open grey circle*). See text for details

Note also that, within `pricefit`, we redefine this internal function, calling it 'cost' and passing the '...' arguments; this makes the code more readable.

During the initialization phase, a matrix with randomly-chosen parameter values, drawn between minimal and maximal bounds is generated (`populationpar`). R's function `runif(npar*npop)` generates npar * npop uniformly distributed random numbers between 0 and 1. The parameters receive the same names (if any) as can be found in the initial parameter set. This is important if the cost function uses the names of these parameters.

The first of these random parameter values (`populationpar[1,]`) is over-written by the initial parameter set, and passed as an argument to the function (`par`).

Then the cost value for each of these parameter sets is estimated and stored in a vector (`populationcost`), and the worst parameter set (with the highest population cost) assessed. R's function `apply` applies a certain function to either the rows (`MARGIN=1`) or columns (`MARGIN=2`) of a matrix. Here the matrix is `populationpar`, and function 'cost' is applied to its rows (`MARGIN=1`).

After the initialization, the algorithm performs a number of hybridization steps, during which a number (default=3) of parameter sets (`selectpar`) is randomly picked from the population, and their mean estimated. R-function `sample` generates these random samples, and R-function `colMeans` takes column-wise means. Another randomly selected parameter set (`mirrorpar`) is used to mirror the mean value; it is checked that the new parameter values do not exceed their imposed ranges. We use R-functions `pmin` and `pmax` to take vector-wise mimina and maxima.

The cost function associated with the new parameter vector is now calculated (`newcost`) and compared with the worst cost function in the population; if it is better, the new parameter set replaces the worst set. Iteration goes on until either the maximum number of iterations (`numiter`) has been reached, or the range in the population cost values is smaller than the fraction of the minimum cost specified by the factor `varleft`. After performing all iterations, the best parameter set is selected.

```
pricefit <- function (
         par,                              # initial par estimates
         minpar=rep(-1e8,length(par)),     # minimal parameter values
         maxpar=rep(1e8,length(par)),      # maximal parameter values
         func,                             # function to minimise
         npop=max(5*length(par),50),       # nr elements in population
         numiter=10000,                    # number of iterations
         centroid = 3,                     # number of points in centroid
         varleft  = 1e-8,                  # relative variation upon stopping
         ...)

{

# Initialization

  cost    <- function (par) func(par,...)
  npar    <- length(par)
  tiny    <- 1e-8
  varleft<-max(tiny,varleft)

populationpar <- matrix(nrow=npop,ncol=npar,byrow=TRUE,
            data= minpar+runif(npar*npop)*rep((maxpar-minpar),npop))
colnames(populationpar)<-names(par)
populationpar[1,]<-par

populationcost <- apply(populationpar,FUN=cost,MARGIN=1)
iworst           <- which.max(populationcost)
worstcost        <- populationcost[iworst]

# Hybridization phase
iter<-0
while (iter<numiter & (max(populationcost)-min(populationcost))
                            >(min(populationcost)*varleft))
```

```
{
  iter<-iter+1

  selectpar <- sample(1:npop,size=centroid)      # for cross-fertilization
  mirrorpar <- sample(1:npop,size=1)             # for mirroring
  newpar   <- colMeans(populationpar[selectpar,])  # centroid
  newpar   <- 2*newpar - populationpar[mirrorpar,] # mirroring

  newpar   <- pmin( pmax(newpar,minpar) ,maxpar)

  newcost <- cost(newpar)

  if (newcost < worstcost)
  {
      populationcost[iworst] <-newcost
      populationpar [iworst,]<-newpar
      iworst    <- which.max(populationcost) # new worst member
      worstcost <- populationcost[iworst]
  }
  } # end j loop

  ibest    <- which.min(populationcost)
  bestpar  <- populationpar[ibest,]
  bestcost <- populationcost[ibest]
  return (list(par = bestpar, cost = bestcost,
              poppar = populationpar, popcost=populationcost))
}
```

All that is left to do is to test the algorithm of Price (1977). We test it on a problem that is particularly hard to fit, i.e. we try to find the amplitude, period and phase of a sine wave. We start by generating 'data points' conform a sine wave with known parameters. The 20 data points are generated at random x-values between 0 and 13 (runif), and a small amount of normally distributed noise (rnorm) with small standard deviation (sd) is added to the y-values. The data are plotted (pch=16 selects filled squares).

Then we define the model cost function, which takes as input the three parameters (par) and calculates the sum of squared residuals of model output (generated with these parameters) and the 'observed' y-values.

The minimum of this function is then sought using 3 methods: p1 is the minimum using the default, mathematical method implemented in R's function optim; p2 is the minimum retrieved by the method of simulated annealing (SANN), also implemented in optim, and p3 is based on the algorithm of Price. The latter also allows to impose bounds on the function; we give high upper bounds for the amplitude and period, and the upper limit of $2*\pi$ for the phase.

Finally the results of the three fitting routines are added; we use R-function curve, with parameter add=TRUE, to do this.

```
amp     <- 6
period <- 5
phase  <- 0.5

x <- runif(20)*13
y <- amp*sin(2*pi*x/period+phase) +rnorm(20,mean=0,sd=0.05)
plot(x,y,pch=16)

cost <- function(par)
{
 with(as.list(par),{
    sum((amplitude*sin(2*pi*x/period+phase)-y)^2)
                })
}

p1 <- optim(par=c(amplitude=1,period=1,phase=1), cost)
p2 <- optim(par=c(amplitude=1,period=1,phase=1), cost,method="SANN")
p3 <- pricefit(par=c(amplitude=1,period=1,phase=1),minpar=c(0,1e-8,0),
            maxpar=c(100,2*pi,100), func=cost,numiter=3000)

curve(p1$par[1]*sin(2*pi*x/p1$par[2]+p1$par[3]),lty=2,add=TRUE)
curve(p2$par[1]*sin(2*pi*x/p2$par[2]+p2$par[3]),lty=3,add=TRUE)
curve(p3$par[1]*sin(2*pi*x/p3$par[2]+p3$par[3]),lty=1,add=TRUE)

legend ("bottomright",lty=c(1,2,3),c("Price","Mathematical","Simulated
annealing"))
```

Due to the randomness in the generation of data points and in the simulated annealing and Price algorithm, every run of this code is different. Figure 4.7 is one output, which shows that the Price algorithm is the only method that has recovered the initial parameter values, whilst both the mathematical and simulated annealing methods have failed. In 50 trials of the fitting, the Price algorithm succeeded in all of them, whilst the simulated annealing algorithm was successful in 11 trials, and the mathematical routine always (!) failed.

4.4.4 Calibration of a Simple Model

In this example we will show, using a model with two parameters, how to visualise the cost landscape of the model, and how to use optimization routines to estimate the 'best' value of the parameters, given a set of observations.

Calibration of a model requires *solution* of the model: one has to obtain an explicit time course of the state variables, in order to compare their values with the observations. We cannot illustrate calibration without solving the model too. We advise readers to skip this section and return to it after reading Chapters 5 and 6, if they have not yet done so.

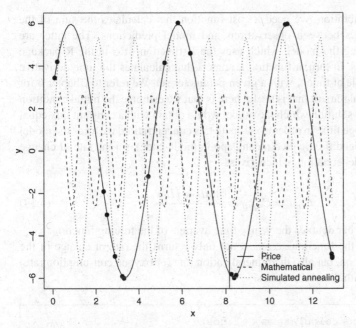

Fig. 4.7 Comparison of best-fit curves, as derived by the Price (pseudo-random search) algorithm (*solid line*), by the method of simulated annealing (*dots*) and by the standard (mathematical) method implemented in R (*dashes*). The fitting points were generated as a sine wave

Our aim is to describe, using a simple model, the rate of sediment carbon mineralization (respiration) in a deep-sea station. For this station observations are available from sediment traps, specifying the rate of carbon deposition over the season. At several occasions, landers have been deployed to measure oxygen consumption by the sediment community. The data are an excerpt from the publication by Sayles et al., 2002.

The simplest model possible for this case is that organic matter decays at a first-order rate. Thus, the dynamics of organic carbon in the sediment is:

$$\frac{dC}{dt} = flux - k \cdot C \tag{4.12}$$

and where k is a parameter (day^{-1}), flux is a forcing function (from sediment trap observations).

For many reasons it can be suspected that sediment traps tend to underestimate the carbon flux to the sediment. We want to account for this effect by assuming that the real flux to the sediment is some multiple of the observed flux in the traps. Thus, we define a dimensionless parameter *mult* that expresses the ratio between the 'real' and the observed flux, which leads to the following model:

$$\frac{dC}{dt} = mult \cdot flux - k \cdot C \tag{4.13}$$

For model calibration, we need a cost function that calculates the sum of the squared differences between observations and model predictions. The latter are obtained by solving the model, which uses the R function `ode` in the R-package `deSolve`. It calls, in turn, a function `minmod` that calculates the time derivative of the state variable at time t, with a given parameter set. We refer to Chapter 6 for details. A technical remark that is useful here is that we estimate the initial condition as the equilibrium solution which the model would obtain with a constant flux equal to the yearly average flux. In order to estimate this concentration, we force the model by this time-averaged flux (called meanDepo in the R-code), set the rate of change to zero in the model equation, which gives:

$$C_{init} = C_{equil} = \frac{mult * \overline{flux}}{k} \tag{4.14}$$

Where the overbar denotes the yearly time average of the forcing function.

Also note that the function `minmod` not only returns the rate of change of the state variable Carbon, but also the mineralization rate (or oxygen consumption rate) for comparison with the observations.

```
# Function to calculate model cost

costf <- function(params)
  {with(as.list(params),
    {
       Carbon     <- meanDepo*mult/k
       outtimes <- as.vector(oxcon$time)
       outmin   <- ode(Carbon,outtimes,minmod,params)
       costt     <- sum((outmin[,3]-oxcon$cons)^2)
       return (costt)
    })
  }

# Function to calculate derivative of state variable

minmod <- function(t,Carbon,parameters)
  {  with (as.list(c(Carbon,parameters)),
    {
        minrate <- k*Carbon
        Depo     <- approx(Flux[,1],Flux[,2], xout=t)$y
        dCarbon <- mult*Depo - minrate
     list(dCarbon,minrate)
    })
  }
```

The dataset consists of 17 observations, clustered at three moments in time, of oxygen consumption of the sediment community. These data are stored in the data frame `oxcon`. The forcing function of carbon deposition is given in the matrix `Flux`. We start the R code by specifying these observational data.

```
#General settings
require(deSolve)

# Define problem and data

Flux <- matrix(ncol=2,byrow=TRUE,data=c(
  1, 0.654, 11, 0.167, 21, 0.060, 41, 0.070,
 73, 0.277, 83, 0.186, 93, 0.140,103, 0.255,
113, 0.231,123, 0.309,133, 1.127,143, 1.923,
153, 1.091,163, 1.001,173, 1.691,183, 1.404,
194, 1.226,204, 0.767,214, 0.893,224, 0.737,
234, 0.772,244, 0.726,254, 0.624,264, 0.439,
274, 0.168,284, 0.280,294, 0.202,304, 0.193,
315, 0.286,325, 0.599,335, 1.889,345, 0.996,
355, 0.681,365, 1.135)))

meanDepo <- mean(approx(Flux[,1],Flux[,2], xout=seq(1,365,by=1))$y)

oxcon<-as.data.frame(matrix(ncol=2,byrow=TRUE,data=c(
 68, 0.387, 69, 0.447, 71, 0.473, 72, 0.515,
189, 1.210,190, 1.056,192, 0.953,193, 1.133,
220, 1.259,221, 1.291,222, 1.204,230, 1.272,
231, 1.168,232, 1.168,311, 0.963,312, 1.075,
313, 1.023)))

names(oxcon)<-c("time","cons")
```

We now have all elements needed to investigate how the model cost varies as a function of the parameters k and *mult*. To do so, we vary both parameters over a range of values and store the model cost for each of them. When doing this, we also find parameter values for which the model cost was minimal.

```
multser    <- seq(1,1.5,by=.05)
numms      <- length(multser)
kseries    <- seq(0.001,0.05,by=0.002)
numks      <- length(kseries)

outcost    <- matrix(nrow=numms,ncol=numks)

   for (m in 1:numms)
   {
   for (i in 1:numks)
    {
     pars <- c(k=kseries[i],mult=multser[m])
     outcost[m,i] <- costf(pars)
    }
   }

minpos<-which(outcost==min(outcost),arr.ind=TRUE)
multm<-multser[minpos[1]]
ki<-kseries[minpos[2]]
```

We now call the Price algorithm, with these parameter values as starting values, to find the optimal set of parameters for cost minimization. Be patient, this may take a while!

```
optpar <- pricefit(par=c(k=ki,mult=multm),minpar=c(0.001,1),
        maxpar=c(0.05,1.5),func=costf,npop=50,numiter=500,
        centroid=3,varleft=1e-8)
```

To demonstrate how the Price algorithm approaches the optimum solution, we call it two more times, with different stopping rules, once when all costs in the population are within 20% of the optimal cost, and once when all costs are within 2.5% of the optimum. We store the results of these calls in two different output lists.

```
optpar20 <- pricefit(par=optpar$par,minpar=c(0.001,1),
        maxpar=c(0.05,1.5),func=costf,npop=50,numiter=500,
        centroid=3,varleft=0.2)
optpar25 <- pricefit(par=optpar$par,minpar=c(0.001,1),
        maxpar=c(0.05,1.5),func=costf,npop=50,numiter=500,
        centroid=3,varleft=0.025)
```

We plot the model output (Fig. 4.8), using the optimal set of data as determined in the first call to the Price algorithm. We call the model solution once more, using this optimal set of parameters and the appropriate initial conditions.

Then we plot the forcing function of the model, stored in Flux, and add to this the model-calculated mineralization rate, and the observed oxygen consumption data.

We then define a new plot, without axes, within which we plot carbon concentration. An axis for this variable is added to the right-hand side of the plot. Finally, a legend is added.

```
windows()
outtimes        <- seq(1,365,by=1)
Carbon          <- meanDepo*optpar$par[2]/optpar$par[1]
names(Carbon) <-"Carbon"
out             <- as.data.frame(ode(Carbon,outtimes,minmod, optpar$par))
names(out)      <- c("time","Carbon","minrate")

par (oma=c(0,0,0,2))
plot(Flux,type="l",xlab="daynr",ylab="mmol/m2/d",
    main="Sediment-detritus model",lwd=2)
lines(out$time,out$minrate,lwd=2,col="darkgrey")
points(oxcon$time,oxcon$cons,pch=25,col="black", bg="darkgray",cex=2)
par(new=TRUE)
plot(out$time,out$Carbon,axes=FALSE,xlab="",ylab="",
    type="l",lty=2)
axis(4)
mtext(side=4,"mmolC/m2",outer=TRUE)
legend("topleft",col=c("black","darkgrey","black"),
leg=c("C flux","C mineralization","C concentration"),
    lwd=c(2,2,1),lty=c(1,1,2))
```

We also plot the model cost landscape (Fig. 4.9), from the grid search we performed varying *k* and *mult*. This is done using a contour plot. We add to this contour plot three more data sets: the results of the calls to the Price algorithm. The call with a stopping rule at 20% of the cost variation resulted in a population of parameter vectors satisfying this condition. These vectors are plotted with small circles. Next we plot, as plusses, the population of parameter vectors when the stopping rule was stricter (at 2.5%), and finally we add the optimal parameter vector as a large black dot. Note that filled contour graphs in R generate several internal coordinate systems, so that plotting additional points within the graph is slightly more complicated than for normal plots.

It can be seen from the plot that the clouds of parameter vectors in the different sets all fall within shapes found in the contour lines as based on the grid search. Within the Price algorithm, the 'population' slowly homes in on the final position, keeping some variation within the population but gradually decreasing this variation.

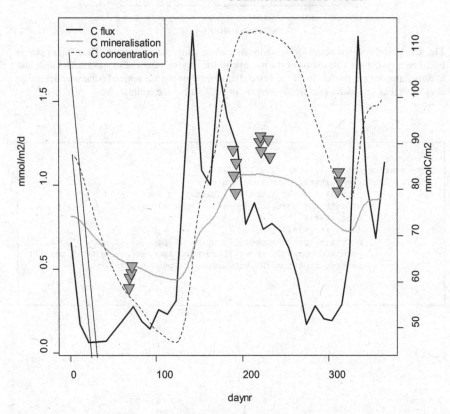

Fig. 4.8 Output of the sediment mineralization model using the optimal parameter set, and also showing the observed data on sediment community oxygen consumption

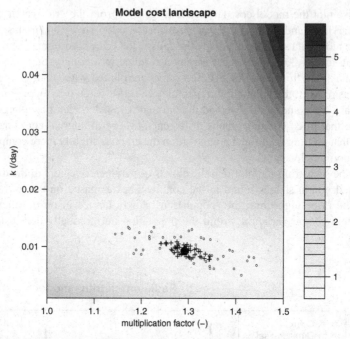

Fig. 4.9 Model cost landscape for the sediment carbon mineralization model. The contour plot is based on a systematic calculation of the model cost for varying values of the parameters mult and k. Superimposed are the results of the Price algorithm, showing the optimal parameter set (*large dot*), as well as sets with a cost within 20% (o) or 2.5% (+) of the optimal cost

```
windows()
pgr<-gray.colors(n=25, start=0.95, end=0.0)
filled.contour(x=multser,y=kseries,z=outcost,
          ylab="k (/day)",xlab="multiplication factor (-)",
          main="Model cost landscape",col=pgr,nlevels=25,
          plot.axes={
          axis(1); axis(2);
          points(optpar20$poppar[,2],optpar20$poppar[,1],pch="o",cex=.5);
          points(optpar25$poppar[,2],optpar25$poppar[,1],pch="+",cex=1);
          points(optpar$par[2],optpar$par[1],pch=16,cex=2)
          }
          )
```

Chapter 5
Model Solution – Analytical Methods

After we have chosen the appropriate mathematical equations (Chapters 2 and 3) and have decided on the values of the parameters (Chapter 4), we may look for a method that will give a mathematical solution to the problem.

For some models, e.g. lattice models, the solution of the model is *algorithmically* determined, i.e. it follows automatically from the microscopic rules defined as the heart of the model. We have seen examples of this in Section 3.6.2 to 3.6.4.

Here, we concentrate on the solution of models formulated as one or as a set of differential equations, which describe how a certain quantity changes in time as a result of the interacting processes (the sources and sinks).

Writing down these differential equations is an intuitive route of capturing our process knowledge of a problem, but scientists are generally more interested in the quantity itself rather than in its rate of change.

To obtain that information, we need to solve the differential equations, that is, to derive the explicit evolution in time or with distance of the state variables. This procedure is based on *integration* of the differential equations, as will be shown in this chapter. Integrating the model requires that *initial conditions* and, if appropriate, *boundary conditions* are specified.

Simple models may allow the calculation of an *analytical* solution, where the solution can be written as an algebraic function of time, space (if appropriate), the parameters and forcing functions. This will be an exact solution, but the number of problems that can be solved analytically is limited. Many analytical solutions can be found in reference works, or using mathematical software tools.

For complex models, it will be necessary to implement a *numerical* solution. Numerical methods will be deferred to Chapter 6. For a proper understanding of what these numerical methods do, it is important to study analytical solutions first.

5.1 An Everyday Life Example

An analytical solution of a differential equation is an equation in which the value of the state variable(s) can be written as an algebraic function of the forcing variables, other state variables and parameters for every point in time (and if appropriate, in space).

Consider the following simple example from everyday life. Someone tells you, not without pride, that his salary has risen by 10% per year over the past three years (he indicates the *rate of change* of his salary). Obviously, you are curious to know if by now he is a rich man – probably because his looks do not confirm the impression that he wants to make on you. In trying to determine what his financial position is now, you face two different problems.

First, the salary obeys the model equation

$$\frac{dSalary}{dt} = 0.1 \cdot Salary \tag{5.1}$$

Somehow, we need to calculate the salary (not the rate of change) at every point in time (t); this is, we need to solve the model, such that we obtain a function:

$$Salary = f(t) \tag{5.2}$$

For the salary problem, we may use our intuition for how this solution will look like. If your salary increases by a fixed percentage every year, it will be a kind of exponential function in time. We know that from bank accounts with fixed interest and other similar examples. However, our intuition also tells us that if the person started off three years ago with a student grant, he will still be a poor man now. In contrast, if he started on a manager salary, he will now be a far-too-rich man. Translated in mathematical terms: we need to know *initial conditions* to fully solve the model.

5.2 Finding an Analytical Solution

Finding the analytical solution is done using *integration*. This is based on the fact that

$$\int d(f(t)) = f(t) + A \tag{5.3}$$

Thus, integration is the 'reverse' of differentiation, and therefore integration is the ideal tool to get rid of the differentials in our differential equation.

There is one problem, however. Integration only yields a solution up to a constant A (called the integration constant), and we need to find ways to estimate the value of this constant A. It is easy to see why this constant A appears in the solution: just remember that $d(f(x)+A) = d(fx)$.

The need for specifying initial conditions is linked with the appearance of constant A: by introducing knowledge of initial conditions, we actually determine the value of A, and the problem is solved.

Summarizing, finding an analytical solution proceeds in two steps (Fig. 5.1 A):

(1) Finding a *general solution* of the differential equation through integration. This is the solution that does not take into account the boundary or initial conditions. The general solution contains integration constants; the number of integration

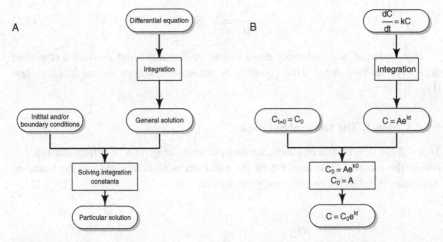

Fig. 5.1 Finding an analytical solution of a differential equation. **A**. General scheme. **B**. Simple example

constants equals the sum of the highest derivative in space and time (see Appendix A.3 for a definition of terms).

(2) The *particular solution* of the model problem is then found by considering the boundary conditions and/or initial conditions. This allows finding values for the integration constants. The necessity for finding values for the integration constants is the mathematical reason for the existence of initial and boundary conditions.

The most pragmatic way of finding a general analytical solution of a (set of) differential equation(s) is to look up the solutions in a book - the lookup technique. This making use of mathematical literature is what most modellers do, thus there is no point in repeating the theory of how to integrate a differential equation. General solutions to many of the differential equations discussed in this book can be found in Appendix B.3.

However, it is worthwhile to explain in some detail how one deals with the initial and boundary conditions, i.e. how these are used to solve for the integration constants and obtain a particular solution. We will explain this procedure via several examples.

5.3 Examples

5.3.1 A Very Simple First-Order Differential Equation

The ecological equivalent of the salary example is a model that describes population growth. We denote with N the density of a population, with λ the instantaneous rate of increase (units of time^{-1}). The differential equation is:

$$\frac{dN}{dt} = \lambda \cdot N \tag{5.4}$$

This equation is a first-order equation, as the highest-order derivative (the time derivative) is first-order. This equation is sometimes known as the Malthus law (Fig. 5.1B).

5.3.1.1 Step 1. The General Solution

To solve the differential equation, we first perform the inverse of differentiating, i.e. taking the integral. The solution of the equation is standard and can be found in Appendix B.3, but it can very easily be derived:

$$\frac{dN}{dt} = \lambda \cdot N$$

$$\frac{dN}{N} = \lambda \cdot dt$$

$$\int \frac{dN}{N} = \int \lambda \cdot dt$$

$$\int d \ln N = \int \lambda \cdot dt$$

$$\ln N = \lambda \cdot t + A'$$

$$N(t) = A \cdot e^{\lambda t} \qquad where \; A = e^{A'} \tag{5.5}$$

In this derivation we have first rearranged terms, then taken integrals of both sides, then used basic integration rules to derive the general solution.

The general solution contains one arbitrary integration constant: A. There is only one constant because the equation is first-order. A second-order equation would have two constants in the general solution (see examples below).

5.3.1.2 Step 2. The Particular Solution

For the particular solution we must find a value for the integration constant A. The specification of the value of N at t_0 ($N = N_{t0}$ at $t = t_0$) makes sure that the time path of N is dully determined and therefore the model fully specified. In the model of the salary, N_0 would be the initial salary, in the population model it would be the initial population density.

The value of A can then be calculated through the initial conditions: for the known density at time $= t0$ we can write:

$$N(t_0) = N_{t0} = A \cdot e^{\lambda t 0} \tag{5.6}$$

Which allows writing the arbitrary constant A as a function of the known initial condition, N_{t0}:

$$A = N_{t0} \cdot e^{-\lambda t0} \tag{5.7}$$

Thus the particular solution to the problem is:

$$N(t) = e^{-\lambda t0} \cdot N_{t0} \cdot e^{\lambda t} \tag{5.8}$$

or

$$N(t) = N_{t0} \cdot e^{\lambda(t-t0)} \tag{5.9}$$

Note: generally, t_0 is taken equal to 0 and the model solution simplifies to:

$$N(t) = N_0 \cdot e^{\lambda t} \tag{5.10}$$

Using this formula we can calculate the value of N for any time t. In the salary example, and with $N_0 = 1000\,€$ and after 3 years, the salary would amount to: $1000*e^{(0.1*3)} = 1350\,€$

5.3.2 The Logistic Equation

We next solve the logistic equation of Verhulst (1838), which is a simple carrying-capacity model from Section 2.7.1.

The differential equation reads:

$$\frac{dN}{dt} = r \cdot N \cdot \left[1 - \frac{N}{K}\right] \tag{5.11}$$

and the general solution of this function can be found in Appendix B.3:

$$N(t) = \frac{K}{1 + A \cdot e^{-rt}} \tag{5.12}$$

where A is an integration constant, that must be derived from the initial condition. With N_0 the initial concentration at time $t = 0$, and with $e^0 = 1$, we obtain:

$$N_0 = \frac{K}{1 + A \cdot e^0} = \frac{K}{1 + A} \tag{5.13}$$

Rearranging:

$$A = \frac{K - N_0}{N_0} \tag{5.14}$$

Thus, the particular solution for the logistic equation with initial density N_0 is:

$$N(t) = \frac{K}{1 + \left[\dfrac{K - N_0}{N_0}\right] e^{-r \cdot t}} \tag{5.15}$$

5.3.3 A Second-Order Differential Equation: Carbon Dynamics in Sediments (*)

As a third, considerably more difficult example, consider the steady-state, 1-dimensional dispersion-advection-reaction equation, expressing the rate of change of carbon (C) as a function of bioturbation, advection, and decay:

$$\frac{\partial C}{\partial t} = 0 = \frac{\partial}{\partial x}\left(D_b \frac{\partial C}{\partial x}\right) - u\frac{\partial C}{\partial x} - kC \tag{5.16}$$

where x is depth in the sediment, D_b is the bioturbation coefficient, u is the advection (sedimentation) rate and k is a first-order decay coefficient.

By 'steady-state', we mean that the concentrations do not change in time. We deal with steady-states in Chapter 7. For now it suffices to note that the 'steady-state' is found by setting $\frac{\partial C}{\partial t} = 0$

The *general solution* of this differential equation can be found in Appendix B.3 of this book and expresses the distribution of carbon (C) as a function of depth (x):

$$C|_x = A \cdot e^{\frac{u-\sqrt{u^2+4\cdot D_b\cdot k}}{2D}\cdot x} + B \cdot e^{\frac{u+\sqrt{u^2+4\cdot D_b\cdot k}}{2D}\cdot x} = A \cdot e^{a\cdot x} + B \cdot e^{b\cdot x} \tag{5.17}$$

where A and B are two integration constants. There are two integration constants as the model is a second-order differential equation (the highest derivative, the diffusive term, is second-order).

To be fully specified this second-order differential equation needs two *boundary conditions* that will be used to solve for the two integration constants. If you wonder why we discuss boundary, rather than initial conditions, remember that we are solving for the 'steady state' of our model. We thus have effectively removed time, but not depth, from our equations. We still need to define though what happens on the edges (boundaries) of the model domain.

For submerged sediments, one natural boundary is the interface with the overlying water (x = 0), where the flux of organic matter to the sediment is specified.

The other boundary condition is generally taken at infinite depth (x = ∞). As we want all the consumption of organic matter to be performed within the sediment column, we specify that the organic matter concentration at infinite depth must be zero (note that we are modelling degradable organic matter only, not the substances that cannot be decomposed in ecologically relevant time scales – our time scale window – and that will eventually be buried in the sediment). Thus the boundary conditions are, at zero depth and infinite depth respectively:

$$Flux = -D_b \left.\frac{\partial C}{\partial x}\right|_{x=0} + wC_{x=0}$$

$$C_{x=\infty} = 0 \tag{5.18}$$

where *Flux* is a known quantity.

These two equations are sufficient to solve for the unknown integration constants A and B, in two steps:

1. We start with the second boundary condition:

$$C_{x=\infty} = A \cdot e^{a \cdot \infty} + B \cdot e^{b \cdot \infty} = 0$$

$$a = \frac{w - \sqrt{u^2 + 4 \cdot D_b \cdot k}}{2D_b} \tag{5.19}$$

$$b = \frac{u + \sqrt{u^2 + 4 \cdot D_b \cdot k}}{2D_b} \tag{5.20}$$

The shape of an exponential equation is depicted in Fig. 5.2, for positive (left) and negative coefficients (right). The function tends to infinity with positive, to 0 with negative coefficients.

With u, D_b and k being positive numbers, the second term $B.e^{(b \cdot \infty)}$ would tend to become infinitely large (as its coefficient b is positive), unless the constant B is zero. The first term ($A.e^{(a \cdot \infty)}$) tends to zero whatever the value of A (as a is a negative value).

Thus, to fulfil the second boundary condition we obtain:

$$B = 0 \tag{5.21}$$

Which simplifies the solution:

$$C|_x = A \cdot e^{a \cdot x} \tag{5.22}$$

2. With B determined, it is now possible to solve for the constant A using the first boundary condition

$$Flux = -D_b \frac{\partial C}{\partial x}\bigg|_{x=0} + wC\bigg|_{x=0} = -D_b \frac{\partial A \cdot e^{a \cdot x}}{\partial x}\bigg|_{x=0} + wA \cdot e^{a \cdot x}\bigg|_{x=0} \tag{5.23}$$

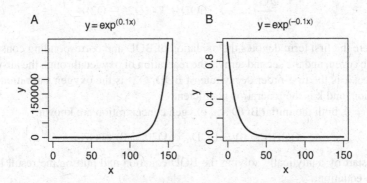

Fig. 5.2 Shape of an exponential curve with positive (**A**) and negative (**B**) coefficient

Based on the fact that $\dfrac{\partial e^{a \cdot x}}{\partial x} = a \cdot e^{a \cdot x}$ and with x = 0, we obtain:

$$Flux = -D_b \cdot a \cdot A \cdot e^{a0} + u \cdot A \cdot e^{a \cdot 0}$$
$$Flux = -D_b \cdot a \cdot A + u \cdot A \tag{5.24}$$

(remember that $e^0 = 1$). This allows calculating A as a function of the known flux:

$$A = \frac{Flux}{-D_b \cdot a + u} \tag{5.25}$$

and therefore C, the organic carbon concentration at any depth x can be expressed as:

$$C(x) = \frac{Flux}{-D_b \cdot a + w} \cdot e^{a \cdot x} \tag{5.26}$$

with

$$a = \frac{u - \sqrt{u^2 + 4 \cdot D_b \cdot k}}{2D_b} \tag{5.27}$$

5.3.4 Coupled BOD and Oxygen Equations (*)

We next solve a model that describes oxygen depletion in an aquatic system that is in contact with the air and that receives sewage. The sewage is expressed by a quantity called 'biochemical oxygen demand' (BOD), which is the amount of oxidizable matter, expressed in oxygen equivalents (mmol $O_2 \, m^{-3}$) (Fig. 5.3A).

There are two coupled differential equations, expressing the rate of change for BOD and oxygen:

$$\frac{d\text{BOD}}{dt} = -\lambda \cdot \text{BOD}$$
$$\frac{d\text{O2}}{dt} = -\lambda \cdot \text{BOD} + k \cdot (O2^* - O2) \tag{5.28}$$

Where the first term denotes the oxidation of BOD and corresponding consumption of oxygen, and the second term is the reaeration of oxygen through the air-water interface; λ is the first-order decay rate of BOD, $O2^*$ is the oxygen saturation concentration, and k is the reaeration coefficient.

At t = 0, both the initial BOD and oxygen concentration are known:

$$\text{BOD}(t = 0) = \text{BOD}_0, \quad \text{O2}(t = 0) = \text{O2}_0 \tag{5.29}$$

We start by analytically solving the BOD equation and putting the result in the oxygen equation.

$$\text{BOD}_t = BOD_0 \cdot e^{-\lambda \cdot t}$$

$$\frac{dO2}{dt} = -\lambda \cdot BOD_0 \cdot e^{-\lambda \cdot t} + k \cdot O2^* - k \cdot O2 \tag{5.30}$$

Note that we have split up the reoxidation term in a constant source term $(k \cdot O2^*)$ and a first-order sink term $(k \cdot O2)$.

The general solution for the differential equation of oxygen can be found in Appendix B.3 (where $Q = k \cdot O2^*$; mind the signs of the coefficients). It is given by:

$$O2(t) = \frac{-BOD_0 \cdot \lambda \cdot e^{-\lambda \cdot t}}{-\lambda + k} + O2^* + A \cdot e^{-k \cdot t} \tag{5.31}$$

where A is an integration constant.

Using the initial condition (at $t = 0$) for oxygen gives:

$$O2_0 = \frac{-BOD_0 \cdot \lambda \cdot e^{-\lambda \cdot 0}}{-\lambda + k} + O2^* + A \cdot e^{-k \cdot 0} = \frac{-BOD_0 \cdot \lambda}{k - \lambda} + O2^* + A$$

$$A = O2_0 - O2^* + \frac{BOD_0 \cdot \lambda}{k - \lambda} \tag{5.32}$$

such that:

$$O2_t = \frac{-BOD_0 \cdot \lambda \cdot e^{-\lambda \cdot t}}{-\lambda + k} + O2^* + O2_0 \cdot e^{-k \cdot t} - O2^* \cdot e^{-k \cdot t} + \frac{BOD_0 \cdot \lambda}{k - \lambda} \cdot e^{-k \cdot t} \tag{5.33}$$

and we obtain the particular analytic solution:

$$\text{BOD}_t = BOD_0 \cdot e^{-\lambda \cdot t}$$

$$O2_t = BOD_0 \cdot \lambda \cdot \frac{e^{-k \cdot t} - e^{-\lambda \cdot t}}{k - \lambda} + O2_0 \cdot e^{-k \cdot t} + O2^* \cdot (1 - e^{-k \cdot t}) \tag{5.34}$$

Note that, in this model, the degradation of BOD is independent from the oxygen concentration. Thus BOD will continue to degrade, even in the absence of oxygen, and this will drive the oxygen concentration to negative values. As this is unrealistic, this model can only be meaningfully applied in those situations where reaeration is large enough to overcome the oxygen demand by the BOD (but see Section 11.3 for an extension of this model).

The result shows that the oxygen concentration initially decreases (when BOD concentration is at its highest), after which it recovers due to aeration; as the BOD concentration approaches 0, oxygen gradually approaches the saturation concentration (Fig. 5.3B).

5.3.5 Multilayer Differential Equations (**)

We finally deal with how to obtain analytical solutions for multiple layer differential equations.

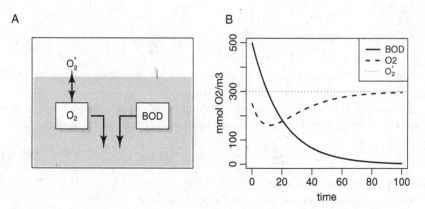

Fig. 5.3 A. Diagram of the O_2-BOD model, **B.** O_2 and BOD concentration as a function of time in the coupled BOD-O_2 model. Parameter values are: $k = 0.1$ d^{-1}, $\lambda = 0.05$ d^{-1}, $O2^* = 300$ mmol m^{-3}, $O2_0 = 250$, $BOD_0 = 500$ mmol m^{-3}

We will solve the sedimentary carbon model introduced in Section 3.5.3.1. This model describes the effect of organisms on organic carbon vertical profiles in marine sediments in the absence of photosynthesis.

Organisms respire the organic matter, which is modelled as a first-order decay rate (rate constant k). In addition, they mix the sediment, with an intensity quantified as a bioturbation coefficient (D_b).

In contrast to previous example from Section 5.3.3, the organisms are present only in an upper sediment layer with thickness L. Below that depth, the sediment is devoid of animal life (but it may still host bacteria!). In this deep layer, bioturbation is absent and only sediment advection occurs.

Thus different carbon (C) dynamics are specified in two layers: layer I from the sediment surface to depth L ([0-L]) and layer II from depth L till infinity ([L-∞]).

Assuming steady-state, the mass balance equation for layer is:

$$\left.\frac{\partial C}{\partial t}\right|^I = 0 = -u \left.\frac{\partial C}{\partial x}\right|^I + D_b \left.\frac{\partial^2 C}{\partial x}\right|^I - k \cdot C^I \qquad (5.35)$$

and the general solution (see Appendix B.3) is:

$$C^I = A_1 \cdot e^{a \cdot x} + B_1 \cdot e^{b \cdot x}$$

$$a = \frac{u - \sqrt{u^2 + 4 \cdot k \cdot D_b}}{2 \cdot D_b} \qquad (x \leq L)$$

$$b = \frac{u + \sqrt{u^2 + 4 \cdot k \cdot D_b}}{2 \cdot D_b} \qquad (5.36)$$

where, as before, D_b is the sediment mixing (bioturbation) coefficient in $cm^2 \ yr^{-1}$, u is the advection rate (cm yr^{-1}) and k is the first-order degradation constant (yr^{-1}).

The mass balance equation and the general solution for layer II are:

$$\left.\frac{\partial C}{\partial t}\right|^{II} = 0 = -u \left.\frac{\partial C}{\partial x}\right|^{II} - k \cdot C^{II} \qquad (x \geq L)$$

$$C^{II} = A_2 \cdot e^{-k/u^x} \tag{5.37}$$

There are 3 integration constants (A1, B1, and A2), therefore we must specify 3 boundary conditions.

At the sediment-water interface $(x = 0)$ we specify the flux of organic matter:

$$Flux = -D_b \left.\frac{\partial C}{\partial x}\right|^{I}_{x=0} + u \cdot C|^{I}_{x=0}$$

$$Flux = -A_1 \cdot a \cdot D_b - B_1 \cdot b \cdot D_b + u \cdot A_1 + u \cdot B_1 \tag{5.38}$$

The two remaining conditions are specified at the interface between the two layers, i.e. at depth L. They are the continuity of concentration and the continuity of flux equation.

By 'continuity' we mean that whether we calculate the flux or concentration at depth L using Eq. (5.36) or Eq. (5.37), we must obtain the same value.

The continuity of concentration at depth L reads:

$$C|^{I}_{x=L} = C|^{II}_{x=L}$$

$$A_1 \cdot e^{a \cdot L} + B_1 \cdot e^{b \cdot L} = A_2 \cdot e^{-k/u \cdot L} \tag{5.39}$$

The continuity of flux at depth L:

$$-D_b \left.\frac{\partial C}{\partial x}\right|^{I}_{x=L} + u \cdot C|^{I}_{x=L} = u \cdot C|^{II}_{x=L}$$

$$- A_1 \cdot a \cdot D_b \cdot e^{a \cdot L} - B_1 \cdot b \cdot D_b \cdot e^{b \cdot L} + u \cdot A_1 \cdot e^{a \cdot L} + u \cdot B_1 \cdot e^{b \cdot L}$$

$$= u \cdot A_2 \cdot e^{-k/u \cdot L} \tag{5.40}$$

This gives three equations in three unknowns, which, with some arithmetic can be solved for the unknowns A_1, B_1, A_2.

The details of the derivation of the solution are omitted since they are somewhat lengthy. Moreover, it is often more efficient NOT to derive the solution by hand, but let the computer do so. We will show how this works, for a slightly different model, in Section 5.4.4. The solution of the multilayer organic carbon model is left as an exercise (Section 5.5.3).

5.4 Case Studies in R

5.4.1 Transient Dispersion-Reaction in One Dimension

Consider organisms whose density increases at a constant rate, and whose random movement causes them to slowly disperse. This dispersal can be modelled as a dispersion process, with dispersion coefficient D_s.

We first assume that the organisms are constrained to move in one dimension only, along a line.

The density of the organism changes according to the one-dimensional dispersion-reaction equation:

$$\frac{\partial N}{\partial t} = \frac{\partial}{\partial x} D_s \frac{\partial N}{\partial x} + k \cdot N \tag{5.41}$$

with boundary and initial conditions:

$$N(x = 0, t = 0) = N_0$$

$$N(x = \pm\infty, t) \text{ is finite} \tag{5.42}$$

To derive the particular analytical solution from the initial and boundary conditions and the general solution is quite complicated, so the particular analytical solution of this equation is given in Appendix B.3.

$$N(x, t) = \frac{N_0}{2\sqrt{\pi D_s \cdot t}} \cdot \exp^{k \cdot t - \frac{x^2}{4D_s \cdot t}} \tag{5.43}$$

With this formula we can now estimate the density at any point in space x and time t.

In the R code below, we estimate the values for N(x,t) on a 1-dimensional grid composed of 50 cells, and extending from -5 to 5 (xx), and for 5 days, with a time increment of 0.025 (tt). We use R's function `outer` to estimate the function value for each (xx,tt) pair. The results of these calculations are stored in a matrix called `grow1D`. We then use R's function `persp` to generate a 3-D surface plot (Fig. 5.4) depicting the density as a function of space and time.

Fig. 5.4 Temporal and spatial
variation of organisms
dispersing and growing on a
line. See text for the R-code
used to generate this figure

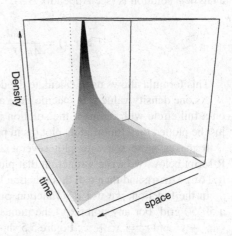

1–D dispersion–reaction

We start by specifying the parameter values:

```
Ds  <- 1      # diffusion coefficient
ini <- 1      # initial condition
k   <- 0.05   # growth rate

grow1D <- outer(xx<-seq(-5,5,length=50),
            tt<-seq(0.1,5,by=0.025),
            FUN = function (x,tt)
                ini/(2*sqrt(pi*Ds*tt))*exp(k*tt-x^2/
                (4*Ds*tt)))

persp(xx,tt,z=grow1D,theta=150,box=TRUE,axes=TRUE,
      xlab="space",ylab="time",zlab="Density")
```

5.4.2 Transient Diffusion-Reaction on a 2-Dimensional Surface

Now, assume the same organisms, but growing on a 2-dimensional (flat) surface, in
concentric circles.

The density of the organism changes according to the one-dimensional diffusive-
reaction equation in cylindrical coordinates and where we assume that the length of
the cylinder is 0.

The equation reads (Section 3.4.5.2):

$$\frac{\partial N}{\partial t} = \frac{1}{r} D_s \frac{\partial}{\partial r} r \frac{\partial N}{\partial r} + k \cdot N \tag{5.44}$$

With similar boundary conditions ($N(0, 0) = N_0$ and $N(t, r = \pm\infty)$ is finite), the analytical solution is (see Appendix B.4):

$$N(r, t) = \frac{N_0}{4\pi D_s \cdot t} \cdot \exp^{k \cdot t - \frac{r^2}{4D_s \cdot t}} \tag{5.45}$$

This formula allows us to calculate the density value at any (r,t) couple.

As one density value at a specific time and radius, $N(r,t)$, represents the density on a full circle with radius r, the solution is essentially 2-dimensional and cannot just be plotted on a time-radius plot (as in previous example).

It makes more sense to plot several 2-D surfaces at different times. In the R-script below, we write a function that plots the density surface at a specific time (plotplane), and then call this function for several time instances.

In the function, we use R's function outer to expand the density values, on a 50*50 grid. For any point(x,y) the radius to the midpoint (0,0) is given by $r = \sqrt{x^2 + y^2}$ and $r^2 = x^2 + y^2$. Figure 5.5 shows the resulting figures.

```
plotplane <- function(time,rmax=5,...)
  {
   val <-outer(xx<-seq(-rmax,rmax,length=50),yy<-xx,
               FUN = function (x,y)
               {
               r2<-x*x+y*y;
               ini/(4*pi*Ds*time)*exp(k*time-r2/(4*Ds*time))
               }
              )
   persp(xx,yy,z=val,theta=150,box=TRUE,axes=TRUE,
         zlab="dens",...)
  }

par(mfrow=c(2,2),mar=c(3,3,3,3))
plotplane(0.1, main= "0.1 day")
plotplane(1  , main= "1 day")
plotplane(2  , main= "2 days")
plotplane(5  , main= "5 days")
```

5.4.3 Steady-State Oxygen Budget in Small Organisms Living in Suboxic Conditions

Oxygen concentrations in aquatic sediments are generally low. Nevertheless, most multi-cellular organisms that live in these sediments rely on oxygen for their metabolism. Many benthic organisms live very near to or even in the anoxic zone. How do these organisms cope with these adverse conditions?

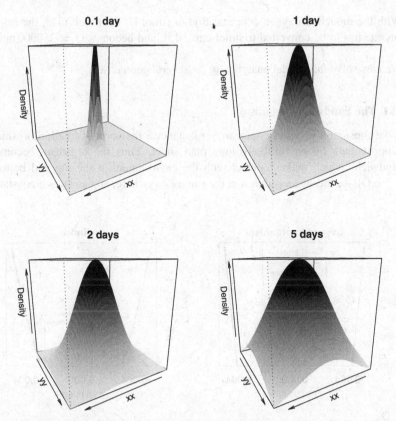

Fig. 5.5 Spatial variation of organisms dispersing and growing on a flat 2-D surface, and for 0.1,1,2 and 5 days. See text for the R-code used to generate this figure

Small organisms living in sediment environments rely on molecular diffusion for their supply of oxygen. As molecular diffusion is only effective over very short distances, it poses a physical constraint on the size and shape of an organism: the animals have to be thin enough for the oxygen to diffuse sufficiently deep into their body.

We will now implement a series of models that reproduce oxygen concentrations in small animals of various shapes and that live under suboxic conditions (at low oxygen concentrations). Assuming steady-state, zero-order consumption (Q) of oxygen (O2) and ignoring advection, the overall equation is:

$$0 = \frac{1}{A} \cdot \frac{\partial}{\partial x} A \cdot D_a \frac{\partial O2}{\partial x} - Q \qquad (5.46)$$

Where A is the surface area, D_a is the diffusion coefficient of oxygen in the tissue of the animal, and Q is the oxygen consumption rate. Typical values for D_a are in the order of $0.4 \, \text{cm}^2 \, \text{d}^{-1}$, whilst $Q \sim 250 \, \mu\text{mole cm}^{-3} \, \text{d}^{-1}$ (e.g. Powell, 1989).

With the modelled oxygen concentration in μmol l^{-1} = nmol cm^{-3}, the respiration rate has to be converted to nmol cm^{-3} d^{-1}, and becomes $Q = 250000$ nmol cm^{-3} d^{-1}.

We now solve this model analytically for several geometries.

5.4.3.1 The Sandwich Organism

First we consider a very flat organism, for instance a flatworm (Turbellaria), which we approximate by an infinitely long plan sheet. Thus the organism becomes a 'sandwich' which makes contact with the environment at the top and bottom (Fig. 5.6.D). As the surface area A is the same everywhere, and as D_a is a constant,

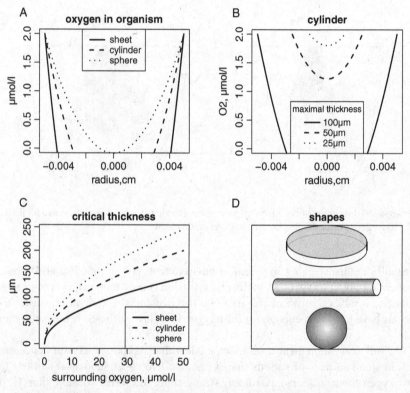

Fig. 5.6 Influence of body shape on the oxygen budget in small organisms. **A.** Oxygen concentration as a function of body radius, for a flat (*solid line*), cylindrical (*large dashes*), and spherical organism (*small dashes*) and for an ambient oxygen concentration of 2 μmol l^{-1}; all organisms are 100 μm thick. **B.** Effect of body radius on internal oxygen concentration in a cylindrical organism and for an ambient oxygen concentration of 2 μmol l^{-1}. **C.** Critical thickness as a function of the ambient oxygen concentration, for the three geometries. When an organism's thickness exceeds the critical thickness, the internal part of its body becomes anoxic. **D.** Shapes considered, from *top* to *bottom*: sheet, cylinder, and sphere. In *grey*: surface through which oxygen is exchanged; the white surface is not penetrable

we can simplify the overall equation

$$0 = D_a \frac{\partial^2 O2}{\partial x^2} - Q \tag{5.47}$$

Which has as solution (see Appendix B.3.2):

$$O2(x) = \frac{Q \cdot x^2}{2D_a} + A \cdot x + B \tag{5.48}$$

and where A and B are integration constants. Now, if W is the width of the organism, then the oxygen concentration at $x = 0$ and $x = W$ has the concentration of the bottom water (BW). Thus:

$$O2(0) = BW = \frac{Q \cdot 0^2}{2D_a} + A \cdot 0 + B \tag{5.49}$$

which gives:

$$B = BW \tag{5.50}$$

and

$$O2(W) = BW = \frac{Q \cdot W^2}{2D_a} + A \cdot W + BW \tag{5.51}$$

such that:

$$A = -\frac{Q \cdot W}{2D_a} \tag{5.52}$$

Thus the equation that specifies the steady-state concentration at any point x ([0,W]) in the flatworm's body is:

$$O2(x) = \frac{Q \cdot x}{2D_a}(x - W) + BW \tag{5.53}$$

Note: as the body is symmetrical around the centre, we obtain the same results if the body is described from the centre to the external world. As this simplifies the problem, this is an often-used technique in solving differential equations. In this case, the spatial coordinate, y, now extends from 0, the centre of the body, to R ($= W/2$). At $y = R$, the bottom water concentration is imposed, while in the organism's centre ($y = 0$), a zero-concentration gradient is imposed $\left(0 = \left.\frac{\partial O2}{\partial y}\right|_{y=0}\right)$. The resulting equation then becomes (for y in [0,R])

$$O2(y) = \frac{Q}{2D_a}(y^2 - R^2) + BW \tag{5.54}$$

which reverts to the same solution as the previous one (take into account that x = y+R and R = W/2).

5.4.3.2 The Cylindrical Organism

Next we consider a cylindrical organism, say a round-worm (a nematode, Fig. 5.6 D). After replacing A with the surface area of a cylinder and simplifying, we obtain:

$$0 = \frac{D_a}{r} \cdot \frac{\partial}{\partial r} r \cdot \frac{\partial O2}{\partial r} - Q \tag{5.55}$$

Where r is the radius, with the origin now defined at the centre of the body.
   We find the general solution in Appendix B.3.3:

$$O2(r) = \frac{Q \cdot r^2}{4D_a} + A \cdot \log_e(r) + B \tag{5.56}$$

One boundary condition prescribes a zero-concentration gradient at the organism's centre (r = 0). This can only be so if A = 0. Prescribing the bottom water oxygen concentration (BW) at r = R, the maximal radius of the organism we obtain: $B = BW - Q \cdot R^2/4D_a$ such that the particular model solution is:

$$O2(r) = \frac{Q \cdot}{4D_a}(r^2 - R^2) + BW \tag{5.57}$$

5.4.3.3 The Spherical Organism

Some organisms (e.g. foraminiferans, ciliates,...) are better approximated by a sphere (Fig. 5.6 D), for which the model becomes:

$$0 = \frac{D_a}{r^2} \frac{d}{dr}(r^2 \frac{dO2}{dr}) + Q \tag{5.58}$$

$$O2(r) = \frac{Q \cdot r^2}{6D_a} - \frac{A}{r} + B \tag{5.59}$$

And with similar boundary conditions as for the cylindrical organism, we obtain:

$$O2(r) = \frac{Q \cdot}{6D_a}(r^2 - R^2) + BW \tag{5.60}$$

5.4.3.4   Implementation in R

Note the similarity in the oxygen concentrations for these various shapes: the formulations are the same, except for some multiplying constant, which is 1/2, 1/4 and 1/6 for the flat, cylindrical and spherical organism respectively.

Figure 5.6 A depicts the oxygen concentration in the various types of organisms, for an oxygen bottom water concentration of $2\,\mu\mathrm{mol\,l^{-1}}$ and a thickness ($2*R$) of the organisms of $100\,\mu\mathrm{m}$. Note the use of cm and days as the model's length and time scale. We calculate concentrations for positive and negative radiuses (rr).

Here is the R-code to generate this plot:

```
BW        <- 2           # mmol/m3
Da        <- 0.5         # cm2/d
R         <- 0.005       # cm, radius
Q         <- 250000      # nM/cm3/d

rr              <- seq(-R,R,length=400)

sandwich <- function(Da,Q,BW,R,r)  BW+Q/(2*Da)*(r^2-R^2)
cylinder <- function(Da,Q,BW,R,r)  BW+Q/(4*Da)*(r^2-R^2)
sphere   <- function(Da,Q,BW,R,r)  BW+Q/(6*Da)*(r^2-R^2)

plot (rr,sandwich(Da,Q,BW,R,rr),lwd=2,lty=1,
      ylim=c(0,BW), type="l",main="oxygen in organism",
        xlab="radius,cm",ylab="µmol/l")
lines(rr,cylinder(Da,Q,BW,R,rr),lwd=2,lty=2)
lines(rr,sphere (Da,Q,BW,R,rr),lwd=2,lty=3)
legend("top",c("sheet","cylinder","sphere"),lty=1:3,
lwd=2)
```

Oxygen is lowest at the centre, and increases to bottom water concentrations at the external surface. Oxygen is depleted at the centre for all shapes; depletion occurs closer to the surface in sandwich-shaped (sheet) organisms. Spherical organisms are best oxidised (Fig. 5.6 A).

Now what happens if the organism becomes thinner? In Fig. 5.6 B this is depicted for a cylindrical organism, 2 and 4 times thinner. Clearly, $50\,\mu\mathrm{m}$ is already thin enough for the organism to remain oxic all over its body.

```
plot  (rr,cylinder(Da,Q,BW,R,rr),type="l",lwd=2,
        xlab="radius,cm",ylab="O2,µmol/l",
        ylim=c(0,2),main="cylinder" )
lines (rr/2,cylinder(Da,Q,BW,R/2,rr/2),lwd=2,lty=2)
lines (rr/4,cylinder(Da,Q,BW,R/4,rr/4),lwd=2,lty=3)

legend("bottom",legend=c("100µm","50µm","25µm"),title="
maximal thickness",lty=1:3,lwd=2)
```

5.4.3.5 The Critical Size

From the previous simulations it is obvious that $100\,\mu$m thick organisms whose metabolism is aerobic will have a hard time surviving at oxygen concentrations as low as $2\,\mu$mol 1^{-1}. However, thinner organisms can cope with these suboxic conditions.

As oxygen concentrations decline, animals should become thinner to ensure that the required aerobic respiratory rate can be maintained. So, what is the maximal thickness that organisms can attain, when living under hypoxic conditions, and for the various shapes?

To resolve this, we calculate, for a certain bottom water oxygen concentration, the thickness at which oxygen in the centre of the animal's body (i.e. at $r=0$) becomes 0. This we call the 'critical thickness', as, for thicker individuals, the resulting oxygen deficiency may impair the oxic metabolism of the organism.

We solve each of the equations above for that maximal radius R at which $(O2(r=0)=0)$.

For the sandwich organism we solve the equation:

$$0 = \frac{Q \cdot}{2D_a}(0 - R^2) + BW \tag{5.61}$$

for the unknown R.

This gives

$$R = \sqrt{\frac{BW \cdot 2D_a}{Q}} \tag{5.62}$$

Similarly, we derive

$$R = \sqrt{\frac{BW \cdot 4D_a}{Q}}, \text{ and } R = \sqrt{\frac{BW \cdot 6D_a}{Q}} \tag{5.63}$$

for the cylindrical, and spherical organisms respectively. Results in Fig. 5.6 C show that sandwich-like organisms have to be thinner than cylindrical, than spherical organisms.

Further extensions of these models can be found in Powell (1989) and Gielen and Kranenbarg (2002).

5.4.4 Analytical Solution of the Non-Local Exchange Sediment Model (***)

We now solve the non-local exchange model (Soetaert et al., 1996a) that was presented in Section 3.5.3.2 and show how to efficiently implement this in R.

This example is much more complex than all examples discussed before. Essentially we include it to demonstrate how we can use matrix algebra to solve for the integration constants in analytical solutions. If you don't consider this too freaky: here it comes!

Recall first the differential equations, specified in two layers ($i = 1, 2$)

$$0 = -u \cdot \frac{\partial C}{\partial x}\bigg|^i + D_b \cdot \frac{\partial^2 C}{\partial x^2}\bigg|^i - \lambda \cdot C|^i \tag{5.64}$$

and the four boundary conditions:

$$Flux_{x=0} = u \cdot C^{i=1}_{x=0} - D_b \cdot \frac{\partial C^{i=1}}{\partial x}\bigg|_{x=0} \tag{5.65}$$

$$C^{i=2}_{x=\infty}\bigg| = 0 \tag{5.66}$$

$$C^{i=1}_{x=L} = C^{i=2}_{x=L} \tag{5.67}$$

$$u \cdot C^{i=1}_{x=L} - D_b \cdot \frac{\partial C^{i=1}}{\partial x}\bigg|_{x=L} + Flux_{x=L} = u \cdot C^{i=2}_{x=L} - D_b \cdot \frac{\partial C^{i=2}}{\partial x}\bigg|_{x=L} \tag{5.68}$$

The general solution for both layers reads:

$$C|^i_x = A_i e^{\frac{u - \sqrt{u^2 + 4 \cdot D_b \cdot k}}{2D} \cdot x} + B_i e^{\frac{u + \sqrt{u^2 + 4 \cdot D_b \cdot k}}{2D} \cdot x} = A_i \cdot e^{a \cdot x} + B_i \cdot e^{b \cdot x} \tag{5.69}$$

and where A_1, B_1, A_2, B_2 are 4 integration constants, 2 for each layer.

In order for the second boundary condition (Eq. 5.66) to be fulfilled, it follows that $B_2 = 0$.

The first boundary condition (Eq. 5.65) leads to an equation in both A_1 and B_1

$$Flux_{x=0} = u \cdot A_1 \cdot e^{a \cdot 0} + u \cdot B_1 e^{b \cdot 0} - D_b \cdot a \cdot A_1 \cdot e^{a \cdot 0} - D_b \cdot b \cdot B_1 \cdot e^{b \cdot 0}$$

$$Flux_{x=0} = u \cdot A_1 + u \cdot B_1 - D_b \cdot a \cdot A_1 - D_b \cdot b \cdot B_1 \tag{5.70}$$

whilst continuity of concentration (Eq. 5.67) and flux (Eq. 5.68) at depth L give:

$$A_1 \cdot e^{a \cdot L} + B_1 \cdot e^{b \cdot L} = A_2 \cdot e^{a \cdot L} \tag{5.71}$$

$$(u - D_b \cdot a)A_1 \cdot e^{a \cdot L} + (u - D_b \cdot b) \cdot B_1 \cdot e^{b \cdot L} + Flux_{x=L}$$

$$= (u - D_b \cdot a) \cdot A_2 \cdot e^{a \cdot L} \tag{5.72}$$

Rearranging and collecting terms, we obtain three linear equations in the three remaining unknowns A_1, B_1, A_2:

$$(u - D_b \cdot a) \cdot A_1 + (u - D_b \cdot b) \cdot B_1 = Flux_{x=0}$$

$$e^{a \cdot L} \cdot A_1 + e^{b \cdot L} \cdot B_1 - e^{a \cdot L} \cdot A_2 = 0$$

$$[(u - D_b \cdot a) \cdot e^{a \cdot L}] \cdot A_1 + [(u - D_b \cdot b) \cdot e^{b \cdot L}] \cdot B_1 - [(u - D_b \cdot a) \cdot e^{a \cdot L}] \cdot A_2$$

$$= -Flux_{x=L} \tag{5.73}$$

It is possible (and maybe not that complex) to solve these equations for the three unknowns by hand. However, it is more convenient to rearrange them in matrix format and let the computer solve the resulting set of linear equations.

Thus we write:

$$\begin{bmatrix} u - D_b \cdot a & u - D_b \cdot b & 0 \\ e^{a \cdot L} & e^{b \cdot L} & -e^{a \cdot L} \\ (u - D_b \cdot a) \cdot e^{a \cdot L} & (u - D_b \cdot b) \cdot e^{b \cdot L} & -[(u - D_b \cdot a) \cdot e^{a \cdot L}] \end{bmatrix} \cdot \begin{bmatrix} A_1 \\ B_1 \\ A_2 \end{bmatrix}$$

$$= \begin{bmatrix} Flux_{x=0} \\ 0 \\ -Flux_{x=L} \end{bmatrix} \tag{5.74}$$

R-code

We implement this model as a function, called `nonlocal`, that takes as input the parameter values (`k`, `u`, `Db`, `depo`, `L`, `injectflux`) and the sediment depths (`sed`) at which the concentration has to be estimated.

The default parameter settings of this function reflect the behaviour of Pb[210] in a typical deep-sea sediment. ($k = 0.031$ yr^{-1}, Db in cm^2 yr^{-1}, u in cm yr^{-1}, *depo* and *injectflux* in dpm cm^{-2} yr^{-1} and L in cm).

First the coefficients a and b, and the exponents at depth L (`expaL`, `expbL`) are calculated. Then the matrix of coefficients (`A`) and the right hand side of the equation (vector `B`) are generated. Note that by default, matrices in R are filled column-wise; it is easier to fill them row-wise (`byrow=TRUE`).

The linear system of equations is solved using R function `solve`; it returns the values of the integration constants A1, B1, A2.

Finally, the concentrations in the various sediment layers are estimated. We use a different equation for those layers where sediment depth is smaller than L (`s1`) and those for which it is larger or equal than L (`s2`), Finally, the concentrations are returned.

```
nonlocal <- function(
         k=0.031,u=0.001,Db=0.1,depo=0.1,
         L=5,injectflux=0.3,sed )
{
 a <- (u/Db - sqrt ( (u/Db)^2 + 4*k / Db))/2.
 b <- (u/Db + sqrt ( (u/Db)^2 + 4*k / Db))/2.
 expaL <- exp(a*L)
 expbL <- exp(b*L)

 A <- matrix(nrow=3,ncol=3,byrow=TRUE,data= c(

   u-Db*a          ,u-Db*b         ,0              ,
   expaL           ,expbL          ,-expaL         ,
 (u-Db*a)*expaL ,(u-Db*b)*expbL ,(Db*a-u)*expaL )
             )
 B  <- c(depo       , 0            ,- injectflux)
 X  <- solve(A,B)

 s1 <- which (sed<L)
 s2 <- which (sed>=L)

 conc <- vector(length=length(sed))
 conc[s1] <- X[1]* exp(a*sed[s1])+X[2]*exp(b*sed[s1])
 conc[s2] <- X[3]* exp(a*sed[s2])
 return(conc)
}
```

Next, we define the depths at which to calculate the Pb^{210} activity, call the model, and plot the results (Fig. 5.7):

```
depth        <- seq(0,15,by=0.01)          # cm
Pb           <- nonlocal(sed=depth)

plot(Pb,depth,ylim=c(15,0),xlab="Pb, dpm/cm2/yr",
ylab="cm",type="l",lwd=2,main="75% injected")
```

5.5 Projects

5.5.1 Organic Matter Sinking Through a Water Column

Consider the aphotic part of an oceanic water column (i.e. the part below the euphotic or illuminated zone). The depth of this water column is 400 m. Organic matter is raining down from the productive euphotic zone, and as it sinks through the water column, it is being degraded.

The organic matter has a constant sinking velocity u of 50 m.d^{-1}.

Fig. 5.7 Result from the non-local exchange model of Section 5.4.4

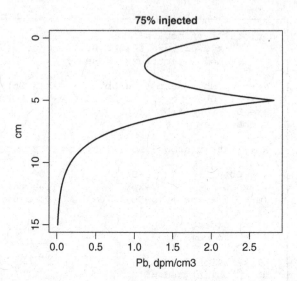

Model the degradation as a first-order process, with a degradation rate k of $0.2\,\mathrm{d}^{-1}$.

The flux of organic matter from the euphotic zone, i.e. the upper boundary for the model, is prescribed as $Flux = 100\,\mathrm{mmolC.m^{-2}.d^{-1}}$. At the lower end of the water column, material leaves the water column to settle on the bottom.

We consider no diffusive mixing due to turbulence or other physical features of the water column.

Tasks:

1. Write the model expressing the rate of change of organic matter as a function of sinking and degradation. Is sinking an advective or a diffusive process?
2. Check the dimensions of your equations.
3. It is relatively easy in this case to calculate the steady state analytically. Put the rate of change in time equal to zero, and solve the resulting differential equation in x by integration. Make a graph of concentration versus water depth.
4. Assuming that the bottom is at 400 m of water depth, what is the deposition flux on the bottom?
5. Find a relationship between the decay rate and the part of the surface flux that settles on the bottom. Find also a relationship between the sinking velocity and the part of the flux deposited on the sediments.

5.5.2 Oxygen Dynamics in the Sediment

Consider the vertical extent of marine or lacustrine sediments, where the origin of the x-axis is positioned at the sediment-water interface and the x-axis is pointing downward (we ignore horizontal gradients here).

Model the oxygen profile in a sediment column with a constant surface of $1\,cm^2$. Oxygen is not produced in the sediment, but consumed at a first-order rate. Typically, the advection rate in sediments is very small, and it can be ignored when modelling relatively fast processes such as oxygen dynamics.

As the surface A is constant, it can be taken out of the derivatives and the equation becomes:

$$0 = \frac{\partial}{\partial x} D_s \frac{\partial O2}{\partial x} - k.O2$$

At the sediment-water interface the concentration of oxygen in the overlying water is specified. Oxygen consumption in the sediment proceeds until all oxygen is exhausted. Thus, the boundary conditions read:

$$O2_{x=0} = O2_0$$
$$O2_{x=\infty} = 0 \tag{5.75}$$

Tasks:

1a. Find the particular solution that describes the oxygen concentration as a function of sediment depth.
1b. Find an analytical formula for the flux of oxygen in the sediment. Recall that:

$$Flux = -D_s \frac{\partial O2}{\partial x}\bigg|_{x=0} + wO2_{x=0} \tag{5.76}$$

and that

$$\frac{\partial e^{ax}}{\partial x} = a \cdot e^{ax} \tag{5.77}$$

1c. Calculate analytically the oxygen penetration depth, which we define as the sediment depth where the oxygen concentration is $0.1\,mmol.m^{-3}$.
2. Typical values of k range between 1 and $1000\,d^{-1}$, whilst the diffusion coefficient of oxygen in the sediment, D_s, is in the order of $1\,cm^2\,d^{-1}$. Make a graph of oxygen concentration versus sediment depth, and with a decay rate k equal to $10\,d^{-1}$ and $D_s = 1\,cm^2\,d^{-1}$. Add two other sediment depth profiles, with k equal to 1 and $100\,d^{-1}$. Tip: use R functions 'plot' and 'lines'.
   Investigate the consequences for the oxygen concentrations.
3. For each of these cases, calculate the flux of oxygen in the sediment and the oxygen penetration depth;
4. Make a graph of oxygen penetration depth in mm, versus oxygen flux, expressed in $mmol\,m^{-2}\,d^{-1}$
   What is the relationship between both; can you explain?

5. Assume the following measurements have been performed, using micro-electrodes:

Sediment depth (cm)	O2 concentration (mmol m^{-3})
0	300
0.1	65.52
0.2	14.31
0.3	3.13
0.4	0.68
0.5	0.59
1	0.27

Estimate the values of the parameters that can explain these observations. Which parameter would you vary in the first place?

Use the fitting procedure from Section 4.4.2 to fit the model to the data.

5.5.3 Carbon Dynamics in the Sediment

In analogy to the implementation in R of the non-local exchange model (Section 5.4.4), implement the two-layer model of carbon dynamics in the sediment from Section 5.3.5.

Chapter 6
Model Solution –Numerical Methods

The analytical solutions discussed in the previous chapter have the advantage of simplicity, as they can be solved using a hand calculator or in a spreadsheet. However, for many real-world problems, the differential equations are too difficult to solve analytically, or it may become impractical to do so. This often arises for instance because non-linear terms are included in the equations, or because the boundary conditions or forcing functions are a variable function of time. When this is the case, a numerical method must be used.

Numerical solutions are approximations where the continuous model equations are approached in discrete steps, both in time, and for spatial models, also in space (Fig. 6.1). The advantage of numerical methods is that there is virtually no limit to the complexity of problems that can be solved, but the price to be paid for this generality is the introduction of a new kind of errors, so-called numerical errors which must be controlled.

Generally, numerical solutions are obtained by writing a computer programme. As numerical models generally proceed by stepping through time, a model calculation is often called a simulation, or we talk about running a model.

6.1 Taylor Expansion

One important mathematical equation, the Taylor expansion, forms the backbone of many numerical methods, so we will discuss this equation first.

Taylor expansion allows evaluating, *for small h*, the unknown function value at some point x + h, when all the derivatives of the function at x are known. The Taylor

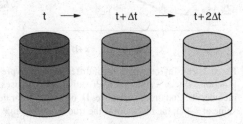

Fig. 6.1 Numerical solutions proceed by subdividing time and space in discrete steps

expansion function is a polynomial, which contains the function value at x (0th order derivative), the slope of the function at x (1st order derivative), the curvature (2nd order derivative), and higher-order derivatives of f(x).

Intuitively, one may compare the derivation of Taylor expansion with forecasting the weather. Assume somebody asks you what the weather will be like tomorrow. The first best guess is to assume that the weather will not change during the day so that tomorrow will have the same weather as today.

Taylor-expansion-wise we assume that the function value did not change in the interval h such that the new value f(x + h) is approximated as (Fig. 6.2A):

$$f(x + h) \approx f(x) \tag{6.1}$$

Of course this is only exact *if* the function f(x) remains constant over the interval from x to x + h. If function f(x) is constant, its first-order derivative $f' = df/dx$ is 0.

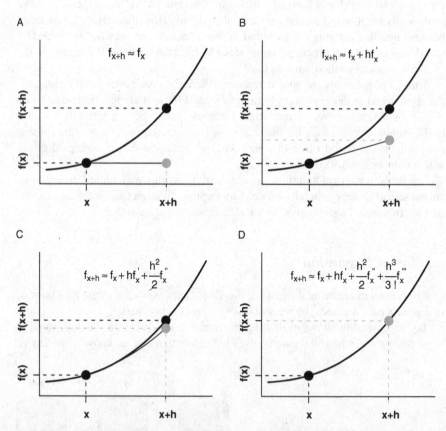

Fig. 6.2 The Taylor expansion of different order to predict a function value at x+h. **A**. uses only the function value at x (0th order derivative). **B**. also uses the rate of change (1st order derivative). **C**. includes the 2nd order derivative. **D**. up to 3rd order derivative. The grey dot denotes the estimated value at x + h; the black dot is the true function value at x and x + h

If the weather *does* change during the day, then our statement that tomorrow will be the same as today does not hold. We might argue that, as today it is getting warmer already (the rate of change of the weather), the weather will be similar as today, but somewhat warmer.

Similarly, in Taylor expansion, a better estimate is given by taking the (known) derivative into account (Fig. 6.2B):

$$f(x + h) \approx f(x) + h \cdot f'(x) \tag{6.2}$$

This formula is exact only if the derivative, $f'(x)$ remains constant in the interval h, i.e. if the second-order derivative, $f''(x)$ is zero.

Even better is to adjust for the change of this derivative (second order derivatives) (Fig. 6.2C).

$$f(x + h) \approx f(x) + h \cdot f'(x) + \frac{h^2}{2} f''(x) \tag{6.3}$$

Repeating this argument, taking ever-higher derivatives into account, leads to the Taylor series:

$$f(x + h) = f(x) + h \cdot f'(x) + \frac{h^2}{2} f''(x) + \frac{h^3}{6} f'''(x) + \cdots \frac{h^n}{n!} f^n(x) + \cdots$$

$$\tag{6.4}$$

and where n! is the factorial (n*(n−1)*(n−2)*... 1) and $f^n(x)$ denotes the nth derivative of f at point x.

Similarly we may write (replacing h with –h) :

$$f(x - h) = f(x) - h \cdot f'(x) + \frac{h^2}{2} f''(x) - \frac{h^3}{6} f'''(x) + \cdots \tag{6.5}$$

In the next sections we will see how Taylor expansion can be used to numerically integrate differential equations in time, or to approximate spatial derivatives.

6.2 Numerical Approximation and Numerical Errors

How is this Taylor expansion used?

Consider a very small h ($<<1$), say h$=0.02$. The powers of h are: $h^2 = 0.0004$, $h^3 = 0.000008$, $h^4 = 0.00000016$,.. $h^8 = 2.5610^{-14}$. The terms $n! = (1 \cdot 2 \cdot \ldots \cdot n)$ by which the powers of h are divided, increase rapidly with n (1, 2, 6, 24, 120, 720,...)

As the terms of the Taylor series become smaller and smaller for higher powers of h, at some point we may consider their value to be negligible.

Numerical methods typically proceed by truncating the Taylor series at some point, arguing that the error that is introduced by doing so is small enough to be unimportant. It is this truncation that is the reason for so-called numerical errors.

We write that a method is n-th order accurate, if the largest power of h included is h^n. Thus, the largest truncation error made at each step is proportional to h^{n+1}. This also means that halving the step h will divide the error by 2^{n+1}, i.e. by 8 for a second order method.

Why should we bother about the (small) errors in numerical solutions? First, of course, it is never pleasing to make errors. Scientists want to be accurate, and this alone is a compelling reason to try and reduce numerical errors to an acceptable level. There is a second reason, however, which relates to the (numerical) *stability* of the solution. When a small error is made in one time step, this error is sometimes compensated by another error in the following time step. For instance, if the first time step leads to a small underestimation of biomass at the end of the step, this may be compensated if this too-small biomass slightly overestimates the rate of increase, such that after two time steps the approximation is still reasonably close to the true solution of the model. If this is the case, we are lucky, and will have no problems of numerical stability. However, the reverse can also happen: small errors in one time step lead to magnification of errors in the next step, and so on... very soon the model solution will be far off-track (see Fig. 6.3A) or will start oscillating wildly (Fig. 6.3B). Particular problems of numerical stability may also arise if small errors lead to slightly negative values for state variables such as concentrations or biomass that, by definition, must be positive (Fig. 6.3B). You would be surprised how fast model algae can grow on negative model nitrate! Such negative values for a state variable may easily lead to growing oscillations in the numerical model solution, which have nothing to do with the real model solution.

One notorious example of truncation error, which is linked to the *spatial* approximation, is called 'numerical dispersion' and will be discussed at the end of this chapter.

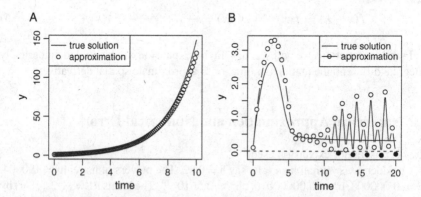

Fig. 6.3 Two examples of numerical errors. **A.** In this (exponential growth) model the error is small at first, but it is successively magnified; after 10 days, the approximated solution clearly deviates from the true solution. **B.** In this model, after 5 days of simulation the numerical solution starts to oscillate with ever increasing magnitude. These oscillations lead to negative values (*black dots*) after 10 days

6.3 Numerical Integration in Time – Basics

When a dynamic differential equation is solved numerically, the method jumps from one point in time to the next:

$$t_0 \Rightarrow t_1 = t_0 + \Delta t \Rightarrow t_2 = t_1 + \Delta t \Rightarrow t_3 \Rightarrow t_4 \Rightarrow \ldots \Rightarrow t_n.$$

and where Δt is the time step. Note that this time step is not necessarily constant; some methods allow it to vary through time. For simplicity of notation, we will however explain the basic methods assuming a constant time step.

The model starts off with the specification of the initial conditions, i.e. the start values of the state variables C, at time t_0 (Fig. 6.4). Then, at each point in time (t), the rates of changes of the state variables, dC/dt are estimated. After this, the numerical integration method uses the values of the state variables plus their rates of changes to estimate the values of the state variables at the next point in time (t+Δt). This procedure is repeated until the total time has been simulated.

To estimate the values at the next time step, an update formula is needed.

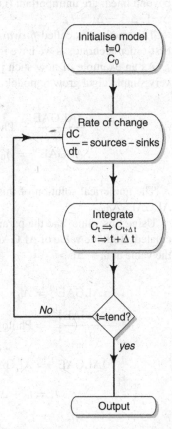

Fig. 6.4 Schematic representation of the numerical integration method

$$C^{t+\Delta t} = f\left(C^t, \frac{\partial C^t}{\partial t}, \Delta t\right) \tag{6.6}$$

6.3.1 Euler Integration

The simplest update formula is to use Taylor equation (Eq. 6.4), with h $=\Delta$t and f(x)=C^t, and to ignore all terms of order higher than one:

$$C^{t+\Delta t} = C^t + \Delta t \cdot \frac{\partial C^t}{\partial t} + O(\Delta t)$$

$$C^{t+\Delta t} \approx C^t + \Delta t \cdot \frac{\partial C^t}{\partial t} \tag{6.7}$$

where $O(\Delta t)$ stands for the error we make by truncating the Taylor series.

This is a reasonable approximation only for small enough Δt, as then the terms beyond linear are unimportant (i.e. for Δt $=0.01$, $\Delta t^2 = 1e^{-4}$, $\Delta t^3 = 1e^{-6}$; whereas for Δt$=1$, $\Delta t^2 = \Delta t^n = 1$).

This method is called *forward differencing* or Euler integration and said to be first-order accurate, as we have ignored all terms of order higher than 1.

As an example of how such procedure might work in practice, let's consider a very simple algal growth model:

$$\frac{d\text{ALGAE}}{dt} = \text{Photosynthesis-Respiration-Grazing}$$

$$\text{ALGAE}^{t0} = A_0 \tag{6.8}$$

The numerical solution of this model starts at t_0 by setting the initial condition: ALGAE=A_0.

Using this value, and the parameters, the rate of change dALGAE/dt at t_0 is calculated. Both the value of ALGAE^{t0} and its rate of change are then used to calculate the value at $t_0 + \Delta t$.

$$\text{ALGAE}^{t0} = A_0$$

$$\left.\frac{d\text{ALGAE}}{dt}\right|^{t0} = \text{Photosynthesis}^{t0} - \text{Respiration}^{t0} - \text{Grazing}^{t0}$$

$$\text{ALGAE}^{t1} = \text{ALGAE}^{t0} + \Delta t \cdot \left.\frac{d\text{ALGAE}}{dt}\right|^{t0}$$

$$t_1 = t_0 + \Delta t \tag{6.9}$$

This procedure is then repeated for successive time steps.

$$\left.\frac{d\text{ALGAE}}{dt}\right|^{tI} = \text{Photosynthesis}^{tI} - \text{Respiration}^{tI} - \text{Grazing}^{tI}$$

$$\text{ALGAE}^{tI+1} = \text{ALGAE}^{tI} + \Delta t \cdot \left.\frac{d\text{ALGAE}}{dt}\right|^{tI} \tag{6.10}$$

Unfortunately the first-order accuracy of the Euler method is not good enough for all applications.

By truncating the Taylor series at order 2, we effectively assume that the rate of change remains constant in the interval Δt. If this is not the case, then the method will not be very accurate (see Fig. 6.5A).

One remedy to obtain better accuracy is reducing the time step Δt. However, when the model's properties are really bad, it is possible that we will need to make the time step so small that the model never gets anywhere!

There are several other solutions to this problem, but before we (briefly) discuss them, we have a look at the criteria to which integration routines should obey.

6.3.2 Criteria for Numerical Integration

There are several criteria that determine the quality of the integrator. Here we discuss accuracy, stability, speed and memory requirements.

Accuracy is a measure for the correctness of the solution. As explained above, the discretization error which is due to the numerical approximation and made at each time step can be quantified by comparing the solution with the Taylor expansion.

$$C(t + \Delta t) = C(t) + \Delta t \times \frac{\partial C}{\partial t} + \frac{1}{2}\Delta t^2 \times \frac{\partial^2 C}{\partial t^2} + \cdots. \tag{6.11}$$

and we write that a method is of nth order, if the truncation error made at each step is proportional to Δt^{n+1}. Generally, higher order methods attain higher accuracy at larger time steps.

The *stability* of a method refers to its potential to lead to increasing oscillations between consecutive solution points (see Fig. 6.3B). Some models are especially prone to this type of problems. Clearly, we want an integration routine to be as accurate and stable as possible!

However, we also want to see the results within a reasonable time frame. This makes *speed* a third, albeit highly subjective, criterion. Speed depends on (1) the size of the time step, Δt, taken and (2) the number of function evaluations needed to advance from time t to time t+Δt. Euler, the computationally most simple of all algorithms requires only one function evaluation, whereas higher-order integration routines (see below) may require more evaluations. However, these methods can often run with a larger time step.

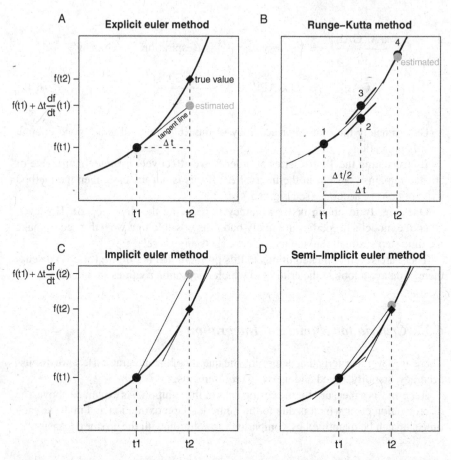

Fig. 6.5 **A**. The explicit Euler integration method. The function value at time t2 is estimated based on the value at time t1 and the rate of change (as denoted by the tangent line). It is assumed that the rate of change remains constant in the time step between t1 and t2. As this is not the case, the estimated value considerably underestimates the true value. **B**. The 4-th order Runge-Kutta method. One function evaluation at the start point (1), two evaluations at the mid-point (2,3), and one evaluation at the end point (4) are combined to provide the final projected value (*grey dot*). **C**. The implicit Euler integration method uses the estimate of the rate of change at the next time step, t2. **D**. The semi-implicit method combines both the explicit and implicit approach

Finally, the more complex the integration routine, the more *memory* it requires. The reason for this is that for every time step more information (basically: higher and higher order derivatives in the Taylor expansion) must be remembered. For some large-scale applications with constituents described in many spatial cells, available memory may become a limiting factor.

A particularly hard-to-integrate type of problems are so-called 'stiff' sets of equations. The problem typically occurs when processes are modelled that have different inherent time scales. When one tries to model dynamics of bacterial population dynamics (typical time scales hours to days), together with a forcing (e.g. organic matter input) with a seasonal or year-to-year variation, then it can be expected that the long-term result will mostly be influenced by the long-term varying factor. Bacterial populations, at any particular moment, will be more or less in equilibrium with the organic matter forcing that varies only slowly. Yet almost all computer resources needed to solve the model will go to the fast process. When using simple Euler integration, e.g., one will probably need a time step of minutes in order to simulate the bacterial dynamics correctly, while the process of real interest varies over months!

To comply with these –sometimes conflicting- requirements, more complex integration methods have been devised. These are briefly introduced in the next chapters.

Basically the improvements include:

- Performing extra function evaluations and interpolating between them. Some Runge-Kutta methods (Section 6.3.3) are based on this principle.
- Performing the simulation with a flexible time step. These integration routines use a finer subdivision of time only when it is actually needed, i.e. when environmental changes are most rapid. The principle is explained using 5th order Runge-Kutta (Section 6.3.4)
- Implicit methods of solution. This is a frequently used remedy against instability (but it comes at the cost of reduced accuracy).

Most modern integration routines somehow combine these features.

6.3.3 Interpolation Methods – 4th Order Runge-Kutta (**)

Runge-Kutta methods use extra evaluations of the differential equations at various positions in time, and interpolate between these values.

Most commonly used is the 4th order Runge-Kutta method, which requires four evaluations per time step Δt (Fig. 6.5 B). The methods' equations are the following:

$$k_1 = \Delta t \cdot f(t_i, \mathbf{C}_i)$$
$$k_2 = \Delta t \cdot f(t_i + \Delta t/2, \mathbf{C}_i + k_1/2)$$
$$k_3 = \Delta t \cdot f(t_i + \Delta t/2, \mathbf{C}_i + k_2/2)$$
$$k_4 = \Delta t \cdot f(t_i + \Delta t, \mathbf{C}_i + k_3)$$
$$\mathbf{C}_{i+1} = \mathbf{C}_i + \frac{1}{6}k_1 + \frac{1}{3}k_2 + \frac{1}{3}k_3 + \frac{1}{6}k_4, \qquad (6.12)$$

where the k_i's are the intermediate evaluations of the differential equations, and the last equation estimates the state variable's value at the next time step by interpolation. This method has a truncation error in the order of Δt^5, i.e. it is a 4th order method.

Similarly as the Euler method, 4th order Runge-Kutta integration uses a fixed time step. For certain applications, using a fixed time step is not a good idea. For instance, a model that is applied over certain periods of time where changes occur very rapidly, and other periods where there is hardly any change at all, the size of (the constant) Δt must be chosen such as to perform well in the period where changes are rapid. This may lead to unnecessary computation during periods where the model is relatively smooth.

6.3.4 Flexible Time Step Methods –5th Order Runge-Kutta (**)

If the solution varies rapidly in one region of the integration interval whilst staying nearly constant in another region, then a method that automatically adapts the time step is much better suited than the constant step size methods (as in previous section). Integration routines that use an adaptive step size, consist of two parts:

– the stepper routine which actually performs the integration, and
– a quality control routine that checks if the step taken is acceptable or not.

The time step is taken as large as possible, but such that the precision of the integration remains within predefined bounds.

The 5th order Runge-Kutta method for instance, combines the robustness of 4th order Runge-Kutta with the elegance of a flexible time step. The idea is to adapt Δt to some desired accuracy Δ_0.

First, a 'reference' error Δ_1 is estimated from the difference between two Runge-Kutta applications, one with a large step Δt_1, the other with two successive steps, with time step half the large stepsize. As in 4th order methods, the error Δ scales as Δt^5 and this is used to estimate the step size step Δt_0 producing the desired accuracy Δ_0 as:

$$\Delta t_0 = \Delta t_1 \cdot \left| \frac{\Delta_0}{\Delta_1} \right|^{0.2} \tag{6.13}$$

6.3.5 Implicit and Semi-Implicit Integration Routines (**)

Implicit methods are less accurate than explicit methods (all the methods discussed before), but are much more stable.

Whereas in explicit methods, the new value of the state variable at time $t+\Delta t$ is calculated entirely in terms of the old value at time t, in *implicit* methods, the derivative is estimated *at the next time step* $(t+\Delta t)$ (Fig. 6.5C) as follows:

$$\mathbf{C}^{t+\Delta t} = \mathbf{C}^t + \Delta t \cdot \left. \frac{\partial \mathbf{C}}{\partial t} \right|^{t+\Delta t} \tag{6.14}$$

As this derivative is a function of the unknown concentration at the next time step, it is itself unknown, and thus integration boils down to finding a solution of the (often nonlinear) set of equations. It is not necessary to know the fine details of how this is done; the important part is that, solving these equations requires the creation and inversion of an (n*n) matrix involving the Jacobian matrix (see

Appendix B.3.4), where n equals the number of state variables. For models with many state variables, this can be quite time-consuming.

In *Semi-implicit methods*, the derivative is approached partly at the current, partly at the next time step (t+Δt) (Fig. 6.5.D) as follows:

$$\frac{\mathbf{C}^{t+\Delta t} - \mathbf{C}^{t}}{\Delta t} = \gamma \cdot \left. \frac{\partial \mathbf{C}}{\partial t} \right|^{t} + (1 - \gamma) \cdot \left. \frac{\partial \mathbf{C}}{\partial t} \right|^{t+\Delta t} \tag{6.15}$$

and where γ is usually 0.5. This method is also known as the Crank-Nicholson scheme.

In explicit Euler and implicit Euler the truncation error is similar (both are 1st order accurate); however, implicit Euler is stable, such that larger time steps can be used. Note however, that, due to the inversion of the (n*n) matrix, performing one time step is much more costly in the implicit method. The semi-implicit method is 2nd order accurate and also stable; its computational demand is comparable to the implicit method.

6.3.6 Which Integrator to Choose?

The good news is that, for those that use R as a modelling platform, a robust integration routine (`ode`) is available (package `deSolve`) that automatically selects the step size such as to achieve both reliability and efficiency. This is *the* method of choice and you can use it without worrying about the accuracy of the model; this is taken care of by the integration routine.

However, there are some model problems that are faster solved with the more simple integration routines. We run a model because we want output at some (regular) time interval. If the time step Δt at which the Euler or 4th order Runge-Kutta method is accurate and stable, is similar to or larger than this output interval, then these simpler integration methods are probably the best choice, as they are much faster. However, before we can be sure that the model output produced by fixed time step methods is sufficiently accurate, we must check whether the constant time step is small enough. Here is a rule of thumb to guide you:

* First run the model with a certain timestep,
* double the timestep and inspect the difference between the two runs.
* If the difference is significant, halve the time step until the results stop changing significantly.
* If insignificant, double the time step until the results start to change significantly.
* To be on the safe side, take a timestep half the one that was just accurate enough.

For other problems, there exist especially-designed integration routines that perform the job optimally. One of these special cases are 1-dimensional reaction-transport equations. We give an example at the end of this chapter, when we model transport and death of marine zooplankton in an estuary.

6.4 Approximating Spatial Derivatives (*)

Numerical integration in time proceeds by jumping from one point in time to the next. Similarly, numerical approximation of spatial derivatives in one dimension calculate the value of the function in a finite number of nodes or layers only: x_1, x_2, x_3, .. (Fig. 6.6A).

To show how this is done, we derive the numerical approximation for the general reaction-transport equation from Section 3.4.3. We start by approximating the equation written in flux-divergence form, after which we find suitable expressions for the advective and dispersive fluxes.

These derivations require that we keep track of positions in space very carefully. In particular, we must always be aware whether a concentration, surface, flux or distance refers to an interface between boxes, or to the centre of the boxes. The symbolism is detailed in Fig. 6.6.

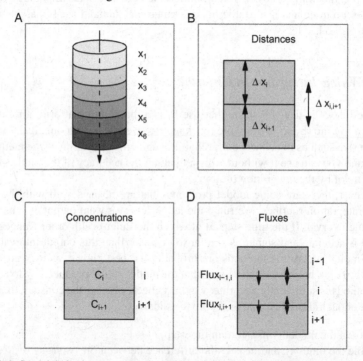

Fig. 6.6 A. Spatial derivatives are approximated by subdividing the spatial domain in a number of grid cells or boxes. Positions (x_i) are defined at the centre of the boxes. **B.** Relevant distances are either between centres of the boxes (e.g. $\Delta x_{i,i+1}$ between centres of boxes i and i+1) or lengths of boxes (Δx_i for length of box i). **C.** Concentrations, densities, etc.. are defined in the centre of the grid cells. **D.** Fluxes are defined on the box interfaces

6.4.1 Approximating the Flux Divergence Equation

As a first step, we numerically approximate the general reaction-transport equation written in the flux-gradient form, and where g is a first-order growth rate:

$$\frac{\partial C}{\partial t} = -\frac{1}{A} \cdot \frac{\partial A \cdot Flux}{\partial x} + g \cdot C \tag{6.16}$$

This formula is now discretized to represent the rate of change of the concentrations in each box i, C_i, as follows:

$$\frac{dC_i}{dt} = -\frac{1}{A_i} \frac{\Delta_i(A \cdot Flux)}{\Delta x_i} + g \cdot C_i \tag{6.17}$$

Where C_i is concentration in box i, Δx_i is the thickness of box i, A_i is the surface area in the middle of box i, and Δ_i denotes that the flux gradient is to be taken around box i.

At this point, it is useful to pinpoint that, when the model domain is divided in a number of discrete cells, we generally specify the concentrations or densities in the *middle* of boxes (Fig. 6.6B), while the fluxes are defined on the box *interfaces* (Fig. 6.6C). Fluxes as much leave a box as enter the next box, so it makes sense to prescribe them in between both boxes.

Keeping this in mind, the flux gradient for box i is then defined as the difference of the fluxes at the interface with the next box (i,i+1) and with the previous box (i−1, i). Thus we write:

$$\frac{dC_i}{dt} = -\frac{A_{i,i+1} \cdot Flux_{i,i+1} - A_{i-1,i} \cdot Flux_{i-1,i}}{A_i \cdot \Delta x_i} + g \cdot C_i \tag{6.18}$$

Here $Flux_{i-1,i}$ is the flux on the interface between cell i−1 and i, and $A_{i-1,i}$ is the surface area on this interface.

Depending on the problem, the surfaces at the interfaces and at the middle of the boxes may all be known (for instance, they may be defined as a continuous function of x). Alternatively, the surfaces at the interfaces may be derived by linear interpolation of the known surfaces at the middle of the boxes.

To make the numerical approximation complete, we need to find a suitable description for the advective and dispersive fluxes. It is simplest to start with the latter.

6.4.2 Approximating Dispersion

The continuous representation of the dispersive flux is:

$$Flux|_{dispersion} = -D \frac{\partial C}{\partial x} \tag{6.19}$$

which states that the flux is directed against the concentration gradient (hence the minus sign), i.e. there will be transport from high to low concentrations. As the flux is defined on the interface, the concentration gradient should also be defined on the interface. Thus it makes sense to approximate this gradient by simply taking the differences of the concentrations in the adjacent boxes, i.e. from box i to box i−1, and divide this by the distance between the centres of both boxes:

$$Flux_{i-1,i}\big|_{dispersion} = -D_{i-1,i} \cdot \frac{C_i - C_{i-1}}{\Delta x_{i-1,i}} \tag{6.20}$$

where $\Delta x_{i-1,i}$ is the 'dispersion distance', i.e. the distance from the centre of box $i-1$ to the centre of box i where the concentrations C_{i-1} and C_i are prescribed. $D_{i-1,i}$ is the value of the dispersion coefficient at the interface (see Fig. 6.6).
This is indeed how dispersive fluxes are approximated.

6.4.3 Approximating Advection

The advective fluxes are simply the product of a concentration times a velocity.

$$Flux\big|_{advection} = u \cdot C \tag{6.21}$$

With the flux defined on the interface, it is most logical to use the concentration on this interface to estimate the flux. However, as we model concentrations in the center of boxes the concentration on the interface is not known, but it can simply be estimated as an average of the concentrations of the two adjacent boxes.

Thus, the logical choice to approximate the advective flux on the interface between box $i-1$ and i is to interpolate spatially:

$$Flux_{i-1,i}\big|_{advection} = u \cdot \frac{C_i + C_{i-1}}{2} \tag{6.22}$$

This is called the 'centred difference' approximation, as it uses the central value of two adjacent concentrations. With equal box sizes, this is the (interpolated) concentration at the boundary between adjacent boxes. With unequal box sizes, this value is situated somewhere else, although the formula would be easily adapted to represent the interpolated concentration at the interface exactly.

Unfortunately, this is often not a good idea! When used to approximate the advective equation, the centred differencing has the undesired effect of producing negative concentrations. In numerical jargon it is said that the scheme is 'non-monotone'.

Why is this so? The flux on the interface of box $i-1$ and i not only enters box i but it also leaves box $i-1$. In the centred difference approximation, the advective flux leaving box $i-1$ is a function of the concentration in box $i-1$, but also of the concentration in the next box, i. Now, assume the case where the concentration in this box (i-1) is zero, while it is positive in box i. Then the centred difference approximation will calculate a net efflux from box $i-1$, although there is nothing present to leave the box! Clearly, at the next time step, concentrations C_{i-1}, will be negative.

Such negative concentrations are biologically unrealistic and should be avoided. Moreover, these negative concentrations in turn often create oscillations, such that the scheme is also 'unstable'.

It is much safer to make the flux leaving box $i-1$ a function of the concentration in the box $i-1$ only. Thus:

$$Flux_{i-1,i}\big|_{advection} = u \cdot C_{i-1} \tag{6.23}$$

This is called a 'backward' approximation. If concentration is 0 in box $i-1$, then this formula predicts that there will be no efflux, hence negative concentration will not occur.

However, the world of numerics is never perfect. Whereas the backward differencing scheme does never produce negative concentrations, it is notorious for its ability to create so-called 'numerical dispersion'. We will demonstrate this phenomenon later.

6.4.4 The Boundaries with the External World

We end this section with some consideration of what to do near the boundaries with the external world. In general, the boundaries will receive particular attention in the numerical solution of a model.

1. In case the boundary condition is prescribed as a *flux*, then life is simple: rather than calculating the flux as a function of concentrations (using Eq. (6.20) or Eq. (6.23)), we just plug in the imposed value in eq. 6.18. For instance, if at the upper boundary, i.e. at the interface with cell 1, a flux is prescribed with a value F, then the numerical approximation for the rate of change of concentration in box 1 is:

$$\frac{dC_1}{dt} = -\frac{A_{1,2} \cdot Flux_{1,2} - A_0 \cdot F}{A_1 \cdot \Delta x_1} + g \cdot C_1 \tag{6.24}$$

Where A_0 is the surface at the interface, and $Flux_{1,2}$ is estimated based on eq. 6.20 and eq. 6.23.

2. In contrast, in case the boundary condition is prescribed as a *concentration* defined on the edge, then it follows that the dispersion distance is only half the thickness of the box, i.e. we obtain:

$$Flux_{0,1}\big|_{advection} = u \cdot C_0^*$$

$$Flux_{0,1}\big|_{dispersion} = -D \cdot \frac{C_1 - C_0^*}{\Delta x_1/2} \tag{6.25}$$

Using these formulas, we have transformed the boundary concentration prescription into a flux prescription, which can then be incorporated into the flux divergence equation:

$$\frac{dC_1}{dt} = -\frac{A_{1,2} \cdot Flux_{1,2} - A_0 \cdot (\left. Flux_{0,1}\right|_{advection} + \left. Flux_{0,1}\right|_{dispersion})}{A_1 \cdot \Delta x_1} + g \cdot C_1$$

3. Boundary conditions may also be prescribed as fluxes dependent on the concentration difference between an external concentration, and the concentration at one of the boundaries of the model (so-called Robin or evaporation boundary conditions):

$$Flux_{0,1} = \alpha(C_e - C_{0,1})$$

In principle, $C_{0,1}$ can be approximated from the known concentration at the centre of the first box C_1, and the concentration gradient in the top part of the model, as was done in the previous case. In practice however, one usually takes the concentration in the upper box (C_1) as a sufficient approximation of the concentration at the interface.

Once the flux at the interface is fully defined it can, again, be plugged into the flux divergence equation.

6.5 Numerical Dispersion (∗∗∗)

When approximating advective terms (Section 6.4.3), the backward differencing scheme (eq. 6.26) did not produce negative concentrations and was stable, but it is only first-order accurate.

$$\left. Flux_{i-1,i}\right|_{advection} = u \cdot C_{i-1} \tag{6.26}$$

The truncation error, which is due to ignoring the 2nd and higher order terms from the Taylor expansion, manifests itself as a process that is called 'numerical dispersion' or numerical dissipation. This is one of the most famous types of numerical errors. It means that, even in the absence of true, physical or biological dispersion, existing sharp gradients will be smoothened because of a numerical artefact.

We illustrate the point using an example similar (but not equal) to the one from the book of Gurney and Nisbet (1998). Individuals of a population grow at a constant rate and without mortality. At the start of the simulation ten individuals are born, and we describe their size as a function of time.

This is a purely 'advective' model, where the advection rate is given by the growth rate, the increase in size of the individuals. If we denote the size of the individuals by x, the specification for this model and the backward differencing approximation is:

$$\frac{\partial N}{\partial t} = -growth \cdot \frac{\partial N}{\partial x}$$

$$\frac{dN_i}{dt} = \frac{growth \cdot N_{i-1} - growth \cdot N_i}{\Delta x}$$

$$N_0 = 10 \qquad\qquad\qquad\qquad\qquad\qquad\qquad (6.27)$$

where N_i is the number of individuals in size class i. If we select the size classes (Δx) such that the organisms move through one size class per day, then, in the real world, after the first 10 days all organisms are of size 10, after 20 days, all organisms are of size 20, etc... In Fig. 6.7 the dashed line represents the true solution after ten days of simulation.

Using the backward differencing scheme, the results are quite different: after 10 steps through time, the square pulse has been smoothened, by the 'numerical dispersion' error, into a bell-shaped curve, with a range more than 10 times the original range! (Fig. 6.7).

Faced with this problem, we can take several directions:

First of all, we can be very pragmatic about it. In the example, the square pulse of individuals moving through the size classes may be the true mathematical solution, but it is not very realistic. In real life, variation in growth rates of individuals will cause some individuals to attain larger size earlier; whilst other organisms will grow slower. Thus, biological variation will also cause a spread of the pulse as time proceeds. One might therefore argue that the smoothened pulse is more realistic than the square one. Moreover, in most applications, there will be a mixture of advective and diffusive terms, and in many cases, the 'numerical' dispersion will be but a fraction of the other dispersion terms. Under these circumstances, we may consider the backward numerical scheme to be sufficiently accurate.

Secondly, we can try to fix the problem by using more complex numerical schemes. These mathematical solutions reduce the effects of numerical dispersion

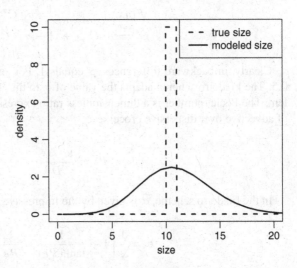

Fig. 6.7 Numerical dispersion. Although this growth model only includes advection, after 10 steps, the true square pulse has been smoothened spectacularly by numerical diffusion

by taking into account the higher order terms of the Taylor series, but most of these schemes are quite complex and still suffer some kind of artificial dispersion. One solution which may be appealing to biologists is provided in the book by Gurney and Nisbet (1998).

Thirdly, as the problem is mainly acute for large boxes, a solution is to increase the number of boxes. However, as this occurs at the expense of increasing computing time, there is a limit to the number of boxes that can be represented in a model.

Finally, we may decide to use centred differences after all, ignoring all warnings about the dangers of negativity. Under certain circumstances, there are perfectly good reasons to do so: as long as concentrations do not increase along the spatial axis, centred differences will never generate negative concentrations, yet they cause very limited numerical dispersion. For instance, sediment organic matter concentrations (nearly) always decrease with depth in the sediment. Thus, models that describe organic matter in a spatially explicit way often approach the advection term by means of centred differences.

6.5.1 Example: Sediment Model

We demonstrate the effect of numerical dispersion in different schemes using a simple model for inert tracers in a sediment, subject to a small dispersion term due to bioturbation (D_b=0.1 cm^{2}y^{-1}) and an advective term due to sediment accretion (u=1 cm y^{-1}). Three numerical schemes are compared: backward differences, centred differences, and a flexible scheme (called the Fiadeiro scheme, Fiadeiro and Veronis (1977)) that tries to optimise in between the other two methods.

For equal box sizes, the three schemes can all conveniently be modelled using a single formula, where the advective flux is expressed as a function of a composite concentration, using a parameter σ:

$$Flux_{i-1,i}\big|_{advection} = u \cdot C^*$$
$$C^* = \sigma \cdot C_{i-1} + (1 - \sigma) \cdot C_i \qquad (6.28)$$

Clearly, in backward differences, σ equals 1. For centred differences, σ equals 0.5. The Fiadeiro scheme adapts the value of σ to the 'Peclet number' of the problem. The Peclet number is a dimensionless ratio expressing the relative importance of advective over dispersive processes:

$$Pe = \frac{u \cdot \Delta x}{D} \qquad (6.29)$$

In the Fiadeiro scheme, σ is given by the impressive formula:

$$\sigma = \frac{1}{2}\left[1 + \frac{1}{\tanh(Pe)} - \frac{1}{Pe}\right] \qquad (6.30)$$

Fig. 6.8 Dependence of the parameter σ on the Peclet number in the Fiadeiro scheme

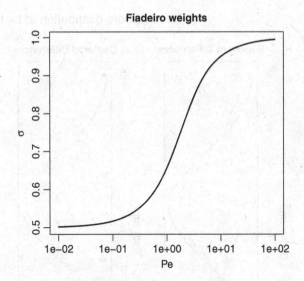

(see Fig. 6.8). Clearly, the scheme switches from centred differences at low Peclet numbers (dispersion – dominated conditions) to backward differences in advection-dominated cases.

Comparison of the results for the passive tracer model (Fig. 6.9) shows that numerical dispersion dominates the results for backward differences, unless the number of boxes is very high (and computation time is accordingly very long). Centred difference results for a very low number of boxes are set somewhat apart, but this may partly be due to the rough specification of the initial conditions in this case. With 30 boxes, the result is almost indistinghuishable from the result with 300 boxes. The Fiadeiro scheme in this case is closer to backward than to centred differences (D is rather low, so Pe is high), but still performs slightly better than the latter. When the dispersion coefficient is raised or the advection is decreased, the Fiadeiro scheme will approach the centred differences solution.

Finally note that in this case, the centred differences did not produce any instability.

6.6 Case Studies in R

In previous chapters we developed a number of models, but did no yet solve them. R-package deSolve contains several numerical methods to solve (systems of) differential equations. Function ode is the most general numerical integration routine from this package and is the method of choice for models whose dynamics are continuous, i.e. there are no sharp jumps in the forcing functions. In case the dynamics is discontinuous, we may run the model in several parts (e.g. Section 6.6.2).

Luminophore distribution at t = 1 year

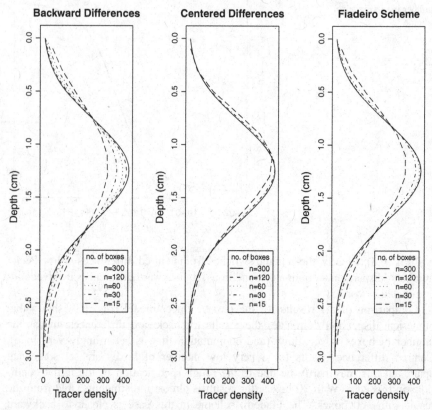

Fig. 6.9 Spatial distribution of inert tracers (e.g. luminophores, fluorescent tracer particles) added as a layer with concentration of 1000 particles cm^{-3} in the top 0.5 cm at t=0. In the model, the particles are subject to dispersion by bioturbation (Db = 0.1 $cm^2 y^{-1}$) and advective movement due to sediment accretion (u = 1 cm y^{-1}). After one year, the peak is situated between 1 and 1.5 cm, but depending on the numerical scheme, considerable numerical dispersion may have taken place. See text for details

6.6.1 Implementing the Enzymatic Reaction Model

We start by implementing the enzymatic reaction model of Section 2.4 in R. The reactions were

$$E + D \underset{k_2}{\overset{k_1}{\leftrightarrow}} I$$

$$I + F \overset{k_3}{\to} E + G \tag{6.31}$$

and the model equations:

$$\frac{d[D]}{dt} = -k_1 \cdot [E] \cdot [D] + k_2 \cdot [I]$$

$$\frac{d[I]}{dt} = k_1 \cdot [E] \cdot [D] - k_2 \cdot [I] - k_3 \cdot [I] \cdot [F]$$

$$\frac{d[E]}{dt} = -k_1 \cdot [E] \cdot [D] + k_2 \cdot [I] + k_3 \cdot [I] \cdot [F]$$

$$\frac{d[F]}{dt} = -k_3 \cdot [I] \cdot [F]$$

$$\frac{d[G]}{dt} = k_3 \cdot [I] \cdot [F] \tag{6.32}$$

There are three model parameters: the rate coefficients k_1, k_2 and k_3 that are defined first. They are stored as a vector and assigned names and values.

```
params<-c(k1=0.01/24,
          k2=0.1/24,
          k3=0.1/24)
```

Similarly the state variable vector that contains the 5 chemical species is created and initial concentrations given.

```
state     <-c(D=100,
              I=10,
              E=1,
              F=1,
              G=0)
```

Next a function ('enzyme') is defined that calculates the rate of change of the state variables. Input to the function is the model time (not used here, but required by the calling routine), and the values of the state variables and the parameters. This function will be called by R's routine that solves the differential equations (ode).

The parameters and state variables are vectors, and their elements can be assessed by indexing (e.g. state[1], state[2], etc...). However, the code is more readable if we use their names instead. To do so, the vectors are treated as a list.

The statement with(as.list(c(state,parameters)),{... }) does this.

The main part of the enzyme model calculates the rate of change of the state variables. At the end of the function, these rates of change are returned, packed as a list.

```
enzyme<-function(t,state,parameters)
{
 with(as.list(c(state,parameters)),{

    dD <- -k1*E*D + k2*I
    dI <-  k1*E*D - k2*I - k3*I*F
    dE <- -k1*E*D + k2*I + k3*I*F
    dF <-                - k3*I*F
    dG <-                  k3*I*F

list(c(dD,dI,dE,dF,dG))
    })
}
```

Now the model can be run. Below we choose to run the model for 300 h, and give output at 0.5 hourly intervals. R's function seq() creates the time sequence.

```
times        <-seq(0,300,0.5)
```

The model is solved using the function ode, which is included in R's package 'deSolve'. This package has to be loaded first, and if necessary installed (see Appendix A.2.6 for how to do this). The function ode takes as input the state variable vector, the times at which output is required, the model function that returns the rate of change and the parameter vector, in that order. It returns a matrix that contains the values of the state variables (columns) at the requested output times.

The output is converted to a data frame and stored in 'out'. Data frames have the advantage, that their columns can be accessed by name, rather than by number. For instance, 'out$D' will take the outputted concentrations of substance D, and so on.

```
require(deSolve)
out <- as.data.frame(ode(state,times,enzyme,params))
```

Finally, the model output is plotted. The figures are arranged in two rows and two columns (mfrow), and the size of the outer upper margin (the third margin) is increased (oma), such as to write a figure heading (mtext) (see Fig. 6.10).

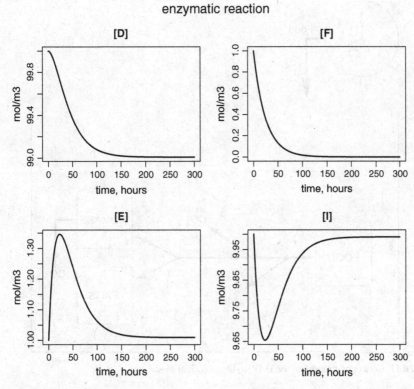

Fig. 6.10 Output of the enzymatic reaction model – for R-code see text

```
par(mfrow=c(2,2),     oma=c(0,0,3,0))

plot (times,out$D,type="l",main="[D]",
     xlab="time,    hours", ylab="mol/m3",lwd=2)
plot (times,out$F,type="l",main="[F]",
     xlab="time,    hours",ylab="mol/m3",lwd=2)
plot (times,out$E,type="l",main="[E]",
     xlab="time,    hours",ylab="mol/m3",lwd=2)
plot (times,out$I,type="l",main="[I]",
     xlab="time,    hours",ylab="mol/m3",lwd=2)
mtext(outer=TRUE,side=3,"enzymatic reaction ",cex=1.5)
```

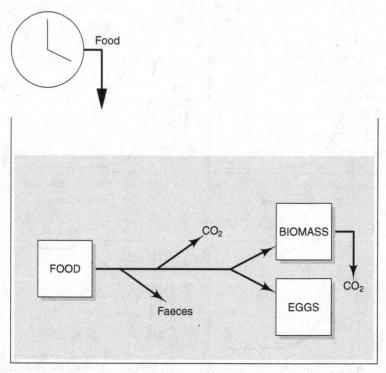

Fig. 6.11 Conceptual model of the DAPHNIA individual model

6.6.2 Growth of a Daphnia Individual

We now discuss a (more complex) model that describes the growth and reproduction of one individual of a zooplankton organism, belonging to the genus *Daphnia* (Cladocera). The model (Andersen, 1997), is based on the energy budget of the organism.

The conceptual model is depicted in Fig. 6.11. The weight of the individual (excluding the eggs), the accumulated egg mass and the food are the state variables. The model is embedded in a transfer culture experiment. This means they are reared in a bottle, and a fixed number of individuals are transferred to new medium at regular intervals.

The implementation of this model is interesting because the biomass of these organisms does not increase continuously. Rather, the individuals grow in-between molts, and the time span between molts depends on temperature. The transfer of organisms to new medium introduces a second temporal discontinuity in their growth.

We use the implementation in R to describe the model in detail.

The model *parameters* fall into three groups.

1. The weight of newborns (neonateWeight), weight at first reproduction, and maximal weight an organism can attain and the duration between moults (instarDuration):

```
neonateWeight       <-   1.1   #µgC
reproductiveWeight  <-   7.5   #µgC
maximumWeight       <-  60.0   #µgC
instarDuration      <-   3.0   #days
```

2. parameters that are used to describe the food ingestion, assimilation, and reproduction and respiration rate

```
ksFood            <-  85.0   #µgC/l
IngestWeight      <-132.0    #µgC
maxIngest         <-   1.05  #/day
assimilEff        <-   0.8   #-

maxReproduction   <-   0.8   #-
respirationRate   <-   0.25  #/day
```

and

3. parameters that describe the experimental setup (time between transfers, the food concentration in the new medium, the number of individuals reared).

```
transferTime      <-   2    # Days
foodInMedium      <- 509    # µgC/l
numberIndividuals <-  32    #   -
```

The three state variables are the individual weight of the organism, the total egg weight and the food in the medium. They are initialized whilst defining them:

```
state     <-c(
INDWEIGHT = neonateWeight ,
EGGWEIGHT = 0                ,
FOOD      = foodInMedium
          )
```

In between moults, the organisms ingest, respire, grow and reproduce. The ingestion rate per unit body weight decreases linearly with body size (variable WeightFactor), and is limited by food availability (half-saturation constant ksFood). Only part of ingested food is assimilated (parameter assimilEff). Assimilated food serves 3 purposes: somatic growth (increase of biomass), reproduction and maintenance (Respiration). The latter is the basal metabolism to support body functions and regeneration of lost material; it is first-order to individual weight.

The net balance between assimilation and respiration, if positive, is used for somatic growth and reproduction.

Reproduction only starts when the individual weight has exceeded the reproductive weight. The fraction allocated to reproduction increases hyperbolically with increasing individual weight. The maximal fraction is less than 1, such that body growth continues after reaching maturity.

Finally the rate of change of individual weight, egg weight and food availability is calculated. For the latter, we take into account the number of individuals (a parameter) that are in the medium. The function returns both the rates of changes and three ordinary variables.

```
model<-function(t,state,parameters)
 {
 with(as.list(c(state)),{ # unpack the state variables

# ingestion, size-dependent and food limited

WeightFactor <- (IngestWeight-INDWEIGHT)/(IngestWeight-neonateWeight)
MaxIngestion <- maxIngest*WeightFactor
Ingestion    <- MaxIngestion*INDWEIGHT*FOOD / (FOOD + ksFood)

Respiration  <- respirationRate * INDWEIGHT

Growth       <- Ingestion*assimilEff - Respiration

if (Growth <=  0. | INDWEIGHT <reproductiveWeight)
   Reproduction <- 0. else
{
  WeightRatio    <- reproductiveWeight/INDWEIGHT
  Reproduction   <- maxReproduction * (1. - WeightRatio^2)
}

# rate of change
dINDWEIGHT <- (1. -Reproduction) * Growth
dEGGWEIGHT <-       Reproduction  * Growth
dFOOD      <- -Ingestion * numberIndividuals

list(c(dINDWEIGHT, dEGGWEIGHT, dFOOD),
     c(Ingestion   = Ingestion,
       Respiration = Respiration,
       Reproduction = Reproduction))

   }) # end of with(as.list(...

} # end of model
```

During moults, increased energetic costs cause the organisms to loose weight. The weight loss is an allometric function of body length; length is an allometric function of individual weight.

```
Moulting    <- function ()

  {
  with(as.list(c(state)),{

    refLoss    <-   0.24 #µgC
    cLoss      <-   3.1  #-

    INDLength    <-  (INDWEIGHT /3.0)^(1/2.6)

    WeightLoss <- refLoss * INDLength^cLoss
    return(INDWEIGHT - WeightLoss)
    })
}
```

Because of the setback of individual weight during moulting and due to the regular transfer of individuals to new medium, running the model proceeds in discrete steps. This is necessary because the integrator ode cannot cope with changes occurring in the values of the state variables: it just requires the values of the rates of changes.

We keep track of the next time at which moulting will occur and the next time at which individuals are transferred to new medium (TimeMoult, TimeTransfer). The integrator proceeds till either of these events occurs (TimeOut), and the state variables are each time reinitialised with the final condition of the integration procedure; this is on the last element of output.

If the organism moults, the individual weight is reset (using the function Moulting, defined above), the egg weight of the instar is reinitialised to 0, and the next time at which moulting (TimeMoult) will occur is calculated.

Transfer of the organisms to new medium, only affects the food concentration (foodInMedium); similar as for moulting, the next time at which transfer will occur (transferTime) is estimated.

```
TimeFrom        <- 0
TimeEnd         <- 40

TimeMoult       <- TimeFrom + instarDuration
TimeTransfer <- TimeFrom + transferTime

Time            <- TimeFrom
Outdt           <- 0.1
out             <- NULL

while (Time < TimeEnd)
{

  TimeOut <- min(TimeMoult,TimeTransfer,TimeEnd)
  times   <- seq(Time,TimeOut,by=Outdt)
  out1    <- as.data.frame(ode(state,times,model,parms=0))
  out     <- rbind(out,out1)

  lout    <- nrow(out1)       # last element of output

  state   <-c(
            INDWEIGHT = out1[lout,"INDWEIGHT"],
            EGGWEIGHT = out1[lout,"EGGWEIGHT"],
            FOOD      = out1[lout,"FOOD"])

  if (Time >= TimeMoult)      # Moulting...
   {
     state[1]     <- Moulting() # New weight individuals
     state[2]     <- 0.          # Eggs = 0
     TimeMoult    <- Time +instarDuration # next moult time
   }

  if (Time >= TimeTransfer) # New medium...
   {
     state[3]     <- foodInMedium
     TimeTransfer <- Time + transferTime  # next transfer time
   }

  Time    <- TimeOut
}
```

Finally we plot the model results (Fig. 6.12), in two rows, two columns (mfrow).
We increase the size of the 3rd outer margin (oma), as we write the main title of the
model in this margin (mtext), enlarged with 50% (cex=1.5).

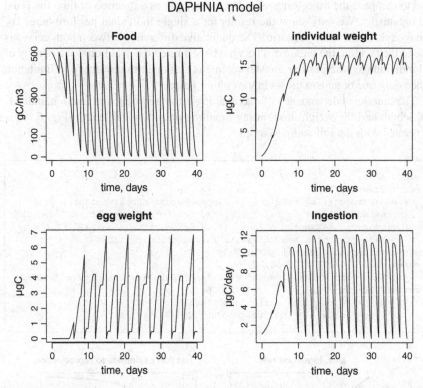

Fig. 6.12 Output generated by the DAPHNIA individual model, for a single individual per culture vessel – see text for R-code

```
par(mfrow=c(2,2),        oma=c(0,0,3,0))

plot (out$time,out$FOOD          ,type="l",main="Food",
      xlab="time, days",ylab="μgC/l")
plot (out$time,out$INDWEIGHT,type="l",main="individual weight",
      xlab="time, days",ylab="μgC")
plot (out$time,out$EGGWEIGHT   ,type="l",main="egg weight",
      xlab="time, days",ylab="μgC")

plot (out$time,out$Ingestion ,type="l",main="Ingestion",
      xlab="time, days",ylab="μgC/ind/day")

mtext(outer=TRUE,side=3,"DAPHNIA model",cex=1.5)
```

It is illustrative to run the Daphnia model for two extreme cases:

1. population density of one organism per litre
2. population density of 50 individuals per litre

and to compare the trajectories of individual weight as a function of time for 1 and 50 organisms. We only show the results for a single individual per litre here. Try the model for the other scenario! The qualitative difference between both curves is due to the fact that the organisms become food limited when cultured at a density of 50 individuals, whereas the transfer regime is sufficient to prevent food limitation when only one organism grows in the culture vessel.

The functional dependency of maximal ingestion, reproduction fraction, individual length and the weight loss during moulting on individual length (Fig. 6.13) is generated with the following R-script:

```
windows()
par (mfrow=c(2,2))
curve(maxIngest*(IngestWeight-x)/(IngestWeight-neonateWeight), 0,60,
    main="Max. ingestion rate",ylab="/d",xlab="ind. weight, µC",lwd=2)
curve(pmax(0., maxReproduction * (1.-(reproductiveWeight/x)^2)),0,60,
    main="fraction assimilate to reproduction ",ylab="-",
    xlab="ind. weight, µC",lwd=2)
curve(((x /3.0))^(1/2.6),0,60,
    main="Individual length",ylab="µm",xlab="ind. weight, µC",lwd=2)
curve(0.24*((x /3.0)^(1/2.6))^3.1,0,60,
    main="Weight loss during moulting",
    ylab="µg",xlab="ind. weight, µC",lwd=2)
```

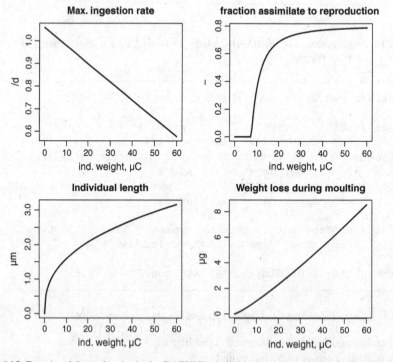

Fig. 6.13 Functional dependencies in the DAPHNIA individual model – see text for R-code

6.6.3 Zero-Dimensional Estuarine Zooplankton Model

Our third example is a very simple zero-dimensional (0-D) model, that describes marine zooplankton entering an estuary through the action of the tides. A significantly more complex and more realistic (1-D) model is implemented in Section 6.6.5. You may refer to this chapter to obtain more information about the rationale of the modelling. We use this simple example to demonstrate the use of the Euler integration method.

The estuarine zooplankton exchanges with the sea at a rate k (d^{-1}), and decays at a constant rate g (d^{-1}). Boundary concentrations (expressed in gram dry weight m^{-3}) in the sea are known and imposed as a forcing function (ZOOsea). The mass balance equation of zooplankton (ZOO) is:

$$\frac{dZOO}{dt} = k \cdot (ZOOsea - ZOO) - g \cdot ZOO \qquad (6.33)$$

Although it is perfectly possible to solve that model with R's function ode, the fact that we do not have control over the exact time at which the value of the forcing function needs to be estimated slows down the computation quite a bit. Here it is much more efficient to use Euler integration instead, as this gives total control over the time step used.

The implementation of the model starts by inputting the zooplankton concentration in the sea (a forcing function), as two vectors, one containing time values, the other the concentrations:

```
fZooTime = c(0, 30,60,90,120,150,180,210,240,270,300,340,367)
fZooConc = c(20,25,30,70,150,110, 30, 60, 50, 30, 10, 20, 20)
```

Then a function (euler) that runs the model from the initial time (start) to the end time (end) and with given time step (delt) and exchange rate and decay rate (k, g) is implemented.

We first form the sequence of time increments which the model will step through (times), and estimate the value of the forcing function at each of these time values (ZOOsea). R's function approx does that; it requires the x and y input data and the output x values (xout). In this case, the data x and y-values are in the fZooTime and fZooConc respectively, while mapping has to be performed on the output times.

We initialise the model with a reasonable guess of zooplankton concentration (5 gDWT/m3)

After declaring a matrix that is to contain the output (out), the model steps through the time sequence (1 : (nt-1)), each time calculating the input and decay rate, and the rate of change, dZOO. After that, the new value of the state variable is estimated by applying the Euler integration formula.

At each time the output is stored in the respective row of out (out[i,]). After finishing the time stepping, we assign column names to matrix out, and the output matrix is returned from the function.

```
euler <-function(start,end,delt,g,k)
  {
  times      <- seq(start,end,delt)
  nt         <- length(times)

  ZOOsea     <- approx(fZooTime,fZooConc, xout=times)$y

  ZOO        <- 5
  out        <- matrix(ncol=3,nrow=nt)

  for (i in 1:(nt-1))
    {
      decay   <- g*ZOO
      input   <- k*(ZOOsea[i] - ZOO)

      out[i,] <- c(times[i],ZOO, input)

      dZOO    <- input - decay

      ZOO     <- ZOO + dZOO *delt
}

colnames(out) <- c("time","ZOO","input")
return(as.data.frame(out))

}
```

Finally, the model is run for one year, the output plotted and a legend added (Fig. 6.14).

```
out <-euler(0,365,0.1,k=0.015,g=0.05)

par (oma=c(0,0,0,2))

plot(fZooTime,fZooConc,type="b",xlab="daynr",
     ylab="gC/m3",pch=15,lwd=2,
     main="Zooplankton model")
lines(out$time,out$ZOO,lwd=2,col="darkgrey")

legend("topright",c("Marine zooplankton","Estuarine zooplankton"),
       pch=c(15,NA),lwd=2,col=c("black","grey"))
```

Zooplankton model

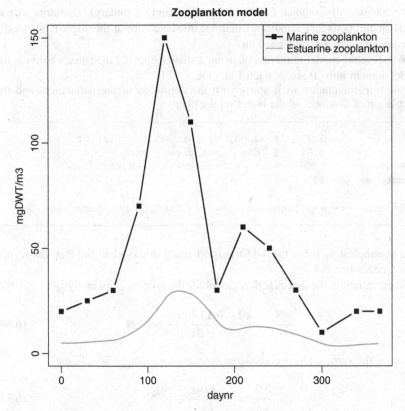

Fig. 6.14 Output generated with the 0-D estuarine zooplankton model – see text for R-code

6.6.4 Aphids on a Row of Plants: Numerical Solution of a Dispersion-Reaction Model

Aphids are serious pests for agriculture as their feeding on the crops reduces the vitality of the plants and makes them less attractive for human consumers. Once they have infected vegetables, they will slowly increase in numbers, spread and infest new plants. We implement and solve a simple model to mimic this behavior. A gardener has planted a row of lattuce, 60 m long. In the middle of the row, a small population of aphids invades the plants.

We assume that the aphids reproduce with a net rate of increase $g=0.01$ d^{-1}. This is a very low value, but it is a *net* rate of increase, i.e. difference between birth and mortality processes, and we assume that the plants are not very good food, so that the aphids just barely survive.

The aphids slowly disperse from their original location. This dispersal is by random movement, and can be modelled as a diffusion-type process with dispersion coefficient $K = 0.3$ m^{2}d^{-1}. At both ends of the row of plants, aphids that happen to move outside will disappear forever (it is an absorbing boundary). Therefore we can set as spatial boundary conditions that densities are zero at the edge of the model domain, i.e. at x= 0m and x= 60m.

We solve the model numerically, using a spatial step of 1 m. Thus, we divide the model domain into 60 boxes, each 1 m wide.

The implementation in R starts with the definition of the parameters and the calculation of distances of the boxes on the plant.

```
D          <- 0.3    # m2/day dispersion coefficient
g          <- 0.01   # /day   net growth rate
delx       <- 1      # m      thickness of boxes
numboxes   <- 60

Distance   <- seq(from=0.5, by=delx, length.out=numboxes)
```

It is simplest to solve the 1-D transport reaction model in the *flux-divergence* form (see Section 6.4.1).

Recapitulating, the flux-reaction model for the process is given by:

$$\frac{\partial N}{\partial t} = -\frac{1}{A} \cdot \frac{\partial A \cdot Flux}{\partial x} + g \cdot N \tag{6.34}$$

which, as the surface (A) is considered constant here, can be simplified:

$$\frac{\partial N}{\partial t} = -\frac{\partial Flux}{\partial x} + g \cdot N \tag{6.35}$$

And where the dispersive flux is given by:

$$Flux_{dispersion} = -D \frac{\partial N}{\partial x} \tag{6.36}$$

With D the dispersion coefficient.

We numerically approximate the rate of change of densities in one grid cell, i, as:

$$\frac{dN_i}{dt} = -\frac{\Delta_i Flux}{\Delta x_i} + g \cdot N_i = -\frac{Flux_{i,i+1} - Flux_{i-1,i}}{\Delta x_i} + g \cdot N_i \tag{6.37}$$

Where N_i is density in box i, Δx_i is the thickness of box i, and Δ_i denotes that the flux gradient is to be taken around box i, i.e. as the difference of the fluxes at the interface with the next box (i,i+1) and with the previous box (i−1,i). $Flux_{i-1,i}$ is the flux on the interface between cell i−1 and i.

The flux on the interface, which is due to dispersion, is given by:

$$Flux_{i-1,i} = -D_{i-1,i} \cdot \frac{N_i - N_{i-1}}{\Delta x_{i-1,i}} \tag{6.38}$$

where $\Delta x_{i-1,i}$ is the dispersion distance, i.e. the distance from the centre of box $i-1$ to the centre of box i, where the densities N_i and N_{i-1} are prescribed. At the external boundaries of the first and last cell, the density is prescribed at the edge, thus it follows that the dispersion distances, $\Delta x_{0,1}$ and $\Delta x_{60,61}$ are only half the thickness of the boxes.

This is how we implement these equations in R: we start by defining the dispersion distances $\Delta x_{i-1,i}$ (deltax), taking into account the fact that they are only half the box thickness at the edges.

To estimate the fluxes (Eq. 6.38), we use R's function diff, which takes the gradient, (i.e. diff (c(1,2,4)) would return a vector with elements $= 2-1$ and $4 - 2$ respectively). We take care to add the boundary densities $(0, \ldots, 0)$.

We then use Eq. (6.37) to calculate the rate of change of densities in each box. The rate of change is returned from the function, as a list.

```
deltax      <- c (0.5,rep(1,numboxes-1),0.5)

Aphid <-function(t,APHIDS,parameters)
{
    Flux       <- -D*diff(c(0,APHIDS,0))/deltax
    dAPHIDS    <- -diff(Flux)/delx + APHIDS*g

    list(dAPHIDS )
}
```

To run the model, we initialise the aphid densities. Density is set to 0 in all boxes, except for the two central ones (30,31), which are 1.

After specifying the times at which we want output, the integrator ode is called; output is stored in matrix 'out'. The first column of this matrix contains time, the other columns the density values at each time. The last line of following code extracts all density values.

```
APHIDS             <- rep(0,times=numboxes)      # ind/m2
APHIDS[30:31]      <- 1
state              <- c(APHIDS=APHIDS)

require(deSolve)
times      <- seq(0,200,by=1)
out        <- ode(state,times,Aphid,parms=0)
DENSITY    <- out[,2:(numboxes +1)]
```

Finally the output is plotted. First we create a temporal-spatial plot of the densities, using function filled.contour. The 3rd outer margin is increased (oma), as the main title will be written there (mtext) (Fig. 6.15).

Then a new window is opened and the initial, the intermediate and the final density versus distance and mean density versus time plotted (Fig. 6.16). For the latter, we take the means of the aphid densities, one for each row (`rowMeans`). The figures are outlined in two rows, two columns (`mfrow`).

```
par(oma=c(0,0,3,0))

filled.contour(x=times,y=Distance,z=DENSITY,color=topo.colors,
      xlab="days", ylab= "Distance on plant, m",main="Density")

mtext(outer=TRUE,side=3,"Aphid model",cex=1.5)

windows()
par(mfrow=c(2,2),oma=c(0,0,3,0))

plot(Distance,DENSITY[1,] ,type="l",lwd=2,
   xlab="Distance, m",ylab="Density", main="initial condition")
plot(Distance,DENSITY[100,],type="l",lwd=2,
   xlab="Distance, m",ylab="Density", main="100 days")
plot(Distance,DENSITY[200,],type="l",lwd=2,
   xlab="Distance, m",ylab="Density", main="200 days")

meanAphid <- rowMeans(out[,2:ncol(out)])
plot(times,meanAphid ,type="l",xlab="time, days",ylab="/m2",
   main="Density versus time")

mtext(outer=TRUE,side=3,"Aphid model",cex=1.5)
```

6.6.5 Fate of Marine Zooplankton in an Estuary (✻ ✻ ✻)

Estuaries are characterized by large salinity gradients, ranging from fresh to marine. The animals that live in these systems need to be adapted to these abiotic conditions. Typically, three different types of zooplankton are found in estuaries: freshwater species that are carried in the estuary through the river flow, endemic brackish-water species, and marine species that enter the estuary from the sea, through the action of the tides. One very well studied estuary is the Scheldt estuary, located in Belgium and the Netherlands. Each year, marine species progressively 'invade' this estuary, moving up the salinity gradient and reaching their maximal upstream position in summer, after which they start retreating back to the sea.

In order to investigate whether the occurrence of these animals is merely due to physical transport (mixing), or whether the marine species are able to increase in the estuary, a model was made that describes the dynamics of marine zooplankton in the Scheldt estuary (Soetaert and Herman, 1994).

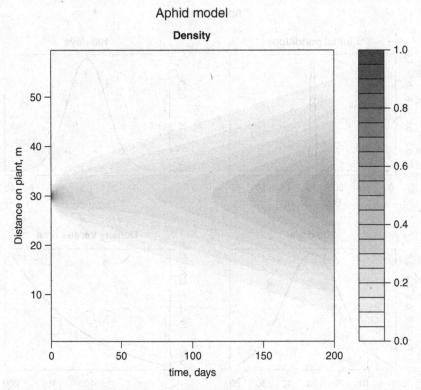

Fig. 6.15 Output generated with the Aphid dispersion-reaction model – see text for R-code

Marine zooplankton (C) is continuously driven to the sea by the freshwater flow (first term), and it is mixed into the estuary by the action of the tides (second term). To keep the model simple, the biological effects (third term) are represented by just one parameter, a net growth rate (g) of the zooplankton. If g is positive, the zooplankton grows in the estuary, if g is negative, there is only decay. The dynamic model reads (see Section 3.4.4):

$$\frac{\partial C}{\partial t} = -\frac{\partial}{A \cdot \partial x}(Q \cdot C) + \frac{\partial}{A \cdot \partial x}\left(A \cdot E \cdot \frac{\partial C}{\partial x}\right) + g \cdot C \tag{6.39}$$

Where C is zooplankton concentration (g dry weight m^{-3}), A is estuarine cross-sectional surface (m^2), Q is flow (m^3 d^{-1}), E is the tidal dispersion coefficient (m^2 d^{-1}) and x is the spatial position along the length axis (m). The spatial extent is subdivided into 100 boxes, extending from the river (box 1) to near the sea (box 100).

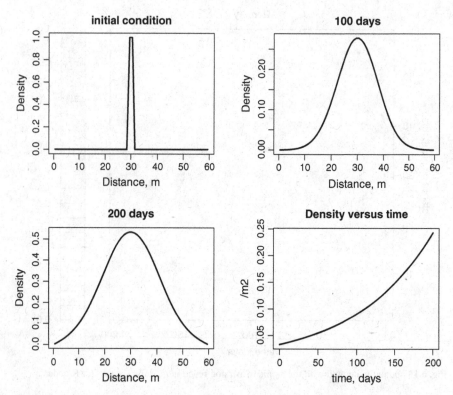

Fig. 6.16 Output generated with the Aphid dispersion-reaction model – see text for R-code

Numerical approximation

This partial differential equation is converted to an ordinary differential equation by approximating the spatial derivatives. It is simplest to start from the 'flux divergence equation':

$$\frac{\partial C}{\partial t} = -\frac{1}{A} \cdot \frac{\partial A \cdot J}{\partial x} + g \cdot C \tag{6.40}$$

And where the *total mass fluxes* $A \cdot J$ are defined as:

$$A \cdot J = Q \cdot C - A \cdot E_x \frac{\partial C}{\partial x} \tag{6.41}$$

To numerically approximate these equations, we keep in mind that fluxes are defined on the box interfaces, while concentrations are defined in the centre of the boxes. Thus, the rate of change of zooplankton concentration in box i can be approximated as:

$$\frac{dC_i}{dt} \approx -\frac{1}{A_i} \cdot \frac{\Delta_i(A \cdot J)}{\Delta x_i} + g \cdot C_i = -\frac{\Delta_i(A \cdot J)}{\Delta V_i} + g \cdot C_i \qquad (6.42)$$

Where we have defined the volume of the box i as:

$$\Delta V_i = A_i \cdot \Delta x_i \qquad (6.43)$$

and where Δ_i denotes that the mass flux gradient is to be taken around box i, i.e. as the difference of mass fluxes at the downstream (i,i+1) and upstream (i − 1,i) interface:

$$\Delta_i(A \cdot J) = (A \cdot J)_{i,i+1} - (A \cdot J)_{i-1,i} \qquad (6.44)$$

These mass fluxes are approximated as:

$$(A \cdot J)_{i-1,i} \approx Q \cdot C_{i-1} - A_{i-1,i} \cdot E_{i-1,i} \cdot \frac{\Delta C}{\Delta x}\Big|_{i-1,i}$$
$$= Q \cdot C_{i-1} - E^*_{i-1,i} \cdot \Delta C_{i-1,i} \qquad (6.45)$$

where we have defined the bulk dispersion coefficient E* (units of $m^3\, d^{-1}$) at the interface between box i − 1 and i as:

$$E^*_{i-1,i} = \frac{A_{i-1,i} \cdot E_{i-1,i}}{\Delta x_{i-1,i}} \qquad (6.46)$$

Note that we use $\Delta x_{i-1,i}$ here, i.e. the 'dispersion distance' from the centre of box i−1 to the centre of box i.

R- implementation

To make the model realistic, the estuarine morphology and physics is roughly patterned to the Scheldt estuary. This estuary is approximately 100 km long (Length). Combining the total length with the number of boxes (nbox), we calculate the distance, from the river (upstream boundary) to the upstream interfaces of each box (IntDist) and to the centre of the boxes (Dist). The cross-sectional area increases in a sigmoid fashion in this estuary, from around $4000\, m^2$ near the river to around $80000\, m^2$ near the estuarine mouth. We need to specify the cross-sectional surfaces both at the interfaces (IntArea) and in the centre of the boxes (Area). The latter is used to estimate the volumes of the boxes (Volume).

The effect of the tides is present up till the upstream boundary, where a sluice separates the estuary from the river. Here the tidal dispersion coefficient is 0 (Eriver); we assume that the dispersion coefficient increases linearly towards the mouth (Esea), where it equals $350\, m^2\, s^{-1}$ (this is only a rough approximation). The tidal dispersion coefficient E ($m^2\, d^{-1}$), is defined at the upstream interface of each box, hence the linear interpolation uses `IntDist`. It is used to estimate the bulk dispersion coefficient (`Estar`, units of m^3 d^{-1}).

```
nbox      <- 100
Length    <- 100000                                # m

dx        <- Length/nbox                           # m

IntDist <- seq(0,by=dx,length.out=nbox+1)   # m
Dist    <- seq(dx/2,by=dx,length.out=nbox)  # m

IntArea <- 4000 + 76000 * IntDist^5 /(IntDist^5+50000^5)   # m2
Area    <- 4000 + 76000 * Dist^5 /(Dist^5+50000^5)         # m2

Volume    <- Area*dx                               # m3

Eriver  <- 0                                  # m2/d
Esea    <- 350*3600*24                        # m2/d
E       <- Eriver + IntDist/Length * Esea     # m2/d

Estar   <- E * IntArea/dx                     # m3/d
```

The freshwater flow Q in this estuary fluctuates relatively smoothly from around $150 \, m^3 \, s^{-1}$ in winter to $50 \, m^3 \, s^{-1}$ in summer; we mimic this as a simple sine wave (parameters `meanFlow`, `ampFlow` and `phaseFlow`).

As the model is one-dimensional, and includes dispersion, we need to specify two boundary conditions, near the river and sea boundaries. Concentration of marine zooplankton in the river is set to 0 (parameter `riverZoo`), whilst at the seaward boundary, true measurements, at monthly intervals are imposed. These measurements are inputted as two vectors: the sampling time (`fZooTime`) and the zooplankton concentration at the sea boundary (`fZooConc`).

```
fZooTime = c(0, 30,60,90,120,150,180,210,240,270,300,340,367)
fZooConc = c(20,25,30,70,150,110, 30, 60, 50, 30, 10, 20, 20)

# the model parameters:
pars     <-    c(riverZoo = 0.0,         # river zooplankton conc
                 g         =-0.05,       # /day   growth rate
                 meanFlow  = 100*3600*24,   # m3/d, mean river flow
                 ampFlow   = 50*3600*24,    # m3/d, amplitude
                 phaseFlow = 1.4)        # -     phase of river flow
```

The model itself is in a function called `Zootran` that estimates the zooplankton rate of change. It starts with estimating the current flow values (`Flow`) and interpolating the time series (fZooTime, fZooConc) to the current simulation time (t); the interpolation uses R's function approx.

Then the mass input fluxes due to advection and dispersion are estimated (Input), and the rate of change calculated as the sum of transport (flux gradient) and growth. The rate of change is returned as a list.

```
require(deSolve)

Zootran <-function(t,Zoo,pars)
{

  with (as.list(pars),{

    Flow    <- meanFlow+ampFlow*sin(2*pi*t/365+phaseFlow)
    seaZoo <- approx(fZooTime, fZooConc, xout=t)$y
    Input  <- +Flow * c(riverZoo, Zoo) +
              -Estar* diff(c(riverZoo, Zoo, seaZoo))
    dZoo   <- -diff(Input)/Volume + g*Zoo
    list(dZoo)

                         })
}
```

To run the model, we provide initial conditions (ZOOP), one value for each box, and a times sequence (one year, daily output). The model is best integrated with one of deSolve's integration routines which is especially designed for solving one-dimensional reaction-transport models (ode.band). We specify that the model includes only one species (nspec).

```
ZOOP  <- rep(0,times=nbox)
times <- 1:365
out   <- ode.band(times=times,y=ZOOP,func=Zootran,parms=pars,nspec=1)
```

Finally, the output is depicted as a filled contour (Fig. 6.17).

```
par(oma=c(0,0,3,0))         # set margin size

filled.contour(x=times,y=Dist/1000,z=out[,-1],
               color= terrain.colors,xlab="time, days",
               ylab= "Distance, km",main="Zooplankton, mg/m3")
mtext(outer=TRUE,side=3,"Marine Zooplankton in the Scheldt",cex=1.5)
```

Marine Zooplankton in the Scheldt

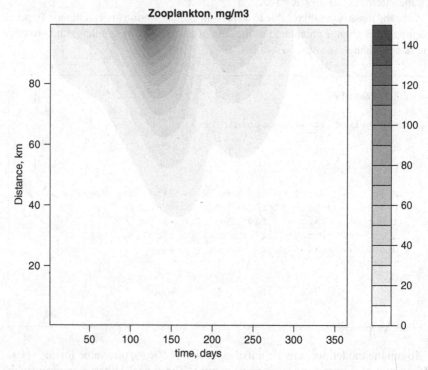

Zooplankton, mg/m3

Fig. 6.17 Output generated with the Estuarine zooplankton transport-reaction model – see text for R-code

6.7 Projects

6.7.1 Numerical Solution of the Autocatalytic Reaction in a Flow-Through Stirred Tank

Implement and solve the model described in Section 3.6.1.

Two chemicals A and B are fed into a flow-through stirred tank where an auto-catalytic reaction occurs between A and B, and that produces substance C.

$$A+B \rightarrow 2B + C$$

The rate of change of the concentrations [A], [B] and [C] is given by:

$$\frac{d[A]}{dt} = d_r \cdot (A_{in} - [A]) - k \cdot [A] \cdot [B]$$

$$\frac{d[B]}{dt} = d_r \cdot (B_{in} - [B]) + k \cdot [A] \cdot [B]$$

$$\frac{d[C]}{dt} = -d_r \cdot [C] + k \cdot [A] \cdot [B] \tag{6.47}$$

Use following parameter settings:

$dr = 0.05 \, \text{s}^{-1}$, $k = 0.05 \, (\text{mmol m}^{-3})^{-1} \, \text{s}^{-1}$, $Ain = 1 \, \text{mmol m}^{-3}$, $Bin = 0.1 \, \text{mmol m}^{-3}$ and for initial conditions for A(t= 0) = 1 mmol m^{-3}, B (t=0) = 0.1 mmol m^{-3} and C(t= 0) = 0.

Run the model for 300 seconds. Use ode to solve the model. Plot the concentration of substances A, B and C as a function of time (results: see Fig. 6.18).

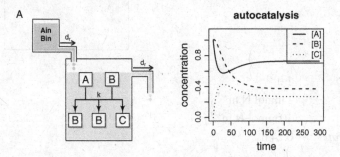

Fig. 6.18 Conceptual model (*left*) and output generated with the autocatalysis in a flow-through stirred tank model (*right*)

6.7.2 Numerical Solution of a Nutrient-Algae Chemostat Model – Euler Integration

The next model is similar to the previous, except that algae are growing in a chemostat, that is poor in nutrients. Culture medium is pumped continuously into the vessel, where it is mixed homogeneously with the existing contents. An identical amount of the existing contents in the vessel is removed by this process. It is assumed that light is in surplus, so that the algal growth is limited only by nutrient availability (Fig. 6.19).

The model equations are:

$$\frac{d\text{PHYTO}}{dt} = pmax \cdot \frac{\text{DIN}}{\text{DIN} + ks} \cdot \text{PHYTO} - dilrate \cdot \text{PHYTO}$$

$$\frac{d\text{DIN}}{dt} = -pmax \cdot \frac{\text{DIN}}{\text{DIN} + ks} \cdot \text{PHYTO} - dilrate \cdot (\text{DIN} - Nin) \tag{6.48}$$

Use the following parameter values:

Fig. 6.19 Schematic
representation of the
chemostat model

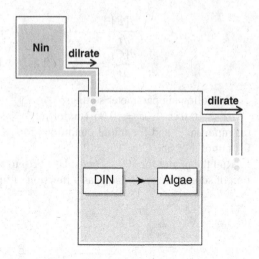

pMax	$1\,d^{-1}$
ks	$1\,\text{mmol N m}^{-3}$
Nin	$10\,\text{mmol N m}^{-3}$
Dilrate	$0.24\,d^{-1}$

And initial conditions:

DIN(t0) $0.1\,\text{mmol N m}^{-3}$

PHYTO(t0) $1\,\text{mmol N m}^{-3}$

Solve the model numerically using Euler integration:

$$C^{t+\Delta t} = C^{t} + \Delta t \cdot \frac{dC^{t}}{dt} \tag{6.49}$$

Tasks:

1. Simulate the dynamics of phytoplankton – DIN for 20 days.
   Use 0.1 day as the time step.
   Start by creating parameters and assigning initial values to the state variables.
   Then loop over time, starting with time = 0 and increasing time each iteration
   with the time step. At each iteration, first calculate the rate of change of DIN
   and the rate of change of algae, and then, update the concentrations using Euler
   integration.
   Make a plot of nutrient and algal concentrations versus time.
   Make a plot of algae versus nutrients.
2. Change the timestep of the model solution.
   First decrease it, then increase it in steps of 0.1 d. always simulate a 20-day
   period.
   What happens to your model solution when you decrease and increase the time
   step?

3. Set the timestep back to 0.1 day. Now increase *pmax* from its 'standard' value of
    $1\,d^{-1}$ in steps of $0.5\,d^{-1}$.
    Watch the stability of your solution. Conclude about the relation between time
    step length and dynamic properties of the model.

6.7.3 Rain of Organic Matter in the Ocean: Numerical Solution of the Advection-Reaction Model

Use the R-code of the aphid model (Section 6.6.4) as a template to implement a
numerical solution of organic matter sinking through the aphotic part of an oceanic
water column. The analytical solution was treated in Section 5.5.1.

The depth of this water column is 400 m. Organic matter is raining down from the
productive euphotic zone. As it sinks through the water column, it is being degraded.
Assume that

*   The organic matter has a constant sinking velocity u of $50\,m.d^{-1}$. The sinking of
    the organic matter is an advective process.
*   Model the degradation as a first-order process, with a degradation rate k of
    $0.2\,d^{-1}$.
*   The flux of organic matter from the euphotic zone, i.e. the upper boundary for
    our model, is prescribed as $Flux = 100\,mmolC.m^{-2}.d^{-1}$. At the lower end of
    the water column, material leaves the water column to settle on the bottom.
*   There is no diffusive mixing.

Run the model for a sufficiently long time, such that the organic matter concentrations stop changing. Compare the final concentration gradient with the analytical
solution.

6.7.4 AQUAPHY Model Implementation

Implement the AQUAPHY algal growth model (Section 2.9.2), under fluctuating
light conditions.

Chapter 7
Stability and Steady-State

Although this book mainly deals with models that are written as a rate of change in time, modellers are often interested in the 'steady-state solution', or equilibrium points, i.e. the conditions where the system does not change anymore.

We may find such equilibrium conditions by two techniques: (1) running the model dynamically for a sufficiently long time, or (2) setting the rate of change equal to 0 and solving the resulting equations. We will introduce these methods in this chapter.

Not all steady-state solutions are equivalent. Some may be quickly abandoned if the system is slightly perturbed; others may function as an attractor of the system, even under a large perturbation. The *stability* of an equilibrium relates to the tendency of the system to return to its position when perturbed.

This chapter essentially deals with two questions: (1) How can we find a steady-state and (2) how stable is the steady-state?

After defining some terms, we first introduce a technique that allows to graphically represent steady-state and stability conditions for models consisting of one or two equations.

Matrix algebra plays an essential role here. The reader is referred to Appendix C to freshen her or his knowledge on the subject, if needed.

7.1 Basics

A dynamic system is at *equilibrium*, when the sources and sinks are in balance. Mathematically, this is when the rate of change with time is zero such that the system does not change, or when the system oscillates in a predictable way with a certain periodicity. The terms 'steady-state' and 'equilibrium' (or better: dynamic equilibrium) are used differently in various disciplines. Here we will use them interchangeably. Also, the 'dynamic equilibrium' that we discuss here should not be confused with the 'chemical equilibrium' that we deal with in the next section.

Stability of an equilibrium relates to the tendency of the system to return to its position when slightly perturbed (Fig. 7.1).

Fig. 7.1 In this landscape, three balls are at rest, representing equilibrium situations. The two dark balls in the valleys are stable: when slightly perturbed, they will return to their original position. The light ball is unstable: perturbing it will cause it to fall in one of the valleys

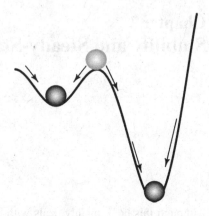

- If the equilibrium point is stable (valley points in the Fig. 7.1), the system will return to this point when it is slightly displaced from the equilibrium, independent of the direction in which it has been displaced from it.
- In contrast, an equilibrium is unstable (top of the hill), if the system diverges away from it, when slightly perturbed.

Many systems of equations have one *global* stable equilibrium: the system will converge to this point, no matter what the starting point is. Often though, there exist multiple *local* stable equilibria (there are two stable equilibria in Fig. 7.1). It depends on the magnitude of the perturbation to which equilibrium the system will evolve.

For models that consist of a limited number of differential equations, the existence and stability properties of the equilibrium can be investigated by graphical methods.

- If the model consists of only one differential equation, we plot the rate of change as a function of the value of the state variable; where this curve intersects the x-axis, the rate of change is 0 and the system is at equilibrium; the sign of the rate of change in its vicinity reflects the stability of the equilibrium.
- For models comprising two differential equations, we plot the lines where the rate of change of each of the state variables is zero, as a function of their values. These lines are called isoclines, and the graph representing the two state-variables is called a phase-plane. Where the isoclines intersect we have an equilibrium point. These graphical methods are referred to as phase-plane analysis.

For more than two equations, the graphical methods become clumsy and we need to resort to mathematical techniques, both for finding the equilibrium point and deriving its stability properties.

7.2 Stability of One First-Order Differential Equation

The following population dynamics model describes density-dependent growth of one species (1st term of the equation) and includes a harvesting term, which is described with Monod kinetics (2nd term).

$$\frac{dN}{dt} = r_i \cdot N \cdot \left(1 - \frac{N}{K}\right) - q \cdot \frac{N}{N + k_s} \tag{7.1}$$

N is species density, r_i is the intrinsic rate of increase, K is carrying capacity, q is the harvesting rate, and k_s is the density at which the harvesting attains half of its maximal value (the half-saturation density).

7.2.1 Equilibrium Points, Stability, Domain of Attraction

We first analyse the **steady-state** behaviour of this model by means of a graphical technique (Fig. 7.2).

If we plot the rate of change (dN/dt) as a function of the density N, the steady-state conditions, by definition, are attained at those density values where the curve crosses the X-axis: at these points, the rate of change is 0. In the example above, there are 3 such points, labelled N0, N1, N2.

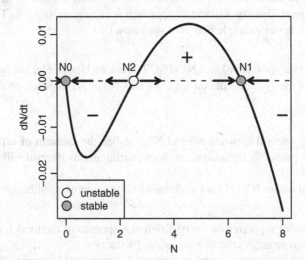

Fig. 7.2 The rate of change (dN/dt) as a function of density (N) for the logistic model with monod-type mortality (harvesting) function. Where the curve intersects the X-axis (*dashed line*) we have an equilibrium point. '+' and '−' denote positive and negative rate of changes; the arrows indicate how the density changes in the vicinity of the equilibrium points. Two types of equilibria are present: the arrows point towards stable equilibrium points (*grey*) and diverge from unstable points (*white*)

If the rate of change (dN/dt) is positive, then, by definition the density will increase with time, while it will decrease when dN/dt is negative.

Thus we can add arrows on the axis that indicate in which direction the solution will change: the arrows point to the right when the rate of change is positive (and density will increase, i.e. move to the right) and to the left otherwise. Now, in the neighbourhood of each equilibrium point we add two arrows, one on each side. For example, on the right of point N2, the rate of change is positive, hence the density will increase and we add an arrow pointing to the right. On the left side of N2, the rate of change is negative, causing a decrease in density, hence we add a left-pointing arrow. In both cases, the solution will move away from N2.

The arrows either point towards the equilibrium point (N0, N1) or point away from it (N2) or show a mixed behaviour.

- When both arrows point to the equilibrium, the equilibrium is **stable**: when perturbing N1 slightly, the solution will again approach it, no matter in which direction it has been displaced.
- When the arrows diverge from the equilibrium point, it is **unstable**: small positive perturbations of N2 will cause divergence towards N1, whilst small decreases of density from N2 will cause divergence towards N0 and extinction of the population.
- It is also possible that the arrows point towards the equilibrium point on one side, and point away from the point on the other side. In this case, the equilibrium point is a **saddle point**. Depending on the direction in which the density is perturbed from this point, density will either approach it or diverge from it. There is no saddle point in our example here (but see below).

The equilibrium points N0 and N1 (Fig. 7.2) are locally stable, but not globally stable: whether the system will converge towards N0 or N1 depends on the starting point.

- The density interval between N0 and N2 is called the **domain of attraction** of equilibrium point N0, because trajectories starting in this interval will ultimately converge to N0.
- The interval where $N > N2$ is the domain of attraction of equilibrium point N1.

As the point N2 separates two very different responses of the model, it is called a '**separatrix**': it separates the two domains of attraction.

Separatrices have some unusual properties. As the rate of change in the vicinity of the separatrix is very low, the system will diverge only slowly from this point. Because of that, small disturbances may cause the system to cross this point and cause the system to flip over to the other stable state. Thus, especially for systems where the separatrix is close to one of the equilibria, flips from one stable state to another may be likely.

7.2.2 Multiple Steady States

Rather than using a graphical technique (previous section), we can also find the steady-state conditions of the previous model by setting the rate of change (dN/dt) equal to 0 and solving for N.

$$0 = N^* \cdot \left[r_i \cdot \left(1 - \frac{N^*}{K} \right) - q \cdot \frac{1}{N^* + k_s} \right] \tag{7.2}$$

(and where the * denotes the equilibrium condition).
   Steady-state is reached when either:

$$N^* = 0 \tag{7.3}$$

or:

$$r_i \cdot \left(1 - \frac{N^*}{K} \right) - q \cdot \frac{1}{N^* + k_s} = 0 \tag{7.4}$$

In the latter case, and rearranging terms, we obtain a quadratic equation in N^*

$$N^{*2} + (k_s - K) \cdot N^* + \frac{qK}{r_i} - k_s \cdot K = 0 \tag{7.5}$$

which has the solutions:

$$N^* = \frac{-(k_s - K) + \sqrt{D}}{2} \quad \text{and} \quad N^* = \frac{-(k_s - K) - \sqrt{D}}{2} \tag{7.6}$$

Where

$$D = (k_s + K)^2 - 4qK/r_i \tag{7.7}$$

Now we can distinguish four cases, as depicted in Fig. 7.3. (for these figures, K, q and ks were set equal to 10, 0.1, 1 and where r_i has different values).

- If D is negative, its square root cannot be taken and the equilibrium points given by Eq. (7.6) do not exist. Thus, there is only one global equilibrium, at $N^* = 0$. In this case the harvesting rate is too large, and the population cannot persist. At equilibrium, the population is completely wiped out. This is depicted in Fig. 7.3 A.
- If D = 0, there are only two equilibrium points: at $N^* = 0$ and at $N^* = 0.5^*$ $(K\text{-}ks)$. The first equilibrium point is globally stable, whilst the other equilibrium point is not; it is a saddle point (Fig. 7.3B). Thus, when initiated at densities higher than N^*, the density will converge to N^*. When starting from lower values the density will diverge from it.

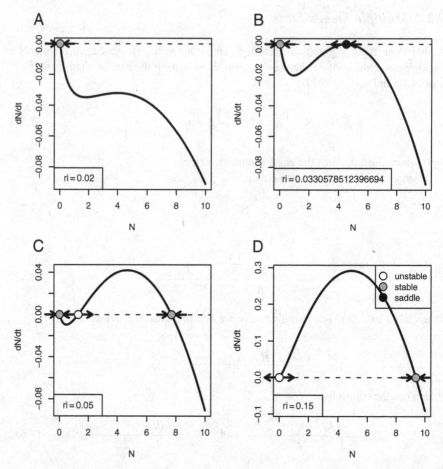

Fig. 7.3 Four different behaviours of the logistic model with monod-type mortality function (see text). Depicted is the rate of change (dN/dt) as a function of density (N). The dashed horizontal line is the X-axis. Parameter values are: K=10, q= 0.1,ks=1, and with variable values for the parameter r_i (as indicted in the *left lower corner*). **A**. One stable equilibrium. **B**. One stable equilibrium, one saddle point. **C**. Two stable, one unstable equilibrium points. **D**. Two equilibrium points, one stable, one unstable

- If $\sqrt{D} < K - ks$ we obtain 3 equilibrium points (Fig. 7.3C), two of which are stable, one of which is unstable.
- If $\sqrt{D} > K - ks$ there are only 2 relevant equilibria, one globally stable, one unstable (Fig. 7.3D).

7.2.3 Bifurcation

From the previous section, it is clear that, by changing one parameter (in this case the rate of increase, r_i), the number of equilibria and their stability properties can change drastically.

Such an abrupt change in the configuration (and stability) of the equilibria is called a **bifurcation**. It brings about a qualitative change in the behaviour of the system.

These changes in equilibrium behaviour can be represented by means of a **bifurcation diagram**, which has on the x-axis the critical parameter (in our case the rate of increase, r_i), and on the y-axis the values and stability properties of the equilibria. Figure 7.4 depicts the bifurcation diagram for our example model and for varying values of r_i; the colour of the equilibrium points reflect their stability properties. The dashed lines give the positions of the values of r_i that were used to generate Fig. 7.3.

We note the following:

- for any value of r_i, the highest equilibrium point is always stable.
- If there are two equilibrium points (at values of $r_i > 0.1$), the equilibrium $N^* = 0$ is always unstable.
- The equilibrium point $N^* = 0$ is always stable when there are 3 equilibrium points.
- The unstable equilibrium points, denoted by the dark symbols are the separatrix for each value of r_i. If a simulation is initiated at density values above

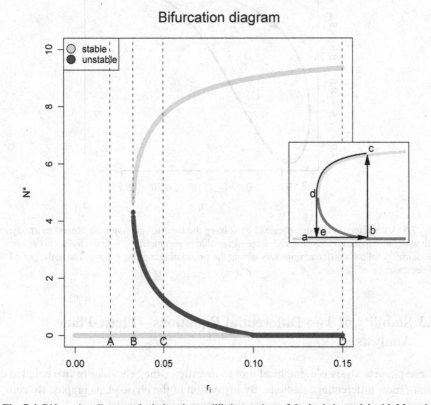

Fig. 7.4 Bifurcation diagram, depicting the equilibrium points of the logistic model with Monod-type mortality function as a function of increased rate of increase (parameter r_i); for the equation and parameter values, see text; A,B,C and D refer to Fig. 7.3

the separatrix, the simulation will converge to the highest stable equilibrium, whereas it will lead to extinction if initiated below the separatrix. The separatrix thus 'separates' regions of different initial conditions.

At point 'a', the species is (locally) extinct and the zero density is a stable equilibrium (Fig. 7.4, inset). As the external conditions improve, the rate of increase (r_i) will rise. As long as r_i remains well below 0.1, any small influx of new organisms will lead to extinction (as long as the density stays below the separatrix, the stable equilibrium is at zero density). Now at values of $r_i > 0.1$ (point 'b'), any influx of organisms will cause an outbreak, i.e. density will increase towards the stable, high equilibrium value (point 'c'). If from that stable point, r_i starts to decline (<0.1), then the densities will remain high, and the path followed will be different, until the rate of increase drops below point 'd', and the population goes extinct ('e') again. The different paths followed by the population under increase and decrease of r_i are called a **hysteresis** effect. A more realistic example of hysteresis is in Fig. 7.5.

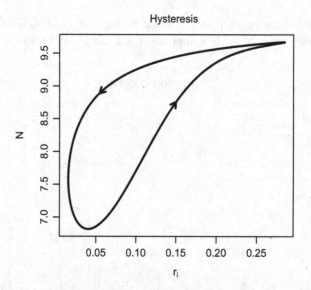

Fig. 7.5 A trajectory of density, obtained by solving the logistic equation with Monod mortality, and with (an imposed) increasing and decreasing value of parameter r_i. Note the hysteresis effect: the densities follow a different trajectory during the period of increasing r_i compared to the period of decreasing r_i

7.3 Stability of Two Differential Equations – Phase-Plane Analysis

Phase-plane analysis is a graphical tool to investigate the behaviour of *two* coupled (first-order) differential equations. By inspection of the phase plane graphs, we can draw general conclusions concerning the dynamics of the model.

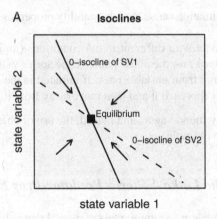

Fig. 7.6 A phase plane graph, depicting two isoclines, one equilibrium point, and four arrows, one in each quadrant

A *phase plane* is an X-Y graph with the values of the first and second state variables plotted on the X and Y-axis respectively (Fig. 7.6). As the state variables are generally positive numbers, the phase plane is restricted to the part where (X>0,Y>0). In what follows, we will use short-hand notation for state variables 1 and 2: SV1, and SV2.

Added to the phase plane is information about the *rate of change* of the two state variables.

1. *The zero isocline* of a state variable is a curve or line, depicting (SV1, SV2) couples at which the rate of change of one of the state variables is 0. Wherever the zero isoclines of the two state variables intersect, we have a point where the rate of change of the two state variables is simultaneously 0, i.e. an equilibrium point. In general there will be more than one equilibrium point, and the analysis allows making inferences about the nature of each equilibrium point.
2. Zero isoclines divide the phase plane into several sections. In each of these sections we may add one *arrow*, where the direction of the arrow is an indication of the sign of the rate of change of the two state variables at a particular point. Recalling that for positive rate of change, the state variable will increase in time, the arrows may point to:

   • North. When the rate of change of state variable 1 (dSV1/dt) is 0, the rate of change of state variable 2 is positive.
   • West. dSV1/dt is negative, dSV2/dt is 0
   • SouthWest: dSV1/dt is negative, dSV2/dt is negative
   • NorthWest: dSV1/dt is negative, dSV2/dt is positive

Etc.

When the arrow in each section points towards an equilibrium point, then the model will converge towards this equilibrium (it is stable). In contrast, when the arrows in each section diverge from an equilibrium point, it is unstable. Often none of these two situations arise, and the stability properties need to be assessed in different ways.

3. We can run the model with different initial conditions (initial SV1-SV2 pairs) and plot the *trajectories* on the phase plane. Trajectories will converge to stable equilibria and diverge from unstable ones. If initiated in the vicinity of a saddle point, they will first approach it and then move away from it.

These concepts may sound vague, and indeed, the use of phase-plane analysis is best illustrated with an example.

7.3.1 Example. The Lotka-Volterra Predator-Prey Equation

The Lotka-Volterra models are a famous class of models that either describe predator-prey interactions or competitive interactions between two species.

A.J. Lotka and V. Volterra formulated the original model in the 1920's almost simultaneously (Lotka, 1925, Volterra, 1926). Since then, there have been many improvements, but the derived equations are still referred to as the Lotka-Volterra models.

The following is an extension of the original Lotka-Volterra model (it includes a logistic term) that solves the temporal evolution of the density (or biomass) of predators and their prey:

$$
\frac{d\text{PREY}}{dt} = r_i \cdot \text{PREY} \cdot \left(1 - \frac{\text{PREY}}{K}\right) - \alpha \cdot \text{PREDATOR} \cdot \text{PREY}
$$
$$
\frac{d\text{PREDATOR}}{dt} = \alpha \cdot \gamma \cdot \text{PREDATOR} \cdot \text{PREY} - m \cdot \text{PREDATOR}
$$
(7.8)

r_i is the rate of increase of prey, K its carrying capacity, α the ingestion rate of the predator, γ the assimilation efficiency and m the mortality rate. In the absence of predators, the prey density increases exponentially at low prey density, moving towards equilibrium (carrying capacity K) (1st term in the prey differential equation).

In the presence of predation, the prey density is reduced proportionally to predator and prey density (assuming a Holling type I functional response) (2nd term in the prey differential equation).

Predator growth is stimulated by the presence of prey (1st term in the predator equation) and experiences a first-order mortality (2nd term).

7.3.1.1 The Phase Plane

In the phase plane, we plot the prey density on the X, the predator density on the Y-axis. Any point in the X-Y plane then represents a pair of prey-predator abundances (Fig. 7.7).

The zero isoclines (lines along which the rate of change of either one of the species is 0) are drawn first.

The rate of change of the predator is zero, and thus predator density will remain constant, when:

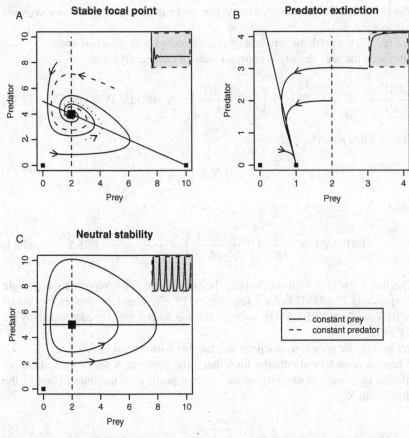

Fig. 7.7 Phase-plane of the Lotka-Volterra predator-prey model. (parameter values: $\alpha = 0.2$, $r_i = 1$, $m = 0.2$, $\gamma = 0.5$, and K variable), with trajectories initiated in the different quadrants. **A.** Stable focal point, $K = 10$. **B.** Predator extinction, $K = 1$, and **C.** Neutral stability, $K = \infty$. This is the original Lotka-Volterra model. The grey inset depicts the temporal change of the prey density for one of the trajectories

$$\frac{d\text{PREDATOR}}{dt} = 0 = \alpha \cdot \gamma \cdot \text{PREDATOR} \cdot \text{PREY} - m \cdot \text{PREDATOR} \quad (7.9)$$

This is when either:

$$\text{PREDATOR} = 0 \quad (7.10)$$

or

$$\text{PREY} = \frac{m}{\alpha \cdot \gamma} \quad (7.11)$$

The first coincides with the X-axis, the second equation is a vertical line crossing X at $m/\alpha \cdot \gamma$.

In Fig. 7.7, the predator zero isoclines are denoted with a dashed line.

Similarly, the prey density is constant (rate of change=0) when:

$$\frac{d\text{PREY}}{dt} = 0 = r_i \cdot \text{PREY} \cdot \left(1 - \frac{\text{PREY}}{K}\right) - \alpha \cdot \text{PREDATOR} \cdot \text{PREY} \quad (7.12)$$

This is when either:

$$\text{PREY} = 0 \quad (7.13)$$

or when

$$\text{PREDATOR} = \frac{r_i}{\alpha} \cdot \left(1 - \frac{\text{PREY}}{K}\right) = \frac{r_i}{\alpha} - \frac{r_i}{\alpha \cdot K} \cdot \text{PREY} \quad (7.14)$$

The first coincides with the Y-axis, the second condition represents a straight line expressing PREDATOR as a function of PREY. This line crosses the Y-axis (set PREY $= 0$) at PREDATOR $= r_i/\alpha$ and the X-axis (zero predator density) at PREY $= K$.

In Fig. 7.7, the prey zero isoclines are denoted with a solid line.

When we consider only the isoclines that differ from the X and Y axes, there are 2 different outcomes, depending on the relative position of the intersection of the isoclines with X.

$$(1) \; \frac{m}{a \cdot \gamma} < K \quad (7.15)$$

$$(2) \; \frac{m}{a \cdot \gamma} > K \quad (7.16)$$

In the first case, the two isoclines intersect, in the second case they don't. These two outcomes relate to different feeding needs of the predator $(a \cdot \gamma)$ relative to its mortality (m) and the carrying capacity of the prey (K). If feeding is inefficient, or mortality too high, the predator goes extinct (case 2). In the other case, both populations can coexist in a stable equilibrium.

Case 1. Stable coexistence

Using the following values for the parameters,

$r_i = 1.$, $\alpha = 0.2$, $\gamma = 0.5$, $m = 0.2$ and $K = 10$

we are in the first case scenario, where $m/a \cdot \gamma < K$ and the isoclines, different from X- and Y-axis, intersect (Fig. 7.7A).

The equilibrium points are determined by the intersection of the isoclines of predator and prey. There are 3 such points (filled square symbols in Fig. 7.7 A):

- The intersection of the X-axis (predator zero isocline) with the Y-axis (prey zero isocline). This is the equilibrium point (0,0) where both predator and prey are absent (leftmost small symbol).
- The intersection of the X-axis (predator zero isocline) with the second prey isocline. This is the equilibrium point $(0, K)$ where there are no predators, only prey (right small symbol).
- The intersection between the two other prey and predator isoclines, the large symbol in Fig. 7.7A.

Now that the equilibrium points have been diagnosed, we examine whether they are stable. By definition, the prey- predator couple that is in the equilibrium point will stay there. But what will happen if it is perturbed, i.e. if there is a sudden small influx of either prey or predator? In case the equilibrium is stable, the system will return to the equilibrium, in case it is unstable, it will diverge.

To investigate that, we add the trajectories of predator/prey couples as they change through time. This means that we solve (integrate) the Lotka-Volterra equation for a sufficiently long time and see where the predator/prey couple ends up.

The various zero-isoclines divide the phase plane in 3 or 4 distinct regions, and for each region, it suffices to calculate the trajectory for one situation only.

From Fig. 7.7 A, it is clear that the prey/predator couple tends towards the central equilibrium point whatever their initial condition, provided that it is not in the two other equilibrium points. Thus this central equilibrium point is stable.

If we perturb the two other equilibrium points on the x-axis (say, we run the model with an initial value of (0.001, 0.001) or $(0.001, K + 0.001)$, then the trajectory of predator/prey couples does not return to the original equilibrium point but converges as well to the central equilibrium. This demonstrates that the two equilibria on the X-axis are unstable.

Case 2. Predator extinction

Setting K equal to 1 is an example of the case where $m/a \cdot \gamma > K$ and the isoclines different from the X- and Y-axis, do NOT intersect (Fig. 7.7 B). Consequently there are only two equilibrium points.

As the carrying capacity of the prey is insufficient to withstand the feeding need of the predator, the predator is driven to extinction: the predator-prey couple is driven to values where prey density = 2 (the carrying capacity), and predator density=0.

The other equilibrium point (0,0) is unstable, as in the previous example.

Case 3. Neutral stability

We now consider the original Lotka-Volterra predator-prey model where the growth of prey is not limited (that is, K is infinitely large in previous example).

$$\frac{d\text{PREY}}{dt} = r_1 \cdot \text{PREY} - \alpha \cdot \text{PREDATOR} \cdot \text{PREY}$$

$$\frac{d\text{PREDATOR}}{dt} = \alpha \cdot \gamma \cdot \text{PREDATOR} \cdot \text{PREY} - m \cdot \text{PREDATOR} \qquad (7.17)$$

In the phase plane, the predator/prey trajectory follows an elliptical path, coming back to where it started. This is called neutral stability. Moreover, the trajectory will be different when started with different initial conditions (Fig. 7.7 C).

This feature of the original Lotka-Volterra model of predator-prey interaction has received much criticism. Wherever you push the system, it will start a new type of oscillation, that will bring it back to the point to which it has been pushed. This is clearly unrealistic behaviour. However, this (too) simple model is also a good starting point to add several mechanisms (carrying capacity limitation for instance, such as in previous examples), because one can immediately see whether such a mechanism brings the system from its pathological neutral stability into a state allowing for a stable equilibrium.

7.4 Multiple Equations

Phase-plane analysis, as discussed in previous section, is a graphical technique that is very well suited for analysing the behaviour of two coupled differential equations. It can be extended to analyse 3 coupled differential equations, but then much of the elegance is lost (one is generally not able to think geometrically in more than 2 dimensions).

In these cases, modellers resort to mathematical methods for finding the equilibrium conditions and apply formal tests of stability. These methods can also be applied to models consisting of 1 or 2 equations. First we deal with mathematical ways of finding steady-state solutions, and then we treat their stability properties.

7.5 Steady-State Solution of Differential Equations (*)

In previous chapters we have shown that a steady-state condition can be found by running the model for a sufficiently long time until it stops changing. Here we discuss a different type of technique which involves root-solving methods.

By setting the rate of change equal to 0, a set of *ordinary* differential equations is transformed to a set of algebraic equations. These can be linear, but most often will be non-linear !

The following set of equations is solved:

$$\frac{d\mathbf{C}}{dt} = f(\mathbf{C}) = 0 \tag{7.18}$$

where $\mathbf{C}$ is the vector with state variables, and dC/dt is the vector with the corresponding rates of changes. The concentrations $\mathbf{C}$ for which $f(\mathbf{C}) = 0$ are called the 'root' of the equations. The root can be found in several ways.

7.5.1 Direct Root Finding: Analytical Solution

If the equations are simple enough, it may be possible to find the steady-state condition by directly solving the (set of) equations.

For instance, the steady-state condition of the logistic equation describing the evolution of population density:

$$\frac{dN}{dt} = r_i \cdot \left(1 - \frac{N}{K}\right) \cdot N \tag{7.19}$$

is found by setting the rate of change $= 0$:

$$0 = r_i \cdot \left(1 - \frac{N}{K}\right) \cdot N \tag{7.20}$$

and solving for N. Thus, steady-state solutions are:
$N = 0$ and $N = K$.

We also used the direct solution technique when we solved the model discussed in Sections 7.2 and 7.3.

Note: if the resulting ordinary equations are linear, then simple matrix inversion will retrieve the root. Numerical approximations of the one-dimensional advection-diffusion equations are linear equations. Therefore, if they are also supplemented with linear reaction equations, matrix inversion may be used to find the steady-state solution.

7.5.2 Iterative Root Finding

Very often, it is not possible to write the solution of the root as an explicit formula. In this case, iterative methods are used.

Basically, one starts with an initial guess of the solution and then uses some kind of update formula to creep closer and closer to the root. The most well-known nonlinear root-finding technique is the Newton-Raphson method.

The model of benthic silicate diagenesis, Section 7.8.4, uses the Newton-Raphson method.

7.5.3 From Partial to Ordinary Differential Equations

If the original, time-dependent, model consists of *partial* differential equations, ignoring the temporal derivative converts them into ordinary differential equations. From there we have two choices:

- The resulting ordinary differential equation can be solved by analytical methods (e.g. it is one of the equations from Appendix B). Several of such solutions have already been discussed in Chapter 5 (analytical solution).

For instance, the steady-state solution of the one-dimensional advection-diffusion equation with first-order reaction (rate λ) and with constant surface:

$$\frac{\partial C}{\partial t} = D\frac{\partial^2 C}{\partial x^2} - u\frac{\partial C}{\partial x} - \lambda \cdot C \tag{7.21}$$

is found by setting the rate of change equal to 0:

$$0 = D\frac{d^2 C}{d x^2} - u\frac{d C}{d x} - \lambda \cdot C \tag{7.22}$$

The solution of this ordinary differential equation can be found in Appendix B.3.2 and is given by:

$$C(x) = A \cdot \exp^{\alpha_1 \cdot x} + B \cdot \exp^{\alpha_2 \cdot x} \tag{7.23}$$

where:

$$\alpha_1 = \frac{u - \sqrt{u^2 + 4D.\lambda}}{2D} \text{ and } \alpha_2 = \frac{u + \sqrt{u^2 + 4D.\lambda}}{2D} \tag{7.24}$$

and where A and B are integration constants to be derived from the boundary conditions.

- In case we cannot find an analytical solution to the ordinary differential equation, we use a numerical approximation of the differential equation. Thus, we try to solve by one of the methods described before (analytical or iterative solution). The estuarine example at the end of this chapter (Section 7.8.5) demonstrates this technique.

7.6 Formal Analysis of Stability (**)

We end this chapter by discussing a formal way to test the stability of the steady state. We will assume that the steady-state conditions (C^*) have been found. The question then is: given this steady state, what happens if the system is slightly perturbed? For instance, if there were a sudden influx or loss of individuals what would happen to the population?

There are several possibilities (Fig. 7.8).

1. The system comes back to the same equilibrium (i.e. the equilibrium is stable). This can happen in two different ways:

   - the perturbation declines smoothly towards the original steady-state. It is a stable equilibrium.
   - the disturbed population overshoots both sides of the equilibrium with decreasing amplitude. It is a stable focal point.

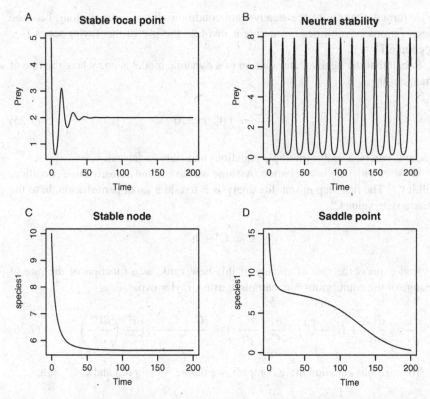

Fig. 7.8 Different trajectories towards the equilibrium. **A** and **B**, generated with the Lotka-Volterra predator-prey model, as in Section 7.3 (Fig. 7.7 A, Fig. 7.7 C). **C** and **D**, Stable equilibrium and saddle point generated with the Lotka-Volterra competition model (see Section 7.8.2)

2. The system oscillates for eternity with certain amplitude. This is called 'neutral stability'

3. The system evolves to a different equilibrium (the original steady-state was unstable). There are two options:

   • the new steady-state is always the same
   • the new steady-state depends on the size of the original perturbation

4. The system increases indefinitely (it is unstable).

   • It increases smoothly
   • It does so with oscillations with ever increasing amplitude

5. The system either diverges from or returns to the equilibrium point, depending on the direction of the perturbation. The point is a saddle point.

A formal analysis of the steady-state condition allows distinguishing between these various possibilities. To derive it involves the use of the Taylor series (see Section 6.1).

Recall that the steady-state solution of a dynamic model is one where the rate of change is 0.

$$\frac{\partial \mathbf{C}}{\partial t} = f(\mathbf{C}, t) = 0 \tag{7.25}$$

where $\mathbf{C}$ can be a vector (multiple equations and state variables).

We start with only one equation. Assume we have found a steady-state condition, call it $\mathbf{C}^*$. The first step in stability analysis is to add a small perturbation, h, to the steady-state value $\mathbf{C}^*$:

$$\mathbf{C} = \mathbf{C}^* + h$$

And express the rate of change of this new value as a function of the rate of change of the equilibrium concentration, using Taylor expansion:

$$\frac{d\mathbf{C}}{dt} = \frac{d\mathbf{C}^*}{dt} + (\mathbf{C} - \mathbf{C}^*) \cdot \frac{\partial}{\partial \mathbf{C}} \left(\frac{d\mathbf{C}^*}{dt} \right) + \frac{(\mathbf{C} - \mathbf{C}^*)^2}{2} \cdot \frac{\partial^2}{\partial \mathbf{C}^2} \left(\frac{d\mathbf{C}^*}{dt} \right) + .. \tag{7.26}$$

To make this equation more compact, we define $\frac{d\mathbf{C}}{dt} = f(\mathbf{C})$ and $\mathbf{C} \text{-} \mathbf{C}^* = h$:

$$f(\mathbf{C}) = f(\mathbf{C}^*) + h \cdot \frac{\partial f(\mathbf{C}^*)}{\partial \mathbf{C}} + \frac{h^2}{2} \cdot \frac{\partial^2 f(\mathbf{C}^*)}{\partial \mathbf{C}^2} + .. \tag{7.27}$$

Now by the definition of steady-state: $f(\mathbf{C}^*) = 0$.

If we assume that h is very small such that terms involving h^2 or higher are so small that they can be ignored, we obtain:

$$f(\mathbf{C}) \approx h \cdot \frac{\partial f(\mathbf{C}^*)}{\partial \mathbf{C}} \tag{7.28}$$

For multiple differential equations, the analysis is similar, but now involves Taylor expansion that takes into account the rate of change associated with the other state variables:

$$f(\mathbf{C}_i) = f(\mathbf{C}_i^*) + \sum_j h_j \cdot \frac{\partial f(\mathbf{C}_i^*)}{\partial \mathbf{C}_j} + .. \tag{7.29}$$

$$f(\mathbf{C}_i) \approx \sum_j h_j \cdot \frac{\partial f(\mathbf{C}_i^*)}{\partial \mathbf{C}_j} \tag{7.30}$$

These equations can be written in matrix notation. We demonstrate this for 2 equations (and where $f(C_i) = dC_i/dt$:

$$
\begin{bmatrix} f(C_1) \\ f(C_2) \end{bmatrix} = \begin{bmatrix} \frac{dC_1}{dt} \\ \frac{dC_2}{dt} \end{bmatrix} = \begin{bmatrix} \frac{\partial f(C_1^*)}{\partial C_1} & \frac{\partial f(C_1^*)}{\partial C_2} \\ \frac{\partial f(C_2^*)}{\partial C_1} & \frac{\partial f(C_2^*)}{\partial C_2} \end{bmatrix} \cdot \begin{bmatrix} h_1 \\ h_2 \end{bmatrix}
\tag{7.31}
$$

Thus: the rate of change of perturbations (left hand side) equals a matrix times the perturbations. Mathematicians refer to the matrix as the Jacobian matrix (see Appendix C.5), while population biologists sometimes refer to the matrix as the community matrix. In what follows, we use the more generic term Jacobian, represented by symbol $\mathbf{M}$.

The question of stability can be translated as: what will the matrix multiplication on the right-hand side of this equation do to the perturbations in the state variable vector? Will it let them grow or shrink, and will it do so in a steady way or in an oscillatory way? This question can be answered by investigating the eigenvalues of the Jacobian matrix ($\mathbf{M}$).

The eigenvalues λ of a matrix $\mathbf{M}$ are numbers such that there exists a vector $\mathbf{v}$:

$$
\mathbf{M.v} = \lambda.\mathbf{v}
\tag{7.32}
$$

In this equation the matrix multiplication (on the left-hand side) equals a scalar multiplication (on the right-hand side). The eigenvalues λ tell something about the 'tendency' of the matrix multiplication to increase or decrease the values in the vector $\mathbf{v}$. Eigenvalues can be real or complex numbers. When they are complex, the non-zero imaginary part indicates a tendency to oscillate.

More formally, the eigenvalues of the Jacobian matrix determine the type of equilibrium:

- the sign of the real part determines whether an equilibrium will be stable ($-$) or unstable ($+$)
- the presence or absence of the imaginary part determines the monotonic change (absent) or oscillatory change (present).

Thus, in the case where there are two eigenvalues (λ_1, λ_2) we distinguish the following cases (Fig. 7.9):

- if both eigenvalues are distinct, real, and positive then the equilibrium will be an **unstable node**. When perturbed, the density will diverge smoothly.
- if both eigenvalues are distinct, real, and negative then the equilibrium will be a **stable node**. After perturbation, the density will come back to the original equilibrium smoothly.
- if both eigenvalues are distinct, real, but of opposite sign, then the equilibrium will be a **saddle point**. In this case, it becomes worthwhile to consider also the eigenvectors: the trajectories in the direction of the eigenvectors are the only paths that lead to the saddle point. The eigenvector associated with the negative eigenvalue determines the path that will stay in the saddle point, whilst the

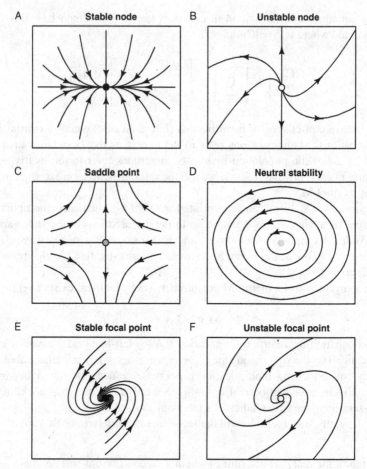

Fig. 7.9 (x,y) trajectories denoting different types of equilibria. All graphs were created with the same coupled differential equations:

$$\frac{dx}{dt} = ax + cy$$

$$\frac{dy}{dt} = by + dx$$

and with the following values for the coefficients a,b,c,d: A. Stable equilibrium: $(-0.1, -0.3, 0, 0)$. B. Unstable equilibrium: $(0.2, 0.2, 0, 0.2)$. C. Saddle point: $(-0.1, 0.1, 0, 0)$. D. Neutral stability: $(0, 0, -0.1, 0.1)$. E. Stable focal point: $(0, -0.1, -0.1, 0)$. F. Unstable focal point: $(0, 0.1, 0.1, 0.1)$. The eigenvalues were respectively: A: $(-0.1, -0.3)$, B: $(0.2, 0.2)$, C: $(0.1, -0.1)$, C: $(0 + 0.1i, 0 - 0.1i)$, D:$(-0.05 + 0.0866i, -0.05 - 0.0866i)$ and E: $(0.05 + 0.0866i, 0.05 - 0.0866i)$

other eigenvector determines the path that will diverge away from the equilibrium point.

- If both eigenvalues are complex (imaginary numbers), with positive real part then the equilibrium will be an **unstable spiral or unstable focus**. When perturbed, the density will diverge with ever increasing oscillations.

- If both eigenvalues are complex, with negative real part, then the equilibrium will be a **stable spiral or stable focus**. When perturbed, the density will return to the equilibrium point, overshooting both sides with decreasing amplitude.
- If the eigenvalues are complex, with 0 real part (it is a purely imaginary number); the system will oscillate around the equilibrium with constant amplitude, the size of which is determined by the size of the original perturbation. In that case we have a **neutrally stable equilibrium**.

7.7 Limit Cycles (***)

The stability analysis from previous chapter only describes the local behaviour, i.e. in the vicinity of the equilibrium point. However, this does not tell us anything about the global behaviour of the model.

Consider the model output depicted in Fig. 7.10A, for instance. This simple model has one single equilibrium point, which is an unstable focus (grey dot). Except when starting in this equilibrium point, all trajectories zoom in on the circular trajectory. This trajectory is called a limit cycle. Trajectories initiated outside of the circle zoom inwards, whilst those initiated within the circle spiral outwards. Once on the circular trajectory, the system oscillates for eternity and stays there. Note: this system is fundamentally different from the neutrally stable cycles that we described in the Lotka-Volterra predator-prey model (Section 7.3, Fig. 7.7C), where we had an infinite continuum of cycles. Neutral stability is best compared with a pendulum in the absence of friction: it will follow a trajectory which depends on the initial condition: the further the pendulum is initially displaced, the larger

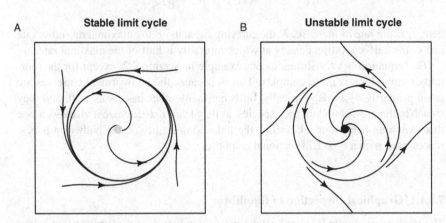

Fig. 7.10 Two limit cycles, created with the following coupled differential equations:
$$\frac{dx}{dt} = ax + cy + ex(x^2 + y^2 - 1)$$
$$\frac{dy}{dt} = by + dx + fy(x^2 + y^2 - 1)$$
A. Unstable focal point ($x = 0, y = 0$) spiralling into a stable limit cycle, generated with the following values for the coefficients: ($a = -1, b = 1, e = -1, f = -1$); the eigenvalues were ($1 + 1i, 1 - 1i$). **B.** Stable focal point ($x = 0, y = 0$) and unstable limit cycle, generated with ($a = -1, b = 1, e = 1, f = 1$) and with eigenvalues ($-1 + 1i, -1 - 1i$)

will be the trajectory. In contrast, a stable limit cycle always converges to the same trajectory, and can therefore be better compared with e.g. a cello string, which will tend towards vibrating with its own natural frequency.

Limit cycles need not be stable. In the model that was used to create Fig. 7.10B, the only equilibrium point is a stable focus (black dot). However, only the trajectories initiated within the circular region are drawn towards this equilibrium, the ones initiated external from the circle move away, whilst the ones starting on the limit cycle stay there for eternity.

7.8 Case Studies in R

7.8.1 Multiple Stable States: the Spruce Budworm Model

The spruce budworm (*Choristoneura fumiferana*) is an insect pest that consumes the leaves of coniferous trees in North America. In some years, their populations explode and the excessive consumption of pine buds and needles causes massive damage to the pineries.

One of the first models where multiple stable-states were demonstrated (Ludwig et al., 1978) was created as a tool to understand the causes of these periodic outbursts. The model describes the density of the spruce budworm (B), as a function of density-dependent growth (first term) and predatory mortality (second term)

$$\frac{d\mathrm{B}}{dt} = r_i \cdot \mathrm{B} \cdot \left(1 - \frac{\mathrm{B}}{K}\right) - \beta \frac{\mathrm{B}^2}{\mathrm{B}^2 + k_s^2} \tag{7.33}$$

here, r_i is the rate of increase, K the carrying capacity, β the maximal mortality rate and k_s the half-saturation density at which mortality is half of the maximal rate.

This equation is very similar as our example in Section 7.2, except for the mortality term, which is more complex. This is because the budworms themselves are eaten primarily by birds. Typically, birds eat many other insects as well, and they switch to the most abundant prey species, as they learn to recognize it visually. Since this switching behaviour takes time, the induced mortality on the budworm is best represented with a type III functional response.

7.8.1.1 Graphical Inspection of Equilibria

The implementation in R starts by defining the parameters and a function that calculates the rate of change dB/dt, for a certain value of B and r_i. Note that, by default, r_i takes on the value of 0.05 in this function.

```
ri   <- 0.05
K    <- 10
beta <- 0.1
ks   <- 1

rate <- function(B, ri=0.05)
        ri*B*(1-B/K)-beta*B^2/(B^2+ks^2)
```

To inspect the behaviour of the model, we plot the rate of change (dB/dt) as a function of the density (B), for a variety of r_i -values (Fig. 7.11A). We use R's plotting function matplot, which plots all columns of a matrix at once. The input matrix has a sequence (seq) of density values as row elements (Bsq), and a sequence of r_i -values as columns (rsq). R's function outer generates this matrix for all combinations of Bsq and rsq. Note that we also add an X-axis (abline(h=0)) and a legend to the plot.

```
Bsq <- seq(0,10,length=500)
rsq <- seq(0.01,0.07,by=0.02)
mat <- outer(X=Bsq,Y=rsq,function(X,Y) rate(B=X,ri=Y))

matplot(Bsq,mat,xlab="B",ylab="dB/dt",
         type="l",lty=1:4,col=1)
abline(h=0,lty=2)
legend("bottomleft",legend=rsq,title="ri",col=1,
        lty=1:4)
```

It is clear from Fig. 7.11A that, depending on the value of r_i, the number of equilibrium points can vary from 1 to 4.

We now select a single case and inspect the number and behaviour of the equilibria more closely:

```
curve(rate(x,0.05),xlab="B", ylab="dB/dt",main="ri=0.05",
       from=0,to=10)
abline(h=0)
```

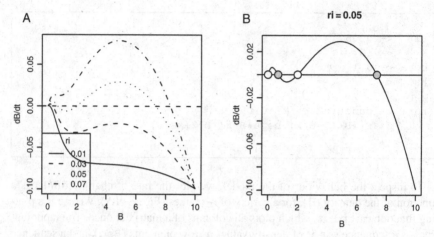

Fig. 7.11 A. The rate of change of spruce budworm density (dB/dt) versus spruce budworm density (B) for varying values of the rate of increase (r_i). **B.** Types of equilibria for r_i =0.05; *grey symbols*: stable equilibria, *white symbols*: unstable equilibria. See text for R-code to generate these figures

7.8.1.2 Estimating the Position and Stability of Equilibrium Points

Whereas it is possible to find an analytical solution for the position of the equilibrium points (see Ludwig et al., 1978), the formula is much more complex as in our example from Section 7.2, so we will use an R-function that uses an iterative procedure instead. This 'approximation' is not as accurate as an analytical formula, but it can easily be applied for different models. R-function uniroot.all (from package rootSolve) finds, for a function f all roots within a certain interval.

In order to estimate the Jacobian 'matrix' for each equilibrium point, we use R-function gradient, also from package rootSolve. With a model consisting of one equation, the Jacobian matrix is just one number:

$$J_{B^*} = \frac{\partial \left(\frac{dB^*}{dt} \right)}{\partial B} \tag{7.34}$$

The eigenvalue of this 'matrix' is simply the number itself. The sign of the eigenvalue refers to the stability of the equilibrium point: a positive sign denotes an unstable, a negative sign a stable equilibrium and 0 is a saddle point.

As we will perform the same procedure, for different values of parameter r_i, when we draw the bifurcation diagram (see below), we write the equilibrium and stability estimation as a function (called equilibrium) that takes as input the value of the parameter r_i, and that returns both the equilibrium points themselves (x) and the type of the equilibrium point (type), as a list. R-function sign returns a value of -1, 0 and $+1$ for negative, zero and positive values respectively. We add 2 such that we obtain a type of 1, 2 and 3 for a stable equilibrium, a saddle point and an unstable equilibrium respectively.

```
require(rootSolve)
equilibrium <- function(ri)
{
  Eq      <- uniroot.all(f=rate,interval=c(0,10),ri=ri)
  eqtype <- vector(length=length(Eq))
  for (i in 1:length(Eq))
   {
     jac         <- gradient(f =rate,x=Eq[i],ri=ri)
      eqtype[i] <- sign(jac)+2
   }
  return(list(x=Eq, type=eqtype))
}
```

We calculate the equilibrium points and their stability and add symbols that refer to the type of equilibrium to the curve that we drew before. We choose as symbols coloured circles (pch=21), twice the default size (cex=2) and with background (bg) set to grey (1), black (2) or white (3) depending on their stability type. The output is given in Fig. 7.11B.

```
eq    <- equilibrium (ri=0.05)
points(x=eq$x,y=rep(0,length(eq$x)),pch=21,cex=2,
       bg=c("grey","black","white")[eq$type]  )
```

7.8.1.3 Drawing the Bifurcation Diagram

We now generate the bifurcation diagram for a sequence of ri -values. We first create an empty plot (type="n"), and then locate, for each ri -value in the sequence, the position of the equilibrium points (eq$x), and their type (eq$type).

Depending on the type of the equilibrium, we add a darkgrey (1, stable equilibrium), black (2, saddle point) or lightgrey (3, unstable equilibrium) symbol.

Finally, at selected points we add arrows depicting how the density will evolve if perturbed (we give the code only for $r_i = 0.05$).

```
rseq <- seq(0.01,0.07,by=0.0001)

plot(0,xlim=range(rseq),ylim=c(0,10),type="n",
xlab="ri",ylab="B*",main=''spruce budworm model'')

for (ri in rseq) {
eq <- equilibrium(ri)

points(rep(ri,length(eq$x)),eq$x,pch=22,
col=c("darkgrey","black","lightgrey")[eq$type],
bg =c("darkgrey","black","lightgrey")[eq$type])
}

equi <-uniroot.all(f=rate,interval=c(0,10),r=0.05)

arrows(0.05,10          ,0.05,equi[4]+0.2,length=0.1 )
arrows(0.05,equi[3]+0.2,0.05,equi[4]-0.2,length=0.1 )
arrows(0.05,equi[3]-0.2,0.05,equi[2]+0.2,length=0.1 )
arrows(0.05,equi[1]+0.2,0.05,equi[2]-0.2,length=0.1 )
```

The results (Fig. 7.12) are roughly similar to the bifurcation diagram in Section 7.2.3. There are either 2, 3 or 4 equilibrium points. In the zone where there are 4, two non-zero densities are stable; their domain of attraction is separated by the separatrix, formed by the intermediate unstable equilibrium points.

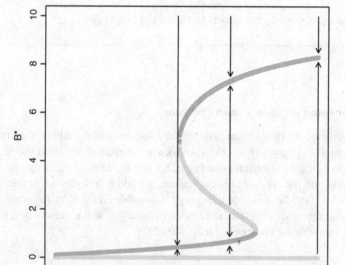

Fig. 7.12 Bifurcation diagram for the spruce budworm model. See text for R-code to generate this figure

When initiated above the separatrix, the simulation will tend to the highest densities (and an outburst will occur); when the simulation starts at densities below the separatrix, it will tend to the lowest stable densities. (this is denoted by the thin arrows on the graph).

7.8.2 Phase-Plane Analysis: The Lotka-Volterra Competition Equations

Consider the following couple of differential equations:

$$\frac{dN_1}{dt} = r_1 \cdot N_1 \cdot \left(1 - \frac{N_1 + \alpha_{12} \cdot N_2}{K_1}\right)$$

$$\frac{dN_2}{dt} = r_2 \cdot N_2 \cdot \left(1 - \frac{N_2 + \alpha_{21} \cdot N_1}{K_2}\right) \tag{7.35}$$

where N_1 and N_2 are the densities of two species that compete for the same resource. Their respective rate of increase and carrying capacity is r_1, r_2, K_1 and K_2; α_{12} and α_{21} are the competitive effect of species 2 on the growth rate of species 1 and vice versa.

This model is used to illustrate the implementation, in R, of the phase-plane and equilibrium analysis.

We start by determining the zero isoclines and equilibrium points.

The zero isoclines of N_1 and N_2 are calculated by setting the rate of change to 0, which gives for N_1:

$$N_1 = 0$$

$$N_1 = K_1 - \alpha_{12} \cdot N_2 \tag{7.36}$$

The former is the Y-axis, the latter is a line that crosses the X-axis at $N_2 = K_1/\alpha_{12}$, the Y-axis at $N_1 = K_1$.

The zero isoclines of N_2 are:

$$N_2 = 0$$

$$N_2 = K_2 - \alpha_{21} \cdot N_1 \tag{7.37}$$

The former line is the X-axis, the latter line crosses the X-axis at $N_1 = K_2/\alpha_{21}$, the Y-axis at $N_2 = K_2$.

R-code

We first define the parameters and implement the Lotka-Volterra model as a function that takes as input the time (t, unused), the species density (N, a vector with 2 values) and the parameter values (not used).

The model returns the rate of change of the two state variables (dN1, dN2), as a list.

```
r1      <- 3
r2      <- 2
K1      <- 1.5
K2      <- 2
alf12 <- 1
alf21 <- 2/
Lotka <-function(t,N,pars)

{
  dN1    <- r1*N[1]*(1-(N[1]+alf12* N[2])/K1)
  dN2    <- r2*N[2]*(1-(N[2]+alf21* N[1])/K2)

  list(c(dN1 , dN2 ))
}
```

Before plotting the isoclines, we first define the intersection of both isoclines with the X- and Y-axis (Ax and Ay for the intersection of the isoclines of N_1 with the X- and Y- axis; and Bx, By for N_2). Then we make sure that both lines will fit the plotting region, by estimating the ranges of the x- and y-axes (xlim, ylim). The first isocline is plotted as a solid line (the default line type), the second isocline is dashed (lty=2).

```
Ax    <- c(0,K2/alf21)
Ay    <- K2 - alf21* Ax
By    <- c(0,K1/alf12)
Bx    <- K1 - alf12* By
xlim <- range(c(Ax, Bx))
ylim <- range(c(Ay, By))

plot  (x=Ax,y=Ay, type="l", lwd=2,    # 1st isocline
       main="Competition phase-plane",
         xlab="N1",ylab="N2",xlim=xlim,ylim=ylim)
lines (Bx,By,lwd=2,lty=2)             # 2nd isocline
```

As we want to add several model trajectories using the same parameter values, but different initial conditions of the state variables, we define a function called 'trajectory' that performs each run and plots both the lines and the arrows.

The model is solved using the integration routine ode, from library deSolve, which is loaded first. We specify the time sequence at which we want output (times), and the initial conditions of the state variables (state). The function ode calls our previously defined model function (Lotka) at each time step and performs the integration; its results are stored in data frame out, which contains the values of the two state variables (columns 2,3) at each time step (column 1). After running the model, we can plot the model trajectory (lines). We also add an arrow, giving the direction of the trajectory (arrows).

After that, all we need to do is call the function 'trajectory' with different initial conditions for N_1 and N_2.

```
library(deSolve)

trajectory <- function(N1,N2)
{
times <-seq(0,30,0.1)
state <-c(N1 = N1, N2 = N2)
out   <-as.data.frame(ode(state,times, Lotka,NULL))

lines (out$N1,out$N2,type="l")
arrows(out$N1[10],out$N2[10],out$N1[11],out$N2[11],
       length=0.1,lwd=2)
}

trajectory (N1=0.05,N2=0.3)
trajectory (0.11,0.3)
trajectory (1.5,1.8)
trajectory (1.0,2.0)
```

At the equilibrium points, both the rate of change of N_1 and N_2 is 0.
This is at the points $(N_1, N_2) = (0, 0), (0 ,K_2), (K_1,0)$
or when $N_1 = \dfrac{K_1 - \alpha_{12}K_2}{1 - \alpha_{12}\alpha_{21}}$ and $N_2 = \dfrac{K_2 - \alpha_{21}K_1}{1 - \alpha_{12}\alpha_{21}}$

```
# 4 equilibrium points
X    <- c(0, 0 , K1, (K1-alf12*K2)/(1-alf12*alf21))
Y    <- c(0, K2, 0 , (K2-alf21*K1)/(1-alf12*alf21))
```

We now proceed with the formal analysis of the stability of these equilibrium points.

The Jacobian (or community matrix) is given by:

$$J_{N_1^*,N_2^*} = \begin{bmatrix} \dfrac{\partial(\dfrac{dN_1^*}{dt})}{\partial N_1} & \dfrac{\partial(\dfrac{dN_1^*}{dt})}{\partial N_2} \\[4ex] \dfrac{\partial(\dfrac{dN_2^*}{dt})}{\partial N_1} & \dfrac{\partial(\dfrac{dN_2^*}{dt})}{\partial N_2} \end{bmatrix} \tag{7.38}$$

and where N^* denotes that it is evaluated at the equilibrium point. The model is simple enough to calculate the Jacobian matrix analytically, but it is easier (and more general) to use a built-in R-function (jacobian.full) that estimates this matrix numerically.

This function is in package `rootSolve` which we attach first.

For each of the 4 equilibrium points, we first calculate the Jacobian and then invoke R's function `eigen` which estimates both the two eigenvalues and the two eigenvectors of this matrix. We require only the eigenvalues ($values). Based on the signs of the eigenvalues, the equilibrium points obtain different background colors. Finally, the equilibrium points are plotted.

```
require(rootSolve)

ei     <- matrix(nrow=4,ncol=2)

for (i in 1:4)
{
 N1 <- X[i]
 N2 <- Y[i]

 # the Jacobian
 Jacob <- jacobian.full(y= c(N1,N2),func=Lotka)
 # eigenvalues
 ei[i,] <- eigen(Jacob)$values

 # colors of symbols
 if (sign(ei[i,1])>0 & sign(ei[i,2])>=0) col <- "white"
 if (sign(ei[i,1])<0 & sign(ei[i,2])<=0) col <- "black"
 if (sign(ei[i,1])* sign(ei[i,2])   <0 ) col <- "grey"

 # equilibrium point plotting
     points(N1,N2,pch=22,cex=2.0,bg=col,col="black")
}
cbind(N1=X,N2=Y,ei)
```

The final statement writes the equilibrium density values together with the eigenvalues: `cbind` binds arrays and vectors columnwise.

The output is:

N1	N2	λ1	λ2
0.0	0	3.000000	2.0000000
0.0	2	-2.000000	-1.0000000
1.5	0	-3.000000	-1.0000000
0.5	1	-2.414214	0.4142135

There is one unstable $(0,0)$ and two stable $((0,2)$ and $(1.5,0))$ equilibrium points. As the two eigenvalues of equilibrium point 4 $(0.5,1)$ are of opposite sign, this is a saddle point. This means that there are two directions that asymptote into this saddle point, whilst all other trajectories end up in either in the first $(0,2)$ or in the second $(1.5, 0)$ stable equilibrium.

The lines that separate these two different behaviours are the stable separatrices; here is how we derive them. The eigenvectors that accompany the eigenvalues of the saddle point give the direction in which the separatrices converge to (linked to the negative eigenvalue) or diverge from (linked to the positive eigenvalue) the saddle point. This direction is linear only very near to the saddle point, but becomes bent further away.

To generate the separatrices, we start from a point, very near to the saddle point, and perturbed in the direction of the eigenvector associated to the negative eigenvalue. The perturbation is performed both in positive and negative direction. From this initial condition, we run the *reverse* of the model.

In the R-code below, first we derive the eigenvector and eigenvalues of the Jacobian that is estimated at the saddle point, and we select the eigenvector corresponding to the negative eigenvalue.

```
eig         <- eigen(Jacob)
vv          <- eig$vector[eig$values<0]
```

The reverse model (revmod) simply outputs the negative of the rate of change, as a vector.

```
revmod <- function(t,N,p) list(-1*unlist(Lotka(t,N,p)))

times   <-seq(0,1.9,0.05)

# first direction
state <- c(N1,N2) + 0.01*vv
out     <- as.data.frame(ode(state,times,revmod,NULL))

lines (out[,2],out[,3],lty=2)
arrows(out[25,2],out[25,3],out[24,2],out[24,3],
       length=0.1,lwd=2)

# second direction
state <- c(N1,N2) - 0.01*vv
times <-seq(0,10,0.05)
out     <-as.data.frame(ode(state,times,revmod,NULL))
lines   (out[,2],out[,3],lty=2)
arrows  (out[11,2],out[11,3],out[10,2],out[10,3],
         length=0.1,lwd=2)
```

We also draw the trajectories diverging out of the saddle point and which are derived by perturbing the equilibrium point in the (positive and negative) direction of the eigenvector associated to the positive eigenvalue.

```
ww    <- eig$vector[eig$values>0]

trajectory (N1+0.05*ww[1],N2+ 0.05*ww[2])
trajectory (N1-0.05*ww[1],N2- 0.05*ww[2])
```

Finally we add a `legend`, on the `right` side of the graph (Fig. 7.13).

```
legend("right",legend=c("isocline N1","isocline
N2","trajectory","separatrice","saddle point",
"stable equilibrium","unstable equilibrium"),
lty=c(2,1,1,2,NA,NA,NA),lwd=c(2,2,1,1,NA,NA,NA),
pch=c(NA,NA,NA,NA,22,22,22),pt.bg=c(NA,NA,NA,NA,"grey","black","white"))
```

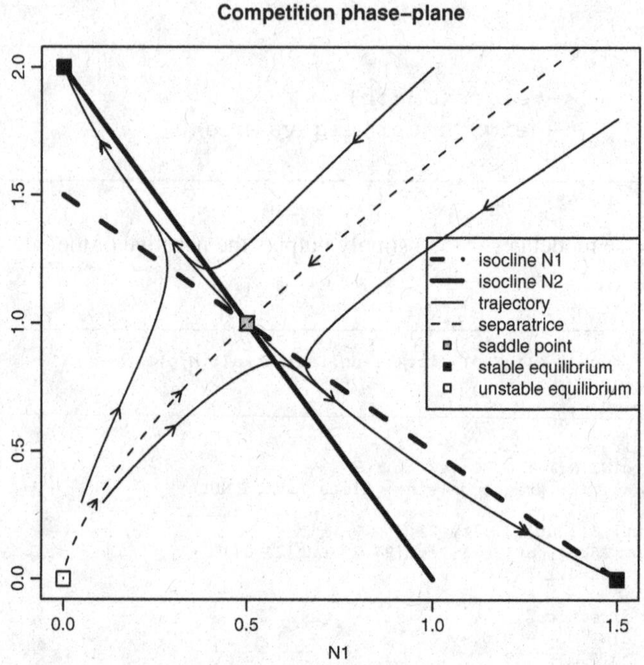

Competition phase–plane

Fig. 7.13 Phase-plane diagram for the Lotka-Volterra competition model. See text for R-code to generate this figure

7.8.3 The Lorenz Equations – Chaos

Chaotic solutions are aperiodic, recurrent solutions that appear random due to their sensitive dependence on the initial conditions. The behaviour is not random though: running the model with the same settings will recreate the same complex patterns. They are, by their nature, unpredictable in the long term.

The Lorenz equations (Lorenz, 1963) were the first chaotic system to be discovered. They are three differential equations that were derived to represent idealised behaviour of the earth's atmosphere.

$$\frac{dx}{dt} = -\frac{8}{3} \cdot x + y \cdot z$$

$$\frac{dy}{dt} = -10 \cdot (y - z)$$

$$\frac{dz}{dt} = -x \cdot y + 28y - z \tag{7.39}$$

It takes about 10 lines of R-code to generate the solutions and plot them; we need to load the integration package (deSolve) and a package that allows generating 3-D scatterplots for the output (Fig. 7.14).

```
require(deSolve)
require(scatterplot3d)

Lorenz<-function(t,state,parameters)
  {
  with(as.list(c(state)), {

    dx      <- -8/3*x+y*z
    dy      <- -10*(y-z)
    dz      <- -x*y+28*y-z

    list(c(dx,dy,dz))                     })

  }

state <-c(x=1, y=1, z=1)
times <-seq(0,100,0.001)
out   <-as.data.frame(ode(state,times,Lorenz,0))

scatterplot3d(out$x,out$y,out$z,type="l",
main="Lorenz butterfly",ylab="",grid=FALSE,box=FALSE)
```

7.8.4 Steady-State Solution of the Silicate Diagenetic Model (**)

The steady-state condition of complex diagenetic models is often solved by means of the iterative Newton-Raphson method. We demonstrate this by implementing the silica diagenetic model, which was first introduced in Section 3.6.5.

For completeness we rewrite the model equations for particulate biogenic silica (BSi, μmol l^{-1} solid) and dissolved silica (DSi, μmol l^{-1} liquid), with the boundary conditions.

$$\frac{\partial BSi}{\partial t} = \frac{1}{1 - \phi_x} \frac{\partial}{\partial x} \left[(1 - \phi_x) \cdot D_b \frac{\partial BSi}{\partial x} \right] - \lambda \cdot BSi \cdot \left(1 - \frac{DSi}{eqSi} \right) \tag{7.40}$$

Lorenz butterfly

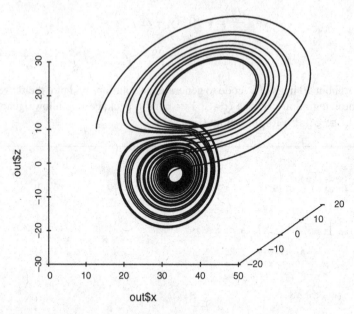

Fig. 7.14 The Lorenz butterfly. See text for R-code

$$Flux_0 = -(1 - \phi_0) \cdot D_b \left. \frac{\partial \mathrm{BSi}}{\partial x} \right|_0$$

$$0 = \left. \frac{\partial \mathrm{BSi}}{\partial x} \right|_\infty \tag{7.41}$$

$$\frac{\partial \mathrm{DSi}}{\partial t} = \frac{1}{\phi_x} \frac{\partial}{\partial x} \left[\phi_x \cdot D_s \frac{\partial \mathrm{DSi}}{\partial x} \right] + \lambda \cdot \mathrm{BSi} \cdot \left(1 - \frac{\mathrm{DSi}}{eq\,Si} \right) \cdot \frac{1 - \phi_x}{\phi_x} \tag{7.42}$$

$$BWDSi = \mathrm{DSi}_0$$

$$0 = \left. \frac{\partial \mathrm{DSi}}{\partial x} \right|_\infty \tag{7.43}$$

Where ϕ is porosity (−), D_b and D_s are the sediment bioturbation rate and the sediment diffusion coefficient respectively (cm^2 d^{-1}), λ is the silicate dissolution rate (d^{-1}), and $EqSi$ the equilibrium dissolved silicate concentration (μmol l^{-1} liquid). In sediments, large gradients occur on very small scales, in the order of mm or cm. Thus, it is convenient (although atypical) to take centimetre (cm) as the length scale in the model. Concentrations are then expressed in nmol cm^{-3} (which is the same as μmol dm^{-3} or μmol l^{-1}).

In R, several functions included in R-package 'rootSolve', allow estimating steady-state conditions. Of these, the function 'steady.1D' is a Newton-Raphson method, especially designed to solve efficiently the steady-state condition of 1-dimensional models that comprise many species (this contrasts to steady.band, which we will use in the next example, and which is best suited for single-species models).

To use these steady-state solvers, the model has to be specified in a similar way as for R's integration routines, i.e. the function is called as: func(time, state, parameters). Generally time will not be used when estimating steady-state.

The model dynamics, implemented in R is given below. A function called SiDiamodel receives as input the time (here not used), the concentrations (Conc, a vector, which contains biogenic, BSi, and dissolved silicate, DSi) and model parameters (not used). It returns the rate of change of the state variables, the vertical profile of Si dissolution rates, and the fluxes at the sediment-water interface and at large depth, as a list.

After retrieving the BSi and DSi concentration from the state variable vector, the fluxes at the interfaces of sediment layers for dissolved (DSiFlux) and biogenic silicon (BSiFlux) are calculated first. Note that, for the deepest layer (N), the zero-gradient boundary condition is imposed by equalling the external concentration to the concentration at the deepest layer (BSi[N], DSi[N]). Also, as the fluxes are defined on the interfaces of layers, the porosity, defined at these interfaces is used (IntPor) to estimate bulk fluxes. The biogenic silicon flux at the sediment-water interface (BSiFlux[1]) is imposed (BSidepo). After calculating the dissolution rate (Dissolution), the rate of change of both state variables (dDSi, dBSi) is estimated as the sum of changes due to transport (the flux gradient), and biogeochemistry (dissolution). Here we convert rates to per unit of liquid or solid, using the porosity defined in the centre of layers (Porosity).

```
SiDiamodel <- function (time=0,Conc,pars=NULL)

{
 BSi<- Conc[1:N]
 DSi<- Conc[(N+1):(2*N)]

 DSiFlux      <- -SedDisp * IntPor    *diff(c(bwDSi ,DSi,DSi[N]))/thick
 BSiFlux      <- -Db       *(1-IntPor)*diff(c(BSi[1],BSi,BSi[N]))/thick
 BSiFlux[1]  <- BSidepo       # upper boundary flux

 Dissolution <- rDissSi * BSi*(1.- DSi/EquilSi )
 Dissolution <- pmax(Dissolution,0)

# Rate of change
 dDSi         <- -diff(DSiFlux)/thick/Porosity       +      # transport
                  Dissolution * (1-Porosity)/Porosity       # biogeochemistry

 dBSi         <- -diff(BSiFlux)/thick/(1-Porosity)    - Dissolution

 return(list(c(dBSi=dBSi,dDSi=dDSi),
             Dissolution=Dissolution,
             DSiSurfFlux =DSiFlux[1],DSIDeepFlux =DSiFlux[N+1],
             BSiDeepFlux =BSiFlux[N+1]))
}
```

We model a sediment column, 10 cm deep, and which is subdivided into thin layers, 0.05 cm thick (parameter thick). To calculate the porosities, we need to define the depth of the upper interfaces of each layer (Intdepth) and the depth at the middle of each layer (Depth). The number of interfaces and layers is then given by Nint and N respectively.

We assume a porosity gradient that changes from 0.9 at the surface (por0) to 0.7 deeper down (pordeep), and with an exponential decay coefficient porcoef (cm^{-1}). We need to calculate porosity at the sediment interfaces (Intpor, defined at sediment depths Intdepth), and in the middle of layers (Porosity, defined at Depth).

Bioturbation is assumed constant (dB0, $cm^2 d^{-1}$) in the upper bioturbated layer (with thickness mixdepth, cm), below which it declines exponentially (decay co-efficient dBcoeff, cm^{-1}). The bioturbation coefficients (Db) are used to calculated fluxes of biogenic silica, which are defined on the layer interfaces, hence it is defined on the interfaces (and uses Intdepth).

```
thick      <- 0.05
Intdepth   <- seq(from=0,to=10,by=thick)
Nint       <- length(Intdepth)
Depth      <- 0.5*(Intdepth[-Nint] +Intdepth[-1])
N          <- length(Depth)

por0       <- 0.9
pordeep    <- 0.7
porcoef    <- 2
Porosity   <- pordeep + (por0-pordeep)*exp(-Depth*porcoef)
IntPor     <- pordeep + (por0-pordeep)*exp(-Intdepth*porcoef)
dB0        <- 1/365
dBcoeff    <- 2
mixdepth   <- 5
Db         <- pmin(dB0,dB0*exp(-(Intdepth-mixdepth)*dBcoeff))
```

After defining the remaining model parameters, we initialise the concentrations with random values (Conc); note that there are 2*N state variables; R-statement runif(2*N) generates 2*N random numbers between 0 and 1. Based on this initial 'guess' of concentrations, the Newton-Raphson method is invoked to solve for the steady-state condition (steady.1D). We specify that there are two species to be solved (DSi, BSi) (nspec=2). Notwithstanding the far-off initial guesses of the concentrations, the model reaches steady-state in very few iterations (typically $4 - 10$).

After the resulting steady-state concentrations are extracted (DSi, BSi), the vertical profiles can be plotted. Note that, by setting the limits of the y-axis ranging from 10 to 0, the y-axis is effectively reversed.

```
SedDisp <- 0.4              # diffusion coefficient, cm2/d
rDissSi <- 0.005            # dissolution rate, /day
EquilSi <- 800              # equilibrium concentration
BSidepo <- 0.2*100          # nmol/cm2/day
bwDSi   <- 150              # µmol/l

Conc    <- runif(2*N)

sol     <- steady.1D (y=Conc, func=SiDiamodel, parms=NULL, nspec=2)
CONC    <- sol$y
Res     <- SiDiamodel(Conc=CONC)


DSi     <- CONC[(N+1):(2*N)]
BSi     <- CONC[1:N]

par(mfrow=c(2,2))
plot(DSi,Depth,ylim=c(10,0),xlab="mmolSi/m3 liquid",main="DSi",type="l",
     lwd=2)
plot(BSi,Depth,ylim=c(10,0),xlab="mmolSi/m3 Solid" , main="BSi",type="l",
     lwd=2)
plot(Porosity,Depth,ylim=c(10,0),xlab="-", main="Porosity",type="l",
     lwd=2)
plot(Db,Intdepth,ylim=c(10,0),xlab="cm2/d", main="Bioturbation",type="l",
     lwd=2)
```

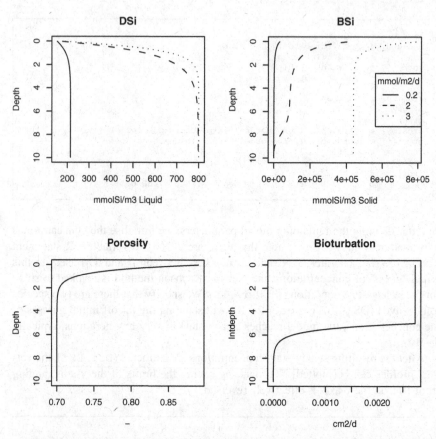

Fig. 7.15 Steady-state solutions of the silicate diagenetic model for different BSi deposition rates. See text for R-code

In Fig. 7.15, the model was run with three different deposition rates, 0.2, 2 and 3 mmol BSi m^{-2} d^{-1}, all other parameters kept constant. Note that, in the highest flux case, a significant part of the biogenic silica is preserved in the sedimentary column.

7.8.5 *Fate of Marine Zooplankton in an Estuary – Equilibrium Condition* (**)

In Section 6.6.5 we calculated the transient dynamics of marine zooplankton in an estuary (Soetaert and Herman. 1994).

We now estimate the steady-state solution, using the same numerical approximation to the spatial derivates:

$$\frac{dC_i}{dt} \approx -\frac{\Delta_i A \cdot J}{\Delta V_i} + g \cdot C_i$$

$$where$$

$$\Delta_i A \cdot J = (A \cdot J)_{i,i+1} - (A \cdot J)_{i-1,i}$$

$$and$$

$$(A \cdot J)_{i-1,i} \approx Q \cdot C_{i-1} - E^*_{i-1,i} \cdot \Delta C_{i-1,i} \tag{7.44}$$

As in previous example we use one of the Newton-Raphson methods from package rootSolve. Function 'steady.band' is especially suited for finding the steady-state condition of single-species 1-dimensional models.

We use the same specification of estuarine morphology and physical conditions, patterned to the Scheldt estuary which was discussed in Section 6.6.5. We will not repeat the R-code here; we just specify the model function (Estuary) and how to solve for the steady-state:

```
require(rootSolve)

Estuary <-function(t,C,pars)
{

  with (as.list(pars),{

    Input <- Q * c(Criver, C) +
              -Estar * diff(c(Criver, C, Csea))
    dC <- -diff(Input)/Volume + rate*C
    return(list(dC))
  })
}

# the model parameters:
pars <- c(Criver = 0.0,       # riverine conc
          Csea  = 100.0,      # seaward conc
          rate  = -0.05,      # /day growth rate
          Q     = 100*3600*24) # m3/d, river flow
```

This model is used to solve for the steady-state salinity profile (in which case, parameter rate =0, and the marine salinity, Csea = 35), and for the steady-state concentration of zooplankton (Csea = 100). We generate multiple profiles of the latter, as a function of the growth rate g, and store the results in columns of matrix Zoo (R-statement cbind(Zoo,st$y) adds the steady-state solution st$y as an extra column to Zoo).

```
# steady-state salinity profile

pars["rate"] <- 0
pars["Csea"] <- 35

sal <- steady.band(y=rep(35,times=nbox),func=Estuary,
                   parms=pars,nspec=1)$y

# steady-state zooplankton profile

gSeq <-seq (-0.05,0.01, by=0.01)
pars["Csea"] <- 100

Zoo  <- NULL

for (g in gSeq)
{
  pars["rate"] <- g
  st <- steady.band(y=rep(0,nbox),func=Estuary, parms=pars,
                    nspec=1 ,pos=TRUE )
  Zoo<- cbind(Zoo,st$y)
}
```

Finally the multiple steady-state profiles of zooplankton are plotted against salinity. R-function `matplot` plots all columns at once (Fig. 7.16).

```
matplot(sal,Zoo,type="l",lwd=1,lty=1:20,col="black",

        xlab="Salinity", ylab="gDWT/m3",main="Zooplankton")

legend("topleft",legend= gSeq,title="g, /day",lty=1:20)
```

7.9 Projects

7.9.1 The Schaefer Model of Fisheries

Consider the logistic model with constant-effort harvesting:

$$\frac{dX}{dt} = r_i \cdot X \cdot \left(1 - \frac{X}{K}\right) - q \cdot E \cdot X \tag{7.45}$$

where X is fish biomass, expressed in tonnes, E is (fishing) effort, in number of (standardised) vessels, and q is the catchability coefficient (vessel^{-1} day^{-1}). The term $q \cdot E \cdot X$ is the yield, in tonnes day^{-1}; both q and E are constant.

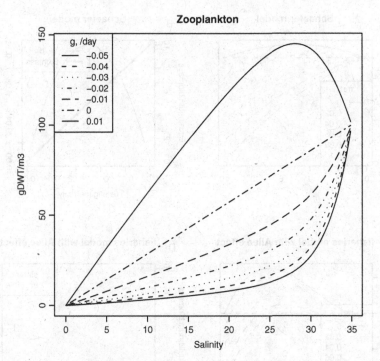

Fig. 7.16 Steady-state zooplankton concentration versus salinity for a well-mixed estuary. The zooplankton 'growth' rate varied from −0.05 (a decrease with 5% per day) to 0.01 d^{-1}. See text for R-code

This model is often called the Schaefer model (Schaefer, 1954) and has been applied to fish populations.

- Plot the rate of change of biomass as a function of biomass and for a series of mortality rates m ranging between 0 and 0.1 d^{-1}, where $m = q \cdot E$. Use values of $r_i = 0.05$ d^{-1}, $K = 10$ tonnes. (the output to generate should look like Fig. 7.17A)
- Derive the equilibrium biomass X* as a function of the parameters. Plot the equilibrium biomasses as a function of increasing values of $m = q \cdot E$. (Fig. 7.17B)
- Also plot the total fisheries yield $q \cdot E \cdot X^*$ at equilibrium and for the sequence of m-values. (Fig. 7.17B).
- At which value of mortality m is the fisheries yield optimal? What is the equilibrium fish biomass at this optimum?
- Now consider the following parameter values which are specific for the Eastern Pacific yellowfin tuna population (Schaefer 1967). $r_i = 2.61$ yr^{-1}, $K = 1.34*108$ kg, $q = 3.8 \, 10^{-5}$ per vessel yr^{-1}. What is the effort level and corresponding fish biomass at which harvest is maximal?

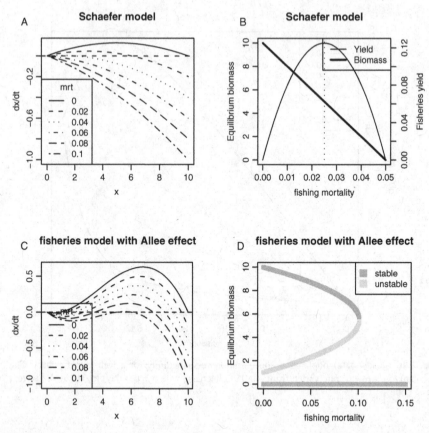

Fig. 7.17 A,B. Output corresponding to project 7.9.1. C.D. Output corresponding to project 7.9.2

7.9.2 A Fisheries Model with Allee Effect

Sometimes, it is more realistic to assume that there exists some kind of minimum density or biomass ($K0 < K$), below which the rate of change is always negative. This can for instance arise because of too low chances of encountering mates at these low densities. Fisheries biologists call this phenomenon 'critical depensation'; biologists call this the 'Allee effect' (Allee, 1931). Here is a model that includes an Allee effect:

$$\frac{dX}{dt} = r \cdot X \cdot \left(\frac{X}{K_0} - 1\right) \cdot \left(1 - \frac{X}{K}\right) - q \cdot E \cdot X \qquad (7.46)$$

Where K_0 is the minimum viable density level

Investigate the equilibrium properties graphically for different values of mortality $m = q \cdot E$ (see Fig. 7.17 C)

- Create a bifurcation diagram for varying values of the mortality. Discuss. (see Fig. 7.17 D)

7.9.3 Ecological-Economical Fisheries Model

How can we better define the fishing effort (E)?

In the Schaefer model (Schaefer, 1954), E was a constant:

$$\frac{dX}{dt} = r \cdot X \cdot \left(1 - \frac{X}{K}\right) - q \cdot E \cdot X \tag{7.47}$$

We now derive a more economically-inspired fisheries model (e.g. Clark, 1990).

Define the price per unit of harvested biomass, p (euros per tonne of fish). The income of the fisheries can be expressed as the product of p and the harvested biomass:

$$RE = p \cdot q \cdot E \cdot X \tag{7.48}$$

where RE stands for 'revenue'.

There is also a cost associated with fishing. Define c, the cost per unit of effort (euros vessel^{-1} year^{-1}), then:

$$TC = c \cdot E \tag{7.49}$$

is the total cost. Note that this term does not contain anything related to fish. This implies that the cost is the same, whether fish are caught or not.

We will model what economists call 'open access fisheries'. Putting it simple, open access fisheries is an (unregulated) fishery where everyone is allowed to harvest freely. Now consider two situations:

In the situation where fish is abundant and RE>>>TC, the fishermen are making a profit. This will attract more fishermen to the fisheries or more boats will be built and the effort will increase: good fishing attracts more investment.

In contrast, when costs exceed the revenue (TC>>>RE), then, on average, the fishermen are loosing money, so that at least some of them will withdraw from the fisheries thus reducing the effort.

Combining both these cases, there is a tendency of the effort to evolve to the level where total revenue (RE) equals total cost (TC), this is to the level where the fishermen are making no profit at all! (a rather depressing thought if you are a fisherman).

Of course, fishing activity may give a quick profit for a few fishermen, but in the long run, this profit will reduce to zero.

One way to put these verbal statements into a model is to make the fishing effort, E, a state variable whose rate of change is a function of total cost and total revenue (Smith, 1969). We can write the following dynamic description for the fisheries effort, E:

$$\frac{d\mathrm{E}}{dt} = k \cdot (p \cdot q \cdot \mathrm{E} \cdot \mathrm{X} - c \cdot \mathrm{E}) \qquad (7.50)$$

where k is a (positive) constant, which governs the speed at which the equilibrium is reached. Note that dE/dt will be negative if RE<TC positive if RE>TC and zero if RE=TC.

Tasks:

Use phase-plane analysis to investigate the equilibrium properties graphically for this model. Derive the stability properties of the equilibria, using the formal derivation described in Section 7.6.

Plot the equilibrium biomass X^* and the equilibrium fishing effort E^* as a function of the cost/gain (c/p) ratio. Discuss.

- Use parameters for the Eastern Pacific yellowfin tuna population (Schaefer 1967). $r_i = 2.61$ yr-1, $K = 1.34*108$ kg, $q = 3.810^{-5}$ per vessel yr^{-1}. What is the equilibrium effort level (E^*) and corresponding fish biomass, assuming that the cost/price ratio is 1527.6.

If you still have the courage, repeat the previous analyses for a model that includes an Allee effect.

7.9.4 Predator-Prey System with Type-II Functional Response

Consider a predator-prey system that has density-dependent growth of the prey and a type II functional response for the predatory grazing.

- Investigate the stability properties of the model, both by phase-plane and by formal analysis. A graphical representation of the results is in Fig. 7.18.
- Similarly as the simpler predator-prey model, this model exhibits 3 different dynamics. Under which conditions can you generate these dynamics?

7.9.5 Succession of Nutrients, Phytoplankton and Zooplankton in a River

In rivers, the growth of phytoplankton is constrained by nutrient input, residence time and zooplankton grazing. Low river discharge may stimulate algal growth by increasing their residence time in the river, but at the same time, reduced river flow lowers the input of nutrients, which may negatively affect algal growth. Moving downstream, one generally observes nutrient depletion in the upstream reaches, co-occurring with algal blooms, followed by zooplankton increases and algal decline. The areas of these peaks shift with river discharge.

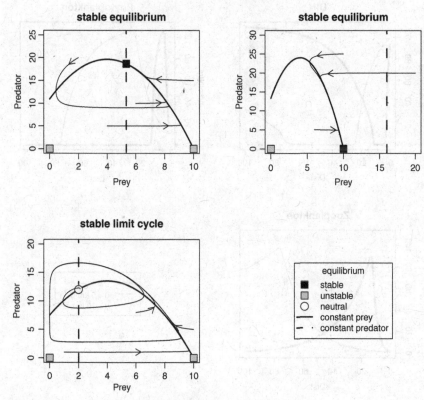

Fig. 7.18 Output corresponding to project 7.9.4

Make a model to investigate the ecosystem response as a function of river flow. The biological component describes Nutrients, Phytoplankton and Zooplankton (NPZ model). The physical component consists of an advective transport term only (it is not an estuary!). The river has constant cross-sectional surface, and is 100 km long. The constitutive equations become:

$$\frac{\partial N}{\partial t} = -u\frac{\partial N}{\partial x} - \mu \cdot \frac{N}{N + k_n}P + g \cdot \frac{P}{P + k_p}Z \cdot (1 - \gamma) + m \cdot Z$$

$$\frac{\partial P}{\partial t} = -u\frac{\partial P}{\partial x} + \mu \cdot \frac{N}{N + k_n}P - g \cdot \frac{P}{P + k_p}Z$$

$$\frac{\partial Z}{\partial t} = -u\frac{\partial Z}{\partial x} + g \cdot \frac{P}{P + k_p}Z \cdot \gamma - m \cdot Z \tag{7.51}$$

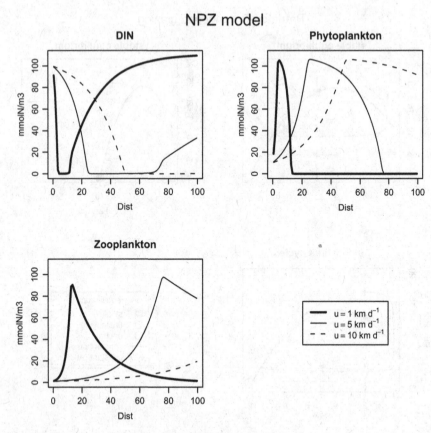

Fig. 7.19 Output corresponding to project 7.9.5

With boundary conditions

$$N_{x=0} = N_0$$

$$P_{x=0} = P_0$$

$$Z_{x=0} = Z_0 \qquad (7.52)$$

Parameter values are:

$\mu = 0.5\,\mathrm{d}^{-1}$, $k_n = 1\,\mathrm{mmol\,N\,m}^{-3}$, $k_p = 1\,\mathrm{mmol\,N\,m}^{-3}$, $g = 0.5\,\mathrm{d}^{-1}$, $m = 0.05\,\mathrm{d}^{-1}$, $\gamma = 0.7$,

and initial conditions $N_0 = 100\,\mathrm{mmol\,m}^{-3}$, $P_0 = 10\,\mathrm{mmol\,m}^{-3}$, $Z_0 = 1\,\mathrm{mmol}$ m^{-3}.

Use the Newton-Raphson method to solve the steady-state condition, and for river flow rates, u of 1, 5, and $10\,\mathrm{km\,d}^{-1}$. Discuss

The results are in Fig. 7.19.

Chapter 8
Multiple Time Scales and Equilibrium Processes

Ecologists and environmental scientists in general, study the environment at various spatial and temporal scales, from micrometers, to square meters to regional or global scales, and from (fractions of a) second, to minutes, to centuries. As temporal and spatial scales change, the nature of key processes also changes. Important processes that need to be represented in great detail at one scale may become less significant at a coarser scale. For instance, in water, turbulence produces more or less homogeneous concentrations at length scales larger than a few centimetres, but at smaller scales, gradients may develop near surfaces that produce or consume the substance. Similarly, what can be considered to stay relatively invariant and is irrelevant at very short time scales may be an important determining process on longer time scales. An example is carbon burial in the sea floor: at a seasonal time scale burial may be but a minor component of the total budget. However, at much longer time scales it is a major factor in the earth's biogeochemistry, because it is a carbon loss term with a very long (geological) recycling time scale.

Not surprisingly, different models are necessary to address similar questions at different scales. Although this section mainly deals with time scales, the temporal and spatial scales are related; long-term processes are generally important at larger spatial scales.

Models that focus on large-scale and long-term processes cannot simply be compiled as the lump-sum of sub-models operating on small temporal and spatial scales. Some kind of simplification is necessary to generate an effective model.

The equilibrium approach is one method to deal with multiple timescales. In many environments, fast and slow reactions occur simultaneously. If the reaction rates differ several orders of magnitude, and if the fast reactions are reversible, it is often possible to assume that the fast reactions are in instantaneous local equilibrium at all times, and only consider the kinetics of the slow reactions. This not only makes the model dynamics simpler but it also avoids spending unnecessary computing time to resolve processes that are not of interest.

We start the chapter with an example that illustrates the equilibrium concept for fast reaction sets. We then add a slower reaction to the model and show how the equilibrium concept can be used to simplify the model. After generalising we give some more examples; finally we present a case study in R.

8.1 Simple Chemical Equilibrium Calculation: Ammonia and Ammonium

Ammonia nitrogen is present in two forms: the ammonium ion (NH_4^+) and ammonia (NH_3). As ammonia can be toxic at sufficiently high levels, it is often desirable to know its concentration.

Ammonium and ammonia are interchangeable via reversible protonation-dissociation reactions, which either consume or produce protons (H^+).

$$NH_4^+ \underset{k^-}{\overset{k^+}{\leftrightarrow}} NH_3 + H^+ \tag{8.1}$$

Rate coefficient k^+ controls the forward (dissociation), k^- the backward (protonation) reaction.

Based on these reactions, and assuming that no other reaction takes place, the rate of change $d[NH_4^+]/dt$ of the concentration of ammonium ($[NH_4^+]$), can be expressed as the rate at which it is produced by protonation minus its consumption rate by dissociation (Section 2.3):

$$\frac{d[NH_4^+]}{dt} = k^- \cdot [NH_3] \cdot [H^+] - k^+ \cdot [NH_4^+]. \tag{8.2}$$

While the rate of change of ammonia ($[NH_3]$) is:

$$\frac{d[NH_3]}{dt} = -k^- \cdot [NH_3] \cdot [H^+] + k^+ \cdot [NH_4^+] \tag{8.3}$$

These protonation-dissociation reactions occur at rates so high that they are often assumed to be instantaneous, such that the concentrations are at steady-state.

At steady-state, the rate of change vanishes giving:

$$\frac{d[NH_4^+]}{dt} = 0, \tag{8.4}$$

$$[NH_3] = \frac{k^+}{k^-} \cdot \frac{[NH_4^+]}{[H^+]} = K_N \cdot \frac{[NH_4^+]}{[H^+]} \tag{8.5}$$

where K_N is the stoichiometric equilibrium constant, expressed in mol (kg solution)$^{-1}$. Its value is $\sim 8 \ 10^{-10}$ and $8 \ 10^{-11} \ mol \ kg^{-1}$ at a temperature of 30°C and 0°C respectively. Note that Eq. (8.5) expresses the well-known equilibrium mass action law of the reaction.

At a certain pH (pH=$-\log_{10}([H^+])$), the contribution of ammonium and of toxic ammonia to the total ammonia concentration (ΣNHx) can be calculated by substituting:

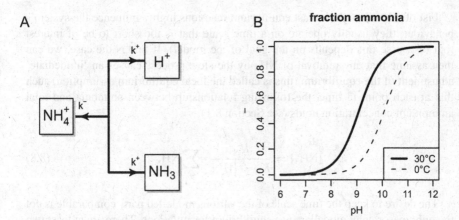

Fig. 8.1 **A**. By dissociation, the ammonium ion produces ammonia and protons. This reversible reaction is so fast, that for most application it is assumed to be at equilibrium. **B**. Ammonia is toxic for many organisms; its relative importance in the aquatic environment increases with pH

$$[NH_3] + [NH_4^+] = \sum NH_x = [NH_4^+] \cdot (\frac{K_N}{[H^+]} + 1)$$

$$[NH_4^+] = \frac{[H^+]}{K_N + [H^+]} \cdot \sum NH_x$$

$$[NH_3] = \frac{K_N}{K_N + [H^+]} \cdot \sum NH_x \tag{8.6}$$

In Fig. 8.1B, the fraction of ammonia as a function of the pH, and for different temperatures is represented.

8.2 Chemical Equilibrium Combined with a Slow Reaction Process

Now assume that, in addition to the fast chemical reactions, another process consumes NH_3 but not NH_4^+, at a first-order rate, with rate-coefficient λ. Compared to the fast consumption and production rates k^- and k^+, the first-order rate λ is several orders of magnitude smaller.

The rate of change of ammonium stays as before (Section 8.1), whereas the rate of change of ammonia now can be written as:

$$\frac{d[NH_3]}{dt} = -k^- \cdot [NH_3] \cdot [H^+] + k^+ \cdot [NH_4^+] - \lambda \cdot [NH_3] \tag{8.7}$$

There are several reasons why it is generally not a good idea to represent all this fine detail in a model.

Fist of all, although the fast equilibrium reactions highly influence the system's behaviour, they usually operate on a time scale that is too short to be of interest (but of course this depends on the goal of the model). If this is the case, we can then assume that any removal of NH_3 by the slow process causes an 'immediate' adjustment of the equilibrium (this is called the local equilibrium assumption) such that at each point in time, the following relationship between ammonia and total ammonium concentration holds (see Section 8.1):

$$[NH_3] = \frac{K_N}{K_N + [H^+]} \cdot \sum NH_x \tag{8.8}$$

The desire to keep the time scale of the various modelled parts comparable is not the only reason why modellers use equilibrium formulations. There are at least two other, equally important reasons: a mathematical and a pragmatic one.

Mathematically, differential equations operating on time scales that differ with several orders of magnitude are referred to as *stiff systems*. These systems are often very hard to solve and require the use of complex solution methods (see Section 6.3.2), making these integrations cumbersome and time-consuming. Therefore, even if we are able to solve the protonation-dissociation reactions dynamically, why waste precious computing time and resources trying to achieve that if we can accurately predict the concentrations based on the local equilibrium assumption?

The other reason is mainly practical: very often the fast process is occurring so fast that its reaction rate cannot even be *measured* with reasonable precision. However, the parameters governing equilibrium are more easily measured. In our example system for instance, it is much simpler to quantify the stoichiometric equilibrium constant K_N, rather than the two rate constants, that govern the reversible reactions (k^+, k^-).

The equilibrium equation above (Eq. 8.8), which was based on the system without the slow reaction, provides a formula that expresses the concentration of ammonia as a function of the total ammonia + ammonium concentration ($\sum NH_x$) at (local) equilibrium, but we still need to derive how this summed quantity $\sum NH_x$ changes in time. Recalling that

$$\frac{d([NH_3] + [NH_4^+])}{dt} = \frac{d[NH_3]}{dt} + \frac{d[NH_4^+]}{dt} = -\lambda \cdot [NH_3] \tag{8.9}$$

We obtain:

$$\frac{d \sum NH_x}{dt} = -\lambda \cdot [NH_3] \tag{8.10}$$

This equation, combined with the equilibrium formulation (Eq. 8.8) fully specifies the model.

8.3 General Approach to Equilibrium Reformulation

We now generalise how the equilibrium re-formulation should be performed. It can be done in four steps.

1. Write the equations of the full dynamic system, assuming that all individual reaction rates are known:

$$\frac{d[NH_3]}{dt} = -k^- \cdot [NH_3] \cdot [H^+] + k^+ \cdot [NH_4^+] - \lambda \cdot [NH_3]$$

$$\frac{d[NH_4^+]}{dt} = k^- \cdot [NH_3] \cdot [H^+] - k^+ \cdot [NH_4^+] \qquad (8.11)$$

2. Compare the magnitude of the reaction rates, to find any reaction sets that are much faster than other reactions. These reactions are assumed to be in instantaneous, local equilibrium. In the example above, these are the protonation-dissociation reactions, involving k^- and k^+ (the shaded area in Fig. 8.2).

3. Choose a suitable summed quantity that is not affected by the fast (equilibrium) reactions. In the example, a suitable summed quantity is $\sum NH_x = [NH_3] + [NH_4^+]$. This now becomes the state variable that will be described. Its rate of change is simply:

$$\frac{d \sum NH_x}{dt} = -\lambda \cdot [NH_3] \qquad (8.12)$$

Thus the fast processes have been removed from the equation.

4. Derive an equilibrium formulation, such that the concentrations of the single species (NH_3 and NH_4^+,) are estimated as a function of the summed quantity ($\sum NH_x$). In order to do so, reduce the full reaction set to the fast reactions only (by removing the slow processes) and set all rates of change equal to zero. Solve the ensuing set of equations by substitution to obtain:

$$[NH_3] = \frac{K_N}{K_N + [H^+]} \cdot \sum NH_x \qquad (8.13)$$

$$[NH_4^+] = \sum NH_x - [NH_3] \qquad (8.14)$$

8.3.1 Enzymatic Equilibrium in a Slow Reaction Process: The Michaelis-Menten Equation (**)

It is somewhat more complex to derive the Michaelis-Menten equation, as an equilibrium formulation of an enzymatic reaction.

Fig. 8.2 Compared to the slow first-order consumption of ammonia (rate λ), the chemical reactions (*shaded area*) occur so fast that they can be considered to operate at quasi-steady-state. This allows to simplify the model significantly (inset)

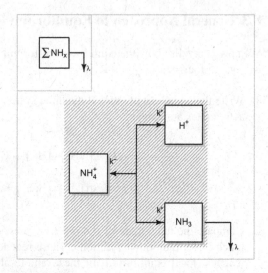

Consider the following reaction: a substance A reacts with enzyme E to form complex EA. This complex either breaks down again into E and A, or it reacts with substance B to form product S, whilst releasing the enzyme.

For the model, we are only interested in the (apparent) reaction of substance A with B and the formation of S (Fig. 8.3, inset). How can this reaction best be formulated?

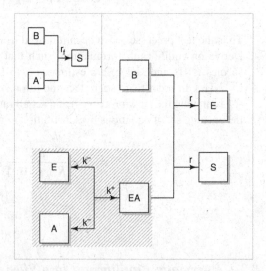

Fig. 8.3 Enzymatic reactions may occur so fast that the production and destruction of the enzyme complex (*in shaded area*) can be considered to operate at quasi-steady-state. This assumption leads to the well-known Michaelis-Menten dynamics (inset)

1. We start by writing the mathematical equations for the full reaction set. These equations are based on the law of mass action (Section 2.3.1), which states that the rate of a chemical reaction involving two reactants is proportional to the

product of their concentration. Thus:

$$\frac{d[A]}{dt} = -k^+ \cdot [A] \cdot [E] + k^- \cdot [EA]$$

$$\frac{d[B]}{dt} = -r \cdot [B] \cdot [EA]$$

$$\frac{d[S]}{dt} = r \cdot [B] \cdot [EA]$$

$$\frac{d[E]}{dt} = r \cdot [B] \cdot [EA] - k^+ \cdot [A] \cdot [E] + k^- \cdot [EA]$$

$$\frac{d[EA]}{dt} = -r \cdot [EA] \cdot [B] + k^+ \cdot [A] \cdot [E] - k^- \cdot [EA] \qquad (8.15)$$

Where r is a second-order reaction rate, units of $[conc]^{-1} \, t^{-1}$ and k^+ and k^- are the second-order, and first-order reaction rates of the reversible enzymatic reaction respectively.

2. The production of the enzyme complex EA (grey area in Fig. 8.3) occurs at a rate much higher than the production of S. Thus we can assume that the consumption of EA through the reaction with B causes a quasi-immediate re-equilibration of EA, by the reversible reaction with A and E.

3. A suitable summed quantity that is not affected by the enzyme kinetics is:

$$\sum A = [A] + [EA] \qquad (8.16)$$

The rate of change of this quantity is:

$$\frac{d \sum A}{dt} = -r \cdot [EA] \cdot [B] \qquad (8.17)$$

which does not include any of the fast reactions.

4. We now consider the reaction set without the slow reactions, and assume local equilibrium (for the fast reactions) by setting the rates of change equal to zero. This set consists of the following three equations, which however all reduce to the same equality:

$$\frac{d[A]}{dt} = -k^+ \cdot [A] \cdot [E] + k^- \cdot [EA] \approx 0$$

$$\frac{d[E]}{dt} = -k^+ \cdot [A] \cdot [E] + k^- \cdot [EA] \approx 0$$

$$\frac{d[EA]}{dt} = +k^+ \cdot [A] \cdot [E] - k^- \cdot [EA] \approx 0 \qquad (8.18)$$

From any of these equations we derive:

$$[EA] \approx \frac{k^+}{k^-} \cdot [A] \times [E]$$

$$\approx \frac{[A]}{K} \cdot \left(\sum E - [EA] \right) \tag{8.19}$$

and

$$[EA] \cdot \left(1 + \frac{[A]}{K} \right) \approx \frac{[A]}{K} \cdot \sum E \tag{8.20}$$

$$[EA] \approx \frac{[A]}{[A] + K} \cdot \sum E \tag{8.21}$$

And where $\sum [E] = [EA] + [E]$ is the total enzyme concentration. Combining Eqs. (8.18) and (8.21) we obtain:

$$\frac{d \sum A}{dt} = -r \cdot \sum E \frac{[A]}{[A] + K} \cdot [B] \tag{8.22}$$

Note that the rate of change of ΣE is zero, i.e. the total concentration of enzyme is constant. Thus, it makes sense to lump this concentration with the rate r from Eq. (8.22). Moreover, we assume that at any moment, the concentration of complex substrate $[EA]$ is small compared to $[A]$ (so that $\sum A \approx [A]$) to arrive finally at:

$$\frac{d[A]}{dt} = -r_f \cdot \frac{[A]}{[A] + K} \cdot [B] \tag{8.23}$$

and where $r_f = r \cdot \sum E$, is the apparent first-order reaction rate (units t^{-1}). Plugging the equilibrium concentration of EA into the formulation for the rate of change of [S] or [B], we obtain:

$$\frac{d[S]}{dt} = r \cdot [B] \frac{[A]}{[A] + K} \cdot \sum E$$

$$\frac{d[S]}{dt} = r_f \cdot \frac{[A]}{[A] + K} \cdot [B]$$

$$\frac{d[B]}{dt} = -r_f \cdot \frac{[A]}{[A] + K} \cdot [B] \tag{8.24}$$

8.3.2 Equilibrium Adsorption in Porous Media (**)

In the sediment, ammonium and phosphate are adsorbed in significant amounts to sediment grains. This adsorption immobilises these substances and therefore needs to be represented in models dealing with the sedimentary N or P cycle. We now derive a model for equilibrium adsorption and diffusion-reaction in porous media.

Fig. 8.4 A diffusing and reacting dissolved substance, C, is adsorbed to sediment grains, S. Assuming equilibrium adsorption, only the dynamics of the dissolved substance needs to be described

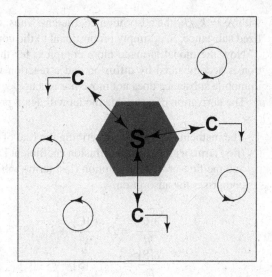

Assume a dissolved substance, which becomes immobilized by adsorption to a solid phase (Fig. 8.4).

Adsorption can be represented by:

$$S \overset{k_1}{\underset{k_2}{\leftrightarrow}} C \qquad (8.25)$$

where S is the immobilized (adsorbed) substance, and C is the dissolved substance, free to diffuse and react. Both substances are expressed as a concentration per unit of bulk sediment (you may want to consult Section 3.4.6 on porous media to refresh your memory on different units). For simplicity, we assume that porosity does not change, neither in time nor in space.

If adsorption reaction kinetics is linear, the rate of change of the concentration of substance C ([C]) and of substance S can be written as:

$$\frac{d[C]}{dt} = k_1 \cdot [S] - k_2 \cdot [C]$$
$$\frac{d[S]}{dt} = -k_1 \cdot [S] + k_2 \cdot [C] \qquad (8.26)$$

which, at equilibrium gives:

$$k_1 \cdot [S] = k_2 \cdot [C] \qquad (8.27)$$

or

$$[S] = K \cdot [C] \qquad (8.28)$$

with $K = k_2/k_1$ the adsorption coefficient. Thus, the concentration of the immobilized substance, S, is simply proportional to the concentration of the free substance.

Now the model is made more complex: for the mobile substance, the adsorption is accompanied by diffusion and a reaction with first-order rate λ, while the immobile substance does not move nor react.

The derivation of a suitable model will again proceed in 4 steps.

1. The mathematical model describing the rate of change of [C] includes diffusion (first term, with sediment diffusion coefficient D_s), adsorption (2nd and 3rd term) and the first-order consumption (last term), while the rate of change of [S] only comprises the adsorption.

$$\frac{\partial [C]}{\partial t} = D_s \frac{\partial^2 [C]}{\partial x^2} + k_1 \cdot [S] - k_2 \cdot [C] - \lambda \cdot [C]$$

$$\frac{\partial [S]}{\partial t} = -k_1 \cdot [S] + k_2 \cdot [C] \qquad (8.29)$$

2. Adsorption is assumed to be in instantaneous, local equilibrium.
3. A suitable summed quantity that is not affected by adsorption is $\sum C = [C]+[S]$, whose rate of change is given by:

$$\frac{\partial \sum C}{\partial t} = \frac{\partial ([C] + [S])}{\partial t} = D_s \frac{\partial^2 [C]}{\partial x^2} - \lambda \cdot [C] \qquad (8.30)$$

this rate of change does not include the fast adsorption terms.
4. We now add the equilibrium formulation $[S] = K \cdot [C]$, and obtain:

$$\frac{\partial ([C] + K \cdot [C])}{\partial t} = (1 + K) \cdot \frac{\partial [C]}{\partial t} = D_s \frac{\partial^2 [C]}{\partial x^2} - \lambda \cdot [C] \qquad (8.31)$$

or

$$\frac{\partial [C]}{\partial t} = D' \frac{\partial^2 [C]}{\partial x^2} - \lambda' \cdot [C] \qquad (8.32)$$

with $D' = \dfrac{D_s}{1 + K}$, the apparent sediment diffusion coefficient and $\lambda' = \dfrac{\lambda}{1 + K}$ the apparent reaction rate.

Form this model equation, it is clear that the effect of the instantaneous adsorption is to slow down the apparent diffusion and reaction. The higher the adsorption, the slower the apparent diffusion and apparent first-order reaction. The factor $1+K$ is often referred to as the retardation factor in soil sciences and hydrology.

8.4 Examples in R

8.4.1 Solving pH in Aquatic Systems

In many aquatic systems, the pH $(= -\log_{10}([H^+]))$ is kept in a narrow range through the action of buffers that release or consume protons (H^+) in water. In natural conditions most of this buffering is due to the dissolved inorganic carbon species: carbon dioxide (CO_2), bicarbonate ion (HCO_3^-) and carbonate ion (CO_3^{2-}).

These substances are subjected to reversible protonation-dissociation reactions, which either consume or produce protons:

- Bicarbonate is formed by dissociation of carbonic acid (H_2CO_3), itself formed by hydration of carbon dioxide:

$$H_2O + CO_2 \underset{k_1^-}{\overset{k_1^+}{\rightleftarrows}} HCO_3^- + H^+ \qquad (\text{reaction A})$$

- The reversible dissociation of bicarbonate to carbonate is given by:

$$HCO_3^- \underset{k_2^-}{\overset{k_2^+}{\rightleftarrows}} CO_3^{2-} + H^+ \qquad (\text{reaction B})$$

(for convenience, we ignore the borate buffer which plays a role in marine waters only, as well as other minor buffer systems).

Similarly as for the ammonium/ammonia model, these acid-base reactions occur sufficiently fast such that they can be considered to operate at equilibrium. The equilibrium equations can be written as:

$$[HCO_3^-] = K_{C1} \cdot \frac{[CO_2]}{[H^+]}$$

$$[CO_3^{2-}] = K_{C2} \cdot \frac{[HCO_3^-]}{[H^+]} \qquad (8.33)$$

Where $[CO_2]$ denotes the concentration of CO_2 and K_{C1} and K_{C2} are the first and second stoichiometric dissociation constant, expressed in $mol\,(kg\,solution)^{-1}$ At a temperature of $25°C$, salinity of 35 and pressure of 1 atm, their values are 0.142 $10^{-5}\,mol\,kg^{-1}$ and $0.12\,10^{-8}\,mol\,kg^{-1}$ respectively.

In these equations, the unknowns are the *four* concentrations:

$$[CO_3^{2-}], [HCO_3^-], [CO_2], [H^+].$$

As there are only *two* equations, we define *two* extra quantities, that are NOT affected by the chemical equilibrium reactions (A and B) but that may be changed by biogeochemical processes.

Thus, we define total dissolved inorganic carbon concentration (DIC) as:

$$DIC = [CO_3^{2-}] + [CO_2] + [HCO_3^-] \tag{8.34}$$

And total alkalinity (TA), which, for our simplified model is defined as:

$$TA = 2[CO_3^{2-}] + [HCO_3^-] - [H^+] \tag{8.35}$$

Note: 'Alkalinity' is a quantity that is commonly measured in aquatic chemistry; however, within the context of this model, it has a subtly different meaning, which is fully determined by the definition given here. 'total alkalinity' in the model thus may not entirely correspond to measured alkalinity in a particular water sample.

It is very important that neither of these quantities (DIC and alkalinity) is affected by the chemical reactions; we will show that this is the case for reaction A; it is similar for the other chemical reaction.

In chemical reaction A, one molecule of DIC (in the form of HCO_3^-) is produced for every molecule of DIC (in the form of CO_2) consumed, and the effect of this reaction on DIC is 0.

For every molecule of bicarbonate (HCO_3^-) produced in reaction A, one proton (H^+) is also produced. Bicarbonate contributes positively, protons contribute negatively to alkalinity, thus the effect of the reaction A on alkalinity is also 0.

Combining these two extra equations with the equilibrium equations, we now have four equations and four unknowns.

It is much more convenient to rewrite the equilibrium concentration of bicarbonate (Eq. 8.36) and of carbonate (Eq. 8.37) as a function of the DIC concentration. This leads to the final 4 equations in the 4 unknowns, which is all we need to solve for the proton concentration (and pH):

$$[HCO_3^-] = \frac{K_{C1} \cdot [H^+]}{[H^+] \cdot [H^+] + K_{C1} \cdot [H^+] + K_{C1} \cdot K_{C2}} \cdot DIC \tag{8.36}$$

$$[CO_3^{2-}] = \frac{K_{C1} \cdot K_{C2}}{[H^+] \cdot [H^+] + K_{C1} \cdot [H^+] + K_{C1} \cdot K_{C2}} \cdot DIC \tag{8.37}$$

$$DIC = [CO_3^{2-}] + [CO_2] + [HCO_3^-] \tag{8.38}$$

$$TA = 2[CO_3^{2-}] + [HCO_3^-] - [H^+] \tag{8.39}$$

Now, in biogeochemical models, both DIC and TA are dynamically described (i.e: their dynamics is expressed as a rate of change), while the pH ($= -\log_{10}([H^+])$) is found at each time step based on equations (Eqs. 8.36–8.39).

Solving for pH can be done in many ways; here is one implementation in R.

The dissociation constants for carbonate (k1, k2) are estimated in R's package seacarb, which has to be loaded first (require).

We then define a function whose root has to be solved (pHfunction). In this function we estimate total alkalinity, based on the guess of pH, the dissociation constants ($k1$, $k2$) and the DIC concentration. The difference of this calculated alkalinity (EstimatedAlk) with the modelled (true) alkalinity is then returned. If pH is correctly estimated, then modelled and estimated alkalinity will be equal, and their difference will be zero. This means that, to find the pH, we need to find the root of this function (the value of pH for which the function value is 0).

Note that the conversion from pH to [H^+] gives the proton concentration in mol kg^{-1}. As the concentrations of the other substances are in μmol kg^{-1}, we convert using a factor 10^6.

It is simplest (but not most efficient) to use R's function uniroot to solve for the root and thus find the pH for a given DIC and total alkalinity concentration. We restrict the region of the pH root in between 0 and 12 (which is more than large enough), and we set the tolerance (tol) to a very small number to increase precision.

```
require(seacarb)
Salinity    <- 0
Temperature <- 20
WDepth      <- 0

k1 <- K1(Salinity,Temperature,WDepth)     # Carbonate k1
k2 <- K2(Salinity,Temperature,WDepth)     # Carbonate k2

pHfunction <- function(pH, k1,k2, DIC, Alkalinity )
{
    H    <- 10^(-pH)
    HCO3 <- H*k1 /(H*(k1+H)  + k1*k2)*DIC
    CO3  <- k1*k2 /(H*(k1+H) + k1*k2)*DIC

    EstimatedAlk  <- (- H) *1.e6  + HCO3 + 2*CO3

    return(EstimatedAlk  - Alkalinity)

}

Alkalinity <- 2200
DIC        <- 2100

sol <- uniroot(pHfunction,lower=0,upper=12, tol=1.e-20,
                k1=k1, k2=k2, DIC=DIC, Alkalinity=Alkalinity)

sol$root
```

8.4.2 A Model of pH Changes Due to Algal Growth (**)

If we want to model the effect of biogeochemical processes on pH, we have to dynamically describe DIC and total alkalinity, and at each time step solve the equilibrium equations for pH.

As an example, we amend an algal growth model with a description for pH.

During growth, algae take up DIC and produce alkalinity. We assume that the algae grow on nitrate, such that the stoichiometry of the reaction is given by (where γ^N is the N:C ratio of algae):

$$CO_2 + \gamma^N NO^-_3 + \gamma^N H^+ + (1 + \gamma^N)H_2O \;-> (CH_2O)(NH_3)_{\gamma N} + (1 + 2\gamma^N)O_2$$

For one mole of algal carbon produced, one mole of DIC (in the form of CO_2) is consumed. Moreover, γ^N moles of protons (H^+) are also consumed. As protons contribute negatively to total alkalinity, their consumption is equivalent to alkalinity production.

In the example below, we model Algae, dissolved inorganic nitrogen (DIN), dissolved inorganic carbon (DIC) and alkalinity as state variables. The algal concentration and DIN is expressed in mmol N m^{-3}. The algae are reared in a closed vessel (batch culture), and their growth is limited by DIN availability (Monod formulation) and light availability (Monod formulation). The light is switched on and off at regular intervals: during one day, algae reside for 12 h in the dark, 12 h in the light.

In addition to the rates of changes, the model also returns the photosynthetically active radiation (PAR), and the pH.

```
model<-function(time,state,parameters)
  {
with(as.list(c(state,parameters)),{

    PAR     <- 0.
    if(time%%24 < dayLength) PAR <- parDay    # switch on light

    Growth <- maxGrowth*DIN/(DIN+ksDIN)*PAR/(PAR+ksPAR)*ALGAE -
              respRate * ALGAE

    dDIN    <- -Growth                    # DIN is consumed
    dDIC    <- -Growth * CNratio          # DIC is consumed ~ CN ratio
    dALGAE <- Growth                      # algae increase by growth
    dALKALINITY <- Growth                 # Alkalinity is produced

    # estimate the pH
    pH <- uniroot(pHfunction,lower=0,upper=12,tol=1.e-20,
         k1=k1,k2=k2, DIC=DIC,Alkalinity=ALKALINITY)$root

    list(c(dDIN,dALGAE,dALKALINITY,dDIC),c(PAR=PAR,pH=pH) )

   })

  }
```

After specifying the model parameters, and estimating the dissociation constants, the model is run for 10 days, producing hourly output (`times`). Finally the output is plotted. In the first plot, the night is indicated by a grey area.

```
# the model parameters:

parms <- c(maxGrowth  =0.125,      #molN/molN/hr
           ksPAR      =100,        #µEinst/m2/s
           ksDIN      =1.0,        #mmolN/m3
           respRate   =0.001,      #/h
           CNratio    =6.5,        #molC/molN
           parDay     =250.,       #µEinst/m2/s
           dayLength  =12.         #hours
           )

Salinity     <- 0
Temperature  <- 20
WDepth       <- 0

# the dissociation constants
k1          <- K1(Salinity,Temperature,WDepth)      # Carbonate k1
k2          <- K2(Salinity,Temperature,WDepth)      # Carbonate k2

#-----------------------#
# the initial conditions: #
#-----------------------#

state    <-c(DIN        =30,       #mmolN/m3
             ALGAE      =0.1,      #mmolN/m3
             ALKALINITY =2200,     #mmol/m3
             DIC        =2100)     #mmolC/m3

#--------------------#
# RUNNING the model:  #
#--------------------#

times <-seq(0,24*10,1)
require(deSolve)
out   <-as.data.frame(ode(state,times,model,parms))

#-----------------------#
# PLOTTING model output: #
#-----------------------#

par(mfrow=c(2,2),  oma=c(0,0,3,0))
plot (out$time,out$ALGAE,type="l",main="Algae",xlab="time, hours",
      ylab="mmol/m3")
polygon(out$time,out$PAR-10,col="lightgrey",border=NA)
box()
lines (out$time,out$ALGAE ,lwd=2 )
```

```
plot (out$time,out$DIN ,type="l",main="DIN" ,xlab="time, hours",
      ylab="mmolN/m3",lwd=2)
plot (out$time,out$DIC ,type="l",main="DIC" ,xlab="time, hours",
      ylab="mmolC/m3",lwd=2)
plot (out$time,out$pH,type="l",main="pH" ,xlab="time, hours",
      ylab="-",lwd=2)

mtext(outer=TRUE,side=3,"Algal growth and pH",cex=1.5)
```

The solution after running the model for 10 days is given in Fig. 8.5.

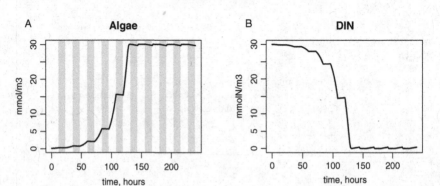

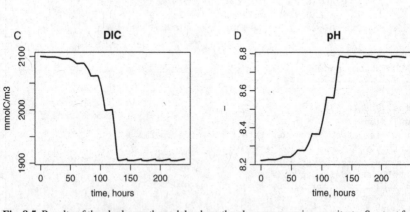

Fig. 8.5 Results of the algal growth model, where the algae are growing on nitrate. See text for the R-code to generate this figure

Chapter 9
Discrete Time Models

Time is an independent variable of dynamic models that can be represented in two fundamentally different ways (Fig. 9.1).

- It can be represented as a *continuum* and the dependent variables (state variables, output variables) assigned a value for each time-point.
- Alternatively, it can be broken up into a sequence of distinct intervals, forming a *discrete* independent variable and the dependent variables assigned a single value for each entire interval.

Continuous time models naturally give way to *differential equations*, which express the change of a variable in time as a continuous function of sources and sinks. Except for the lattice and cellular automaton models of Chapter 3, all the models discussed till now were continuous time models.

In contrast, discrete time models represent time in fixed steps. The mathematical representation of such a model is as a *difference equation*. Discrete time models are mainly used in population dynamics, where time for instance is considered to jump from one generation to the next, from one stage to the next, or from one age class to the next. It is then assumed that one snapshot every generation, stage or age class gives a good representation of population dynamics.

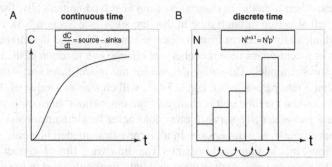

Fig. 9.1 Time can be represented in two fundamentally different ways: continuous (**A**) or discrete (**B**). The mathematical model is then either represented as a differential (**A**) or as a difference (**B**) equation

K. Soetaert, P.M.J. Herman, *A Practical Guide to Ecological Modelling*,
© Springer Science+Business Media B.V. 2009

In this chapter we start by deriving several discrete equivalents of the most simple population model that uses logistic population dynamics. Then we deal with more elaborate models that explicitly include species interactions (host-parasitoids) or life history phenomena (age-structured models). In the latter type of models, matrix algebra is used to estimate the rate of population growth or decline, the population's stable age structure, and their reproductive value.

Matrix multiplication and eigenvalue analysis are the main tools required for this chapter. You may consult the appendices to refresh your knowledge on the subject.

9.1 Difference Equations

Difference equations are a set of update rules that describe the state of the system at time t+1, as a function of the state at time t.

Assume we have a model in which we describe one single variable, the density of a population of animals. The update rule of density from one year to another can be written as a function of recruitment and survival of the animals.

$$\text{DENSITY}_{t+1} = recruitment + survival \cdot \text{DENSITY}_t \qquad (9.1)$$

Here *recruitment* and *survival* are parameters; *recruitment*, which is the number of individuals born during the time step between t and $t + 1$, has the same units as the state variable DENSITY. Parameter *survival*, the probability of survival in the time step is a dimensionless quantity.

Remark that the value of both parameters, *recruitment* and *survival*, depends on the time unit chosen. In the above definition, a time step always has the duration of one time unit. If we were to double the time step (and therefore the time unit), then the value of *recruitment* would double, whereas the probability to survive, i.e. the value of *survival* would decrease. This dependence of parameters on the chosen time frame is a fundamental characteristic of discrete time models. The parameters are then called *finite rate parameters*, which sets them apart from the *instantaneous rate parameters* of the continuous time models.

Note that, when a model in continuous time is solved numerically, the solving algorithm will also step through time in discrete steps (see Section 6.3). However, this is fundamentally different from a discrete time model: the instantaneous rate parameters of a continuous time model do not change as a function of the time *step* chosen to solve a model. They may depend on the model time *unit*, but only in the sense that a rate parameter of, e.g. $0.5\,\text{d}^{-1}$, will change to a value of $0.5/24 = 0.020833\,\text{h}^{-1}$ when the time unit is changed from day to hour. In a continuous-time model, a rate parameter like survival rate could never be a dimensionless quantity. It should have time^{-1} as a dimension. In a discrete-time model, however, as in the example above, this is quite well possible. The dimension time^{-1} cannot occur in such models, because time is not explicit and only implicitly used in the definition of one time step.

9.2 Discrete Logistic Models

In Section 2.7.1 we discussed the logistic model of population growth (Verhulst, 1838). The equation that describes density dependence in a single-species population is:

$$\frac{dN}{dt} = r \cdot \left(1 - \frac{N}{K}\right) \cdot N \qquad (9.2)$$

Whereas the logistic model (Eq. 9.2) is formulated as a continuous equation, its discrete analogues are more famous.

The logistic equation can be approximated in several ways:

- First, rewrite eq. 9.2 as a simple exponential growth formula:

$$\frac{dN}{dt} = \gamma \cdot N$$

$$\gamma = r \cdot \left(1 - \frac{N}{K}\right) \qquad (9.3)$$

Where the quantity γ can be considered the density-dependent reproductive rate of the population (it is called 'density-dependent' because it changes with the values of N). Now, assuming that γ can be considered constant over a time step, we obtain the Ricker equation (1954) by integration:

$$N_{t+1} = N_t \cdot e^{\gamma \cdot 1} \qquad (9.4)$$

$$N_{t+1} = N_t \cdot e^{r \cdot \left(1 - \frac{N}{K}\right)} \qquad (9.5)$$

Which can also be written as:

$$N_{t+1} = N_t \cdot e^r \cdot e^{-r \frac{N_t}{K}}$$

$$N_{t+1} = b \cdot N_t \cdot e^{-cN_t} \qquad (9.6)$$

- Another way to discretise the equation is to write the differential, i.e. the left hand side of Eq. (9.2) in a discrete way:

$$\frac{\Delta N}{\Delta t} = r \cdot N \cdot \left(1 - \frac{N}{K}\right)$$

$$N_{t+\Delta t} - N_t = r \cdot N_t \cdot \left(1 - \frac{N_t}{K}\right) \Delta t \qquad (9.7)$$

rearranging, we obtain:

$$N_{t+1} = N_t \cdot (1+r) - \frac{r \cdot (N_t)^2}{K} \qquad (9.8)$$

for $\Delta t = 1$.

- Mathematicians often rescale this difference equation:

$$x_t = \frac{r}{1+r} \cdot \frac{N_t}{K}$$
and
$$\mu = 1 + r \qquad (9.9)$$

And obtain another equation, the logistic map:

$$x_{t+1} = \mu \cdot x_t \cdot (1 - x_t) \qquad (9.10)$$

9.3 Host-Parasitoid Interactions.

Discrete time models are ideally suited to describe the population dynamics of insects, which are characterised by discrete generations. The modelling of the interactions between a host and its parasitoid was initiated by Nicholson and Bailey (1935). Here we investigate a slightly more complex model (Rogers, 1972), but we rewrite the equations such that they are compatible with the remainder of this book.

An adult female parasitoid searches for a host on which to deposit her eggs. After hatching, the larval parasitoids grow by consuming their host, and eventually killing it before they pupate (Fig. 9.2).

The next generation of parasitoids emerges from hosts that have been parasitized, whilst hosts free of parasitoids will produce their own progeny. The fraction of hosts parasitized depends on the rate of encounter of the two species, and thus on both their densities.

If we denote the fraction of hosts NOT parasitized as $f(H_t, P_t)$, we can write a general formulation for the dynamics of hosts (H) and parasitoid (P) as:

$$H_{t+1} = \lambda_t \cdot H_t \cdot f(H_t, P_t)$$
$$P_{t+1} = c \cdot H_t \cdot [1 - f(H_t, P_t)] \qquad (9.11)$$

and where λ_t is the reproductive rate of the host, and c is the number of eggs deposited on the host. Now we make following assumptions:

- Similar as in previous section, we adopt for the host's growth rate:

$$\lambda_t = \exp\left(r \cdot \left(1 - \frac{H_t}{K}\right)\right) \qquad (9.12)$$

where K is the carrying capacity of the host

Fig. 9.2 Schematic representation of the host-parasitoid interactions

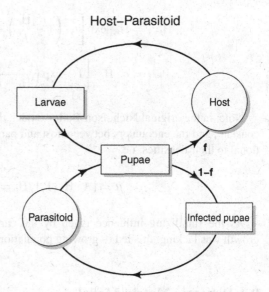

- The egg deposition by parasitoids on the host can be modelled as a Poisson distribution (see Appendix D.3).
- Only the first encounter of parasitoid with the host is significant; a second encounter will not increase the number of eggs hatched from the host.

The Poisson distribution describes the probability that n events will occur in a time interval:

$$p(n) = \frac{e^{-\mu}\mu^n}{n!} \tag{9.13}$$

The likelihood for the host to escape, i.e. to experience zero encounters is then given by:

$$f(H_t, P_t) = p(0) = \frac{e^{-\mu}\mu^0}{0!} = e^{-\mu} \tag{9.14}$$

where μ is the *average* number of hosts parasitized in the time interval. We still need to specify this number. Hereto we assume that the total number of hosts parasitized obeys a Monod-like response. Recalling that the average number = total number of hosts parasitized / host density, we can now write for μ:

$$\mu = \left[A \cdot \frac{H_t}{H_t + k_s} \cdot P_t \right] / H_t = \frac{A}{H_t + k_s} \cdot P_t \tag{9.15}$$

Combining all, the discrete-time dynamics of host (H) and parasitoid (P) is given by:

$$H_{t+1} = H_t \cdot \exp\left[r \cdot \left(1 - \frac{H_t}{K}\right) - \frac{A}{H_t + k_s} \cdot P_t\right]$$

$$P_{t+1} = c \cdot H_t \cdot \left[1 - \exp\left(\frac{A}{H_t + k_s} \cdot P_t\right)\right] \tag{9.16}$$

Note: in the original Nicholson-Bailey model, the growth rate was assumed to be constant, and the encounters between host and parasite were assumed to be proportional to their densities, i.e.

$$\mu = [A \cdot H_t \cdot P_t] / H_t = A \cdot P_t \tag{9.17}$$

As the stabilizing influence given by the carrying capacity in the parasitoid growth was lacking, this led to growing population oscillations in the 2 species.

9.4 Dynamic Matrix Models

In dynamic matrix models the linear dynamics are represented as a matrix equation. Here we will contend ourselves by giving some examples where we use matrices to solve population models that describe individuals of different ages, stages or size structures.

9.4.1 Example: Age Structured Population Model

Leslie matrix models are a special case of discrete time models where a population is divided into age classes, which have the same length as the time step. We will consider a population model consisting of long-lived individuals with overlapping generations. This means that individuals may delay reproduction for one or more years and they may survive after reproduction to reproduce again.

Assume a species with following characteristics: individuals start breeding at the age of 3 years, after which they breed once a year. They live to a maximum of 5 years. Age-specific fecundity and survival parameters are density-independent; they are constant through time.

The duration of an age class is equal to the time step. Thus, individuals in age class i at time $t + 1$, belonged to age class $i - 1$ at the previous time step t. They will enter age class i at time $t + 1$ only if they survived during the previous interval – this is given by their finite survival probability $s_{i-1,i}$

[Number of individuals in age class i at time $t + 1$] = [Number of individuals in age class $i - 1$ at time t] * [fraction surviving from $i - 1$ to i]

$$N_i^{t+1} = N_{i-1}^t \cdot s_{i-1,i} \tag{9.18}$$

The number of offspring (which have age 0) is determined by the fecundity of individuals in class 3–5 within one year: one individual of age class 3 generates f_3 offspring, etc...

[Number of individuals in age class 0 *at time t*] $= N_0^t$

$$N_0^t = N_3^t \cdot f_3 + N_4^t \cdot f_4 + N_5^t \cdot f_5 \qquad (9.19)$$

Rather than explicitly dealing with age 0, we can combine both the fecundity of individuals in age class 3 – 5 with the survival from age class 0 – 1, to determine the number of individuals surviving to the first age class at *time t+1*:

$$N_1^{t+1} = N_0^t \cdot s_{0,1} = \left(N_3^t \cdot f_3 + N_4^t \cdot f_4 + N_5^t \cdot f_5 \right) \cdot s_{0,1} \qquad (9.20)$$

This gives a model reduced with one age class. The life-cycle graph for both the original and the reduced model is depicted in Fig. 9.3. The original model is also referred to as a post-breeding model, because N represents the abundance of each age class directly after reproduction. The reduced model is referred to as a pre-breeding model because here N represents the abundance just before reproduction.

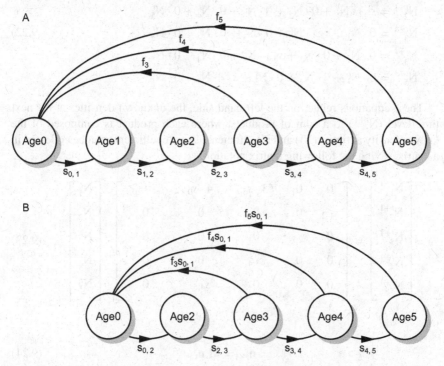

Fig. 9.3 An age-structured population model: **A.** six age classes with their transition probabilities $s_{i,i+1}$ and fecundities f_i. **B.** The age-specific fecundities and survival of the 0th age class have been combined and a reduced model composed of five age classes is obtained

In populations where the maximum age is not very strict, i.e. some animals may survive for quite a long period, it can be useful to represent in the highest age class all animals of that age *and older*. In our example, if some animals would survive beyond the age of 5, age class N_5 would represent all animals aged 5 and older. Doing so can reduce the number of age classes, while maintaining sufficient details about those age classes that matter most.

Obviously, if the highest age class also includes all animals older than the nominal age, one must build into the model a way for the animals to survive. Thus, the dynamic equation for the oldest age class (here 5 as an example) becomes:

$$N_5^{t+1} = s_{4,5} \cdot N_4^t + s_{5,5} \cdot N_5^t \tag{9.21}$$

9.4.2 Matrix Notation

The difference equations of the reduced, pre-breeding version of the population model can be written in a more general way as:

$$
\begin{aligned}
N_1^{t+1} &= 0 \cdot N_1^t + 0 \cdot N_2^t + f_3 \cdot s_{0,1} \cdot N_3^t + f_4 \cdot s_{0,1} \cdot N_4^t + f_5 \cdot s_{0,1} \cdot N_5^t \\
N_2^{t+1} &= s_{1,2} \cdot N_1^t + 0 \cdot N_2^t + 0 \cdot N_3^t + 0 \cdot N_4^t + 0 \cdot N_5^t \\
N_3^{t+1} &= 0 \cdot N_1^t + s_{2,3} \cdot N_2^t + 0 \cdot N_3^t + 0 \cdot N_4^t + 0 \cdot N_5^t \\
N_4^{t+1} &= 0 \cdot N_1^t + 0 \cdot N_2^t + s_{3,4} \cdot N_3^t + 0 \cdot N_4^t + 0 \cdot N_5^t \\
N_5^{t+1} &= 0 \cdot N_1^t + 0 \cdot N_2^t + 0 \cdot N_3^t + s_{4,5} \cdot N_4^t + 0 \cdot N_5^t
\end{aligned}
\tag{9.22}
$$

These equations relate, on the left hand side, the unknown densities at the next time level (N_i^{t+1}) to a sum of products, where each product is composed of the known density at time t (N_i^t) and a coefficient. These coefficients can be collected in a matrix, giving the following matrix notation:

$$
\begin{bmatrix} N_1^{t+1} \\ N_2^{t+1} \\ N_3^{t+1} \\ N_4^{t+1} \\ N_5^{t+1} \end{bmatrix}
=
\begin{bmatrix}
0 & 0 & f_3 \cdot s_{0,1} & f_4 \cdot s_{0,1} & f_5 \cdot s_{0,1} \\
s_{1,2} & 0 & 0 & 0 & 0 \\
0 & s_{2,3} & 0 & 0 & 0 \\
0 & 0 & s_{3,4} & 0 & 0 \\
0 & 0 & 0 & s_{4,5} & 0
\end{bmatrix}
\cdot
\begin{bmatrix} N_1^t \\ N_2^t \\ N_3^t \\ N_4^t \\ N_5^t \end{bmatrix}
\tag{9.23}
$$

or

$$\mathbf{n_{t+1}} = \mathbf{M} . \mathbf{n_t} \tag{9.24}$$

where $\mathbf{n_t}$ and $\mathbf{n_{t+1}}$ are the population structure vectors at time t and t+1 respectively, and $\mathbf{M}$ is a matrix with the fecundity and survival values.

You can find a short reminder of how matrix multiplication works in Appendix C.

To solve the model for a certain time period, it suffices to perform the matrix multiplication a number of times. For instance, with the initial population density given (n_0), the population structure after 5 years is given by:

$$n_5 = M.n4 = M.(M.n3) = M.M.M.M.M.n_0 = M^5.n_0 \qquad (9.25)$$

In the next chapters we discuss three interesting characteristics from such a population model:

(1) the finite rate of increase λ, the rate at which the population as a whole increases
(2) the stable age distribution
(3) the reproductive value

9.4.3 Stable Age Distribution and Rate of Increase

As the fecundity and survival parameters are density independent, such model generally either leads to population explosion or decline when applied for sufficiently long time, similarly as the exponential growth model (Section 5.3.1). It is interesting to estimate the ultimate rate of increase (or decrease) of the population and the age distribution to which it converges. We can estimate these quantities in two ways: by brute force or by mathematical techniques.

- The *brute force technique* to estimate these quantities consists in performing successive matrix multiplications $n_{t+1} = M.n_t$, rescaling the age distribution after each step, and repeating this till the age distribution stops changing. Rescaling consists simply of dividing all elements of the vector n by the sum of vector n, such that, after rescaling, the sum of all these elements equals 1. When the age distribution vector stops changing, the rescaled vector n_{t+1}, contains the stable age distribution and the rate at which total population density increases is the rate of increase. This rate of increase can be assessed as follows: If $N = \sum_i N_i$ is total density (not rescaled), then the difference model can be written as: $N^{t+1} = N^t \cdot \lambda$ from which the finite rate of increase, λ can simply be assessed as the proportion between total density at two successive time steps.
- In *mathematical* terms, we want to find a *value* λ and a vector n_t^* (with the stable age distribution) such that:

$$M.n_t^* = \lambda.n_t^* \qquad (9.26)$$

- As λ is a value, the terms in n_t^* will be changed in proportion at the next time step. Thus their ratio over time (n_{t+1}^*/n_t^*) will stay constant such that n_t^* is the stable age distribution.

Mathematically, λ is known as the dominant root or eigenvalue of the matrix M and $\mathbf{n}_t^*$ is the corresponding eigenvector.

Note: in general a p*p matrix has p eigenvalues, some of which are complex numbers, i.e. they have a real and imaginary part. The dominant eigenvalue is the one with the largest magnitude of the real part.

Thus, to find the finite rate of increase and the stable age distribution of a population model it suffices to find the dominant eigenvalue and corresponding eigenvector.

9.4.4 The Reproductive Value

For a certain age class, the reproductive value is proportional to the number of offspring that an animal of this age is expected to produce during the remainder of its lifetime. It is a relative measure, as the reproductive value of newborns is defined equal to 1. Reproductive value is affected by survival and by fecundity. For a newborn, it reflects the probability to survive until reproductive age, and the expected number of offspring produced once reproductive age is attained. Therefore, reproductive value can be larger than 1, because newborns may have a very low probability to survive until reproductive age, but this probability increases with age as an animal survives longer. Reproductive value generally peaks at the age of first reproduction.

For management purposes, this is an important quantity as it marks the age that is most valuable in terms of future reproduction. In view of sustainable use of resources it is best to harvest individuals with lower reproductive value, whereas translocation of individuals to new territory will be more effective if they have reached their highest reproductive value. Similar as for the rate of increase and stable age distribution, the reproductive value can be derived by brute force and mathematical techniques.

The brute force technique consists of performing successive matrix multiplications of populations with different initial conditions. Consider four different populations, where the life cycle consists of four life stages. The first population initially comprises one individual in the first stage, no individuals in the other stages; the second population initially has one individual only in the second stage, and so on. Now each population steps through time until the stable age distribution has been reached. At this point in time, all populations will increase at the same rate. However, their total densities will not necessarily be the same, but reflect differences in reproduction capacity of the original life stage. Scaling the total densities with the density of stage 1 gives the reproductive value of each stage.

Mathematically speaking, the 'reproductive value' is the left-hand eigenvector of the matrix. Why this is so, is rather complex to explain, so we will NOT derive this formulation here – however, both methods of computation are compared in Section 9.5.4; this should convince you that the left-hand eigenvector indeed extracts the desired quantity.

9.5 Case Studies in R

9.5.1 Bifurcations in the Discrete Logistic Model

All the discrete approximations to the logistic model, discussed in Section 9.2 are famous for the complex behaviour that they can generate.

For the Ricker equation, the number of equilibrium conditions periodically doubles as r increases and eventually the model becomes chaotic. The doubling can be summarised as a bifurcation diagram, where the equilibrium condition is plotted as a function of increasing rates of natural increase (r).

Figure 9.4 shows the bifurcation diagram for the Ricker equation, which was generated in the following R-script, where we have assumed that $K = 1$.

```
ricker   <- function(N,r) N*exp(r*(1-N))
rseq     <- seq(1.5,4,0.001) # sequence of r-values

plot(0,0,xlim=range(rseq),ylim=c(0,5),type="n",
     xlab="r",ylab="Nt",main="discrete logistic model")

for ( r in rseq)
 {
  N  <- runif(1)
   for (i in 1:200) N <- ricker(N,r) # spinup
for (i in 1:200){N <- ricker(N,r)
                points(r,N,pch=".",cex=1.5)}
 }
```

Thus, for a sequence of r-values (rseq), the Ricker model is first iterated 200 times, starting with a random value of N (runif(1) generates one uniformly distributed random value between 0 and 1); then the next 200 iterates are added as points to the bifurcation diagram; cex=1.5 enlarges the size of these points with 50%.

9.5.2 Bifurcations in the Host-Parasitoid Model

We reconsider the two-species model that describes the interactions between a host (H) and its parasitoid (P) from Section 9.3.

The discrete-time dynamics were given by:

$$H_{t+1} = H_t \cdot \exp\left[r \cdot \left(1 - \frac{H_t}{K}\right) - \frac{A}{k_s + H_t} \cdot P_t\right]$$

$$P_{t+1} = H_t \cdot \left[1 - \exp\left(\frac{A}{k_s + H_t} \cdot P_t\right)\right] \tag{9.27}$$

discrete logistic model

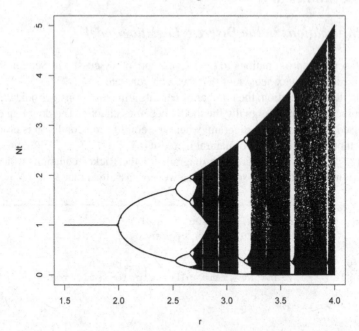

Fig. 9.4 Bifurcation diagram of the Ricker logistic model. See text for R-code to generate this figure

The behaviour of this model was investigated by Kaitala et al. (2000) and it is a particularly nice illustration of the complexities that a simple model can generate.

The bifurcation diagram can be created in a similar way as in the previous example. However, as the dynamics now comprise two state variables and the equations are more complex, the iterations are performed in a function ('Parasite') that takes as input a vector with parasite and host densities at time t and returns the updated densities, as a vector.

```
rH <- 2.82    # rate of increase
A  <- 100     # attack rate
ks <- 1       # half-saturation density

Parasite <- function(P_H,ks)
{
P   <-P_H[1] ; H <- P_H[2]
f   <- A*P/(ks+H)

return(c(H*(1-exp(-f)),
         H * exp(rH*(1-H)-f)))
}
```

We start by running the model for two cases, $k_s = 22$ and $k_s = 20$ (Fig. 9.5 A,B).

```
plottraject<-function(ks)
{
out <- matrix(nrow=50,ncol=2,NA)
P_H <- c(0.5,0.5)
for (i in 1:100) P_H <-Parasite(P_H,ks)
for (i in 1:50) {P_H <-Parasite(P_H,ks); out[i,]<-P_H }

plot (out[,1],type="l",ylim=range(out),lwd=2,xlab="t",
       ylab="Population", main=paste("ks=",ks))
lines(out[,2],lty=2)
}
plottraject(25)
plottraject(20)
```

Note that while the model shows a four-timestep periodicity for $k_s = 25$, it displays chaotic behaviour for k_s values equal to 20.

To make the bifurcation diagram, function Parasite is called several times, first during the spin-up phase and then during the plotting phase. The bifurcation diagram is created with the same initial conditions.

```
ksSeq <- seq(15,35,0.05) # sequence of ks-values
plot(0,0,xlim=range(ksSeq),ylim=c(0.,2),xlab="ks",
     ylab="Nt",main="Bifurcation diagram")

for ( ks in ksSeq)
{
P_H <- c(0.5,0.5)
for (i in 1:500) P_H <-Parasite(P_H,ks) # spinup
for (i in 1:300) {
                  P_H <-Parasite(P_H,ks)
                  points(ks,P_H[2],pch=".",cex=1.5)
                  }
}
```

9.5.3 Attractors in the Host-Parasitoid Model

Finally, as in the paper by Kaitala et al. (2000), a closer look is taken at the region around $k_s = 23.09$.

Here, there are two alternative attractors, with period 4 and 16 respectively. Moreover, to which attractor the simulation converges depends on the initial conditions of host and predators. That such strange behaviour may lead to complex fractal patterns

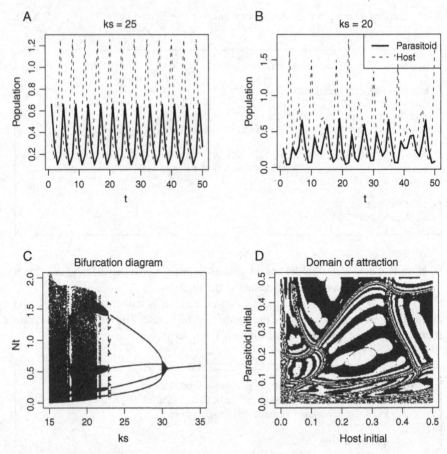

Fig. 9.5 The host-parasitoid model. **A,B**. Two trajectories for half-saturation constants of 25 and 20. **C**. Bifurcation diagram, **D**. Domain of attraction. See Sections 9.5.2 and 9.5.3 for the R-code to generate these figures

becomes clear when we run the next R-script, where we vary the initial condition of host and parasites between 0.002 and 0.5 (xlim,ylim).

The model is run for all possible combinations of initial host and parasite densities, outlined on a grid with mesh size 0.002; R-function expand.grid creates these combinations. After an initialization phase, 20 model output steps are kept, and the frequency in this time series assessed. Note that double precision numbers are rarely equal, so the calculation of the number of 'unique' values is done after truncating the number to 1 significant digit (trunc(PP*10)). Finally a small rectangle (size 0.002) is drawn on the plot, and coloured black if the attractor has frequency four, white otherwise. Depending on the mesh-size, this simulation may take quite some time! Results can be found in Fig. 9.5 D.

```
ks    <- 23.09
dz    <- 0.002
xlim <- c(0.002,0.5)
ylim <- c(0.002,0.5)

Initial <- expand.grid(P = seq(xlim[1],xlim[2],dz),
                       H = seq(ylim[1],ylim[2],dz))
plot(0,0,xlim=xlim,ylim=ylim,ylab="Parasitoid initial",
     xlab="Host initial",
     type="n",main="Domain of attraction")

PP    <- vector(length=20)

for ( ii in 1:nrow(Initial))
{
ini <- Initial[ii,]
P_H <- unlist(ini)
for (i in 1:100) P_H<-Parasite (P_H,ks)   # spinup
for (i in 1:20) {P_H <-Parasite(P_H,ks);
                  PP[i] <- P_H[1]}

Freq <- length(unique(trunc(PP*10)))
ifelse (Freq == 4,col<-"black",col<-"white")
rect(ini$P-dz/2,ini$H-dz/2,
     ini$P+dz/2,ini$H+dz/2,col=col,border=col)
}
```

9.5.4 Population Dynamics of Teasel

We now discuss the implementation in R of a matrix model that describes the population dynamics of teasel (*Dipsacus sylvestris*), a European perennial weed. Its life cycle can be described by six stages (Caswell, 2001; Fig. 9.6): two stages of dormant seeds (DS), small, medium and large rosettes (R) and flowering plants (F).

We start by inputting the model data in R, assigning names to the stages and entering the transition probabilities (per year) into a matrix called M. Note that we enter the transition matrix M row-wise. As this is contrary to the default storage of matrix values in R (which is column-wise), we have to specify this explicitly (byrow=TRUE).

```
Stagenames <-c("dormant seeds 1yr","dormant seeds 2yr",
"rosettes<2.5cm","rosettes2.5-18.9cm","rosettes>19cm",
"flowering plants")

NStages <- length(Stagenames)

M <- matrix(nrow=NStages,ncol=NStages,byrow=TRUE,data =
c(
    0,      0,      0,      0,      0,      322.38,
    0.966,  0,      0,      0,      0,      0     ,
    0.013,  0.01,   0.125,  0,      0,      3.448 ,
    0.007,  0,      0.125,  0.238,  0,      30.170,
    0.008,  0,      0,      0.245,  0.167,  0.862 ,
    0,      0,      0,      0.023,  0.75,   0    ) )

rownames(M) <- Stagenames
colnames(M) <- 1:NStages
M
```

Fig. 9.6 Schematic representation of the population dynamics model of Dipsacus sylvestris, a perennial weed

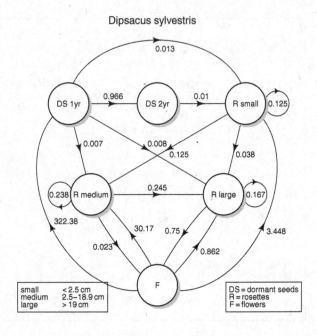

The last statement prints matrix M to the screen:

```
                             1      2      3      4      5       6

dormant seeds 1yr       0.000  0.00  0.000  0.000  0.000  322.380
dormant seeds 2yr       0.966  0.00  0.000  0.000  0.000    0.000
rosettes<2.5cm          0.013  0.01  0.125  0.000  0.000    3.448
rosettes2.5-18.9cm      0.007  0.00  0.125  0.238  0.000   30.170
rosettes>19cm           0.008  0.00  0.000  0.245  0.167    0.862
flowering plants        0.000  0.00  0.000  0.023  0.750    0.000
```

From this transition matrix, we see that the probability of transition in one year between the two types of dormant seeds (from stage 1 to 2) is 0.966. In one year, one flowering plant (column 6) produces several seeds, 323 of which become dormant (row 1), and 3.5 of which germinate and produce small rosettes (row3). All flowering plants die (the probability of survival, M[6,6] = 0).

9.5.4.1 The Rate of Increase and the Stable Stage Distribution

To estimate the finite rate of increase and the stable stage distribution, the model is run for 50 steps, and the model output stored in a matrix called 'out'. We create this matrix first.

The initial population vector (Population) consists of 100 small dormant seeds. This initial condition forms the first row of our output matrix. Then, for each step the population vector is replaced by the matrix multiplication (%*% in R) between the transition matrix M and the population vector. The new population structure is added, as a row, to the output vector.

```
numsteps        <- 50
Population      <- rep(0,times=NStages)
Population[1]   <- 100
out             <- matrix(nrow=numsteps+1,ncol=NStages,data=0)
out[1,]         <- Population

for (i in 1:numsteps)
{
 Population <- M %*%Population
 out[i+1,]  <- t(Population)
}
```

At each time step, the relative stage distribution can be estimated by dividing stage densities by the summed densities. R-function rowSums makes the sums for each row. We can then visualise how the stage structure changes in time using R-function matplot.

```
p    <-  out/rowSums(out)
matplot(p,type="l",lty=1:6,lwd=2,main="stage dist",col="black")
legend("topright",legend=Stagenames,lty=1:6,lwd=2)
```

After an initial period where the stage distribution changes wildly, it converges to the stable stage distribution (Fig. 9.7).

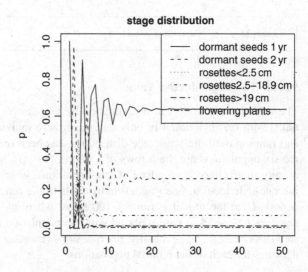

Fig. 9.7 Relative stage distribution versus time for the population dynamics model of *Dipsacus sylvestris*; see text for R-code to generate this figure

The last row of p now contains the final relative stage distribution, while the ratio of total density in out gives an estimate of the finite rate of increase:

```
p[nrow(p),]
sum(out[50,])/sum(out[49,])
```

the output looks like this:

```
[1]  0.63771 0.26395 0.01215 0.06931 0.01224 0.00462

[1]  2.333881
```

To *mathematically* estimate the stable stage distribution, and the intrinsic rate of increase r, we calculate the eigenvalues and eigenvectors of the transition matrix M.

R's function eigen calculates eigenvalues (and sorts them, largest first) and the right eigenvectors; function Re() takes the real part. The stable stage distribution is estimated as follows:

```
e <-eigen(M)
Stablestage <- Re(e$vectors[,1])
(Stablestage <- Stablestage/sum(Stablestage))
```

Note that the stable stage distribution is standardized to the total density (sum (StableStage)). By bracketing the statement, it is executed and printed to screen.

The rate of increase is the largest (real part of the) eigenvalue:

```
(lambda    <- Re(e$values[1]))
```

9.5.4.2 The Reproductive Value

The reproductive value can be estimated by propagating 6 populations, each initiated with one individual in only one stage, zero individuals at the other stages, and running until the stable age distribution has been reached. In the code below, the six populations are the 6 rows of a matrix (rPop). R-statement diag creates a unity matrix (see Appendix C for a definition), which is equivalent to assigning one individual to one stage, 0 individuals to the others, and this for all stages (rows). Then the model is run for 100 steps, where at each time the population densities (rPop[j,]) are replaced with the population densities at the next time step (M%*%rPop[j,]). Finally, the row sums (rowSums) give the ultimate total densities for each of the 6 initial populations.

Standardised on the value of the first stage (`rr[1]`), we obtain the reproductive value.

```
rPop      <- diag(nrow=NStages,ncol=NStages)

for (i in 1:100)
   for (j in 1:NStages) rPop [j,] <- M %*% rPop [j,]

rr<-rowSums(rPop)
rr/rr[1]
```

```
[1]   1.00000000   0.00924182 2.14583970 37.71322106
252.63301061 725.85847839
```

The reproductive value can also be calculated from the left eigenvector of matrix M, which is just the right eigenvector of the transpose of M. Note that this value is a-dimensionalised against the first left eigenvector value, such that the reproductive value of the first stage $= 1$.

```
lefte      <- eigen(t(M))
reprodVal <-Re(lefte$vectors[,1])
reprodVal <-reprodVal/reprodVal[1]
```

Finally, all is written to the screen:

```
lambda
data.frame("stage"=Stagenames,"Stable"=Stablestage,
"Rep rod"=reprodVal)
```

The output looks like:

```
[1]  2.32188
```

And, for the stable stage distribution and reproductive value:

```
                 stage        Stable            Reprod
1   dormant seeds 1yr 0.636901968   1.00000000
2   dormant seeds 2yr 0.264978062   0.00924182
3       rosettes<2.5cm 0.012174560   2.14583970
4 rosettes2.5-18.9cm 0.069281759  37.71322106
5       rosettes>19cm 0.012076487 252.63301062
6    flowering plants 0.004587164 725.85847839
```

You may check that this is (almost) the same as the estimates obtained by the brute force methods.

9.6 Projects

9.6.1 Bifurcation of the Logistic Map

Given the bifurcation example from Section 9.5.1, it should be simple to draw the bifurcation diagram for the logistic map (Section 9.2).

9.6.2 Equilibrium Dynamics of a Simple Age-Class Matrix Model

Investigate the steady-state (equilibrium) dynamics of the following, simple model describing a population of long-lived individuals with 4 age classes (from Gotelli, 2001) (See Fig. 9.8). Start by inputting the transition matrix. Then use matrix algebra (eigenvalues, eigenvectors) to estimate the finite rate of increase and stable age distribution and reproductive value.

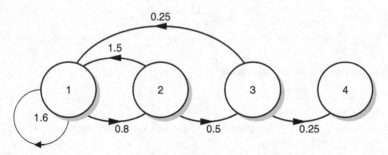

Fig. 9.8 Schematic representation of the age class model of Section 9.6.2

9.6.3 Equilibrium Dynamics of a US Population

The fecundity and transition probabilities for the US population in 1966 are given for 5-year age classes in Fig. 9.9. The example is from Keyfitz and Flieger (1971), in Caswell (2001). If you are surprised to see a reproductive contribution by the 5–10 year old age class, we share this surprise with you, but we take the data as they are. After all, these were the wild sixties!

Estimate the finite rate of increase and the stable age distribution and reproductive value of each age class.

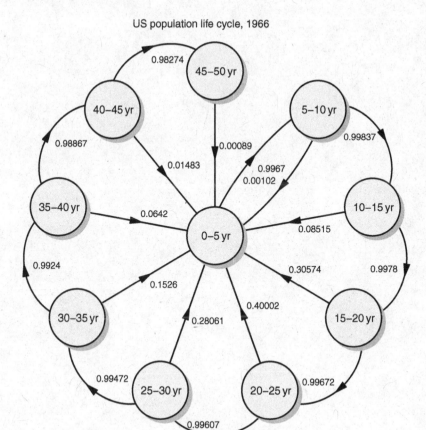

Fig. 9.9 Schematic representation of the US population life cycle from 1966

Answer: rate of increase $= 1.049753$, age class characteristics:

```
   class Stable age Reproductive value
1    2.5 0.12519582 1.0000000000
2    7.5 0.11886860 1.0532286982
3   12.5 0.11305025 1.1064134855
4   17.5 0.10745531 1.0786840286
5   22.5 0.10202672 0.8293320512
6   27.5 0.09680920 0.4724304967
7   32.5 0.09173400 0.2164683044
8   37.5 0.08672214 0.0752098562
9   42.5 0.08167595 0.0149208285
10  47.5 0.07646200 0.0008478185
```

Chapter 10
Dynamic Programming

Throughout their lifetime, individual animals make choices which affect their survival or successful reproduction. Examples include: where, when and how often to feed, when to migrate and the path to follow, whether to allocate food to growth or to reserves etc. All these choices depend on external conditions (food availability, predation risk, temperature, . . .), but also on internal conditions, such as the developmental stage, or the energy reserves of the animal.

Ecologists argue that, as evolution has shaped the organisms for a long time, they will behave optimally, i.e. they function such that some measure of 'fitness' is optimised. This fitness can for instance be ultimate reproductive output, or survival rate until reproductive age. Typically, fitness is the end result of a sequence of decisions in time. A single decision can therefore not be evaluated in isolation, but should be framed into this sequence.

Finding the optimal set of decisions is the field of dynamic programming applications. In ecology, these applications allow us to reconstruct how and why organisms behave/function as they do in their natural environments, assuming that they do it optimally.

Formulating a dynamic programming model starts by specifying different states that an organism can be in at a particular moment in time. This state can e.g. be characterized by the organism's weight, health etc. It further specifies the different decisions the organism can make, and attaches to each decision a reward (positive contribution to fitness) and a cost (negative contribution to fitness). The optimal set of actions is then compiled by stepping backwards in time. This is, we start with the final conditions and their associated fitness, and we try to find out how to get there.

The book by Clark and Mangel (2000) is entirely devoted to the application of dynamic programming in ecology. We start with a rudimentary version of their simplest example in order to explain how to construct and solve dynamic programming problems. We then end with the full example which we also implement in R.

10.1 Sequential Decisions

This chapter essentially deals with making decisions one after another. The combination of all the decisions made is called a *policy*, and the quality of the policy can be measured by the ultimate *fitness* reached.

A sequential decision problem is composed of the following components:

- The problem is divided into a number of *steps* and a *decision* is taken at each step. Choosing a decision generates a *reward* and/or incurs a *cost*.
- At each step, the condition of the organisms is represented by a *state*. There are a finite number of possible *states*. The decision *transforms* one state into a (usually different) state at the next step.

Consider an individual that has a choice out of two patches of land, one for foraging, and one for reproduction. Feeding increases the biomass of the organism, whilst reproduction decreases it. There is a minimal biomass threshold below which the organism dies. There is also a maximal biomass the organism can attain. Dynamic programming is used to estimate optimal patch selection that maximises the organism's total reproductive success over 20 steps.

In this problem, the *states* correspond to all biomass classes (denoted by X) between the minimal and maximal biomass. One *step* is one day during which the animals forage or reproduce. The *decision* is which patch to visit. The *cost* is the metabolic cost and the reproductive cost if the animal visits the reproductive patch, whilst the *gain* is the increase in biomass if the animal feeds.

By taking into account the costs and gains for the patches, the biomass at the new time step ($X(T+1)$) can be estimated as:

$$\boxed{X(T+1) = X(T) + \text{Gain} - \text{Cost}} \qquad (10.1)$$

Thus, if the organism visits the feeding patch:

$$X(T + 1)_{\text{feed}} = X(T) + \text{Feeding gain} - \text{Metabolic cost} \qquad (10.2)$$

Whilst, for animals visiting the reproductive patch:

$$X(T + 1)_{\text{repr}} = X(T) - \text{Reprod. cost} - \text{Metabolic cost} \qquad (10.3)$$

10.2 Finding the Optimal Solution

At any time step, the animal consults its current condition (biomass) and, based on that, decides what to do next. Because the animal's behaviour has been shaped by natural selection, it will take the action that maximises the ultimate (long-term) fitness. This is not trivial: an act that maximises the short-term reward is not necessarily the best on the long-term! The aim of dynamic programming is exactly to find the combination of actions which *maximises* the ultimate fitness of these organisms.

Theoretically, the best set of actions that optimises fitness at the end of the model run can be found in several ways:

- One, not very bright, solution is to start with arbitrary values of states (biomasses), and propagating the model through time, at each time step taking arbitrary decisions, and, after 20 steps, calculating final fitness. If we repeat this infinitely often, we can select the set of actions that leads to the maximal fitness. Clearly, this is not very efficient.
- Another, equally dumb, strategy might be to calculate fitness for every possible combination of actions and then selecting the trajectory which gives maximal fitness after 20 steps. However, even the smallest of problems gives rise to an enormous number of possible strategies, such that the problem is very hard to solve, even on the most powerful computer. Note that for both forward strategies, it is not possible to decide what the optimal strategy is, based solely on the immediate reward: the ultimate fitness reward of the trajectory can only be calculated at the very end.
- In dynamic programming applications, one starts with the possible final conditions and tries to find out how to get there. Thus, the model works backward, solving for each previous time step one after another. This backward recursion is much more effective than forward recursion, although the latter may seem more natural. (Note that the term 'dynamic programming' refers to the method of solution, it is not connected to programming at all).

From these three solution methods, dynamic programming is by far the most efficient way to find the optimal solution of sequential decision models. However, its major difficulty is to define the *final conditions* of the model, i.e. the fitness values associated to any state that the organism can eventually reach.

10.3 A Simple Example

We will apply this algorithm to a simple patch selection model, where at each time step an animal can choose between a visit to the feeding patch, or a visit to the reproductive patch. The fitness to maximise is the total reproductive capacity. The settings of this simple example are as follows:

- the fitness increase of visiting the reproductive patch is 1; there is no immediate fitness gain in visiting the feeding patch.
- the metabolic cost of an animal during its visit to either the feeding or reproductive patch in a time step is given the value 1
- feeding gain while visiting the feeding patch is 2
- reproductive cost, while visiting the reproductive patch, is 1
- no feeding is possible while visiting the reproductive patch
- a visit to the feeding patch is more dangerous than a visit to the reproductive patch. The probability to survive a visit to the feeding patch is 0.9, while it is 0.95 for the reproductive patch

- animals die when their biomass goes below a value of 5
- at the end of the simulation, all animals die (this is a semelparous population) and thus their final fitness, independent of their biomass, is zero

The characteristics of the two patches are summarised in the table below:

	Feeding Patch	Reproductive Patch
Probability of survival	0.9	0.95
Metabolic Cost	1	1
Reproductive Cost	0	1
Feeding gain	2	0
Reproductive (fitness) gain	0	4

The dynamic programming algorithm works as follows (Fig. 10.1):

- We start by specifying the final fitness for all possible states X at the end, i.e. after the last time step. We denote this final fitness as $f[X, T_{end}]$. It is very simple, since all animals die at this moment: it is zero for all X:

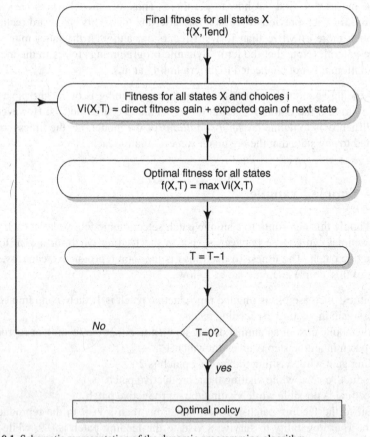

Fig. 10.1 Schematic representation of the dynamic programming algorithm

$$f[X, T_{end}] = 0 \text{ for all } X$$

- During the last time step, $T_{end} - 1$, all animals will obviously try to visit the reproductive patch if they can afford to do so: they have fitness to gain by repro-ducing, but not by feeding, and they will die after this step anyway. An animal visiting the reproductive patch has a metabolic cost of 1, and a reproductive cost of 1. It can only afford to visit the patch if its biomass is 7 or higher, otherwise it will die (as its biomass will have decreased below the threshold of 5) during its visit to the reproductive patch (in which case we assume no reproduction has taken place). Formally, we estimate the fitness value for all biomass classes X at time $T=T_{end} - 1$, and for all possible choices. We denote the fitness value for choice i (i=1..n) of this organism as $V_i(X,T)$. This fitness value is the sum of immediate fitness gain and the expected (optimal) fitness of the next state.

$V_i(X, T) =$ immediate fitness gain $+$ expected optimal fitness of next state

$$(10.4)$$

The expected optimal fitness of the next state is the fitness of this state, calcu-lated earlier, and weighted for the animal's probability to survive until that next state. Thus, if we denote immediate fitness gain in decision i as G_i, expected biomass at the next state in decision i as $X(T+1)_i$, optimal fitness in the next state of this biomass as $f(X(T+1),T+1)$, and probability of survival in decision i as p_i, we have:

$$V_i(X, T) = G_i + p_i \cdot f[X(T + 1), T + 1] + (1 - p_i) \cdot 0$$
$$V_i(X, T) = G_i + p_i \cdot f[X(T + 1), T + 1] \qquad (10.5)$$

(Note: we multiply $(1-p_i)$ with 0 as the fitness of a dead organism is clearly 0).

In our simplified example, there are only two possible choices. $V_1(X,T)$ is the fitness value for biomass class X at time T when selecting the food patch, and $V_2(X,T)$ the fitness value when visiting the reproductive patch.

In the former case, there is no immediate fitness gain and the fitness value at $T_{end} - 1$ is:

$$V_1(X, T_{end} - 1) = p_1 \cdot f[X(T_{end})_1, T_{end}] \qquad (10.6)$$

Now, $X(T_{end})_1$, the biomass at the next time step is simply the current biomass $(X(T_{end} - 1))$ plus gains minus losses:

$$V_1(X, T_{end} - 1) = p_1 \cdot f[X(T_{end} - 1) + \text{Feeding gain - Metabolic cost}, T_{end}] \qquad (10.7)$$

In our example, feeding gain is 2 and metabolic cost is 1. An animal thus gains one biomass unit when it feeds: $X(T+1)_{feed}=X(T)+1$ and consequently,

$$V_1(X, T_{end} - 1) = 0.9 \cdot f[X + 1, T_{end}]$$

but since $f[X,T_{end}]=0$ for all X, $V_1(X,T_{end}-1)=0$ for all X

In the second case, reproduction is the immediate fitness reward. This reproduction fitness gain G_2 has the value 1 for all $X\geq7$, and the value 0 for smaller X. Biomass $X(T+1)$ under this decision is given as $X(T)$-Reproductive cost-Metabolic cost. Both costs have the value 1, thus biomass will decrease by 2.

$$V_2(X, T_{end} - 1) = G_2 + p_2 \cdot f[X(T_{end})_2, T_{end}]$$
$$V_2(X, T_{end} - 1) = 1 + 0.95 \cdot f[X(T_{end} - 1) - 2, T_{end}] \text{ for } X \geq 7$$
$$V_2(X, T_{end} - 1) = 0 + 0.95 \cdot f[X(T_{end} - 1) - 2, T_{end}] \text{ for } X < 7 \qquad (10.8)$$

Since $f[X,T_{end}]=0$ for all X, $V_2(X,T_{end} - 1)$ can either be 1 (for larger organisms) or 0.

- If we now pick the patch with the highest fitness, we have the optimal choice, i.e. whether to feed or to reproduce, and the optimal fitness (f) at time step $T_{end} - 1$ for biomass X:

$$f[X, T_{end} - 1] = \max\left(V_1(X, T_{end} - 1), V_2(X, T_{end} - 1)\right) \cdot \qquad (10.9)$$

The optimal choice for all animals with $X\geq7$ will be to reproduce, and they will have fitness 1. Animals with smaller biomass only have the choice to feed and have fitness 0.

- At this point in the procedure, we have the values for optimal fitness of all states X at time $T_{end} - 1$. We can now repeat the above procedure for time step $T_{end} - 2$. The resulting values for V_1 and V_2 are:

$$V_1(X, T_{end} - 2) = 0.9 \cdot f[X + 1, T_{end} - 1] = 0.9 \qquad \text{for } X \geq 6$$
$$V_1(X, T_{end} - 2) = 0 \qquad \text{for } X < 6$$
$$V_2(X, T_{end} - 2) = 1 + 0.95 \cdot f[X - 2, T_{end} - 1] = 1.95 \quad \text{for } X \geq 9$$
$$V_2(X, T_{end} - 2) = 1 \qquad \text{for } 7 \leq X < 9$$
$$V_2(X, T_{end} - 2) = 0 \qquad \text{for } x < 7$$

Animals with biomass smaller than 7 cannot reproduce and have to feed. All other animals maximize their fitness by choosing the reproductive patch. Taking $f[X,T]$ as the maximum of V_1 and V_2 we find:

$$f[X, T_{end} - 2] = 1.95 \qquad \text{for } X \geq 9$$
$$f[X, T_{end} - 2] = 1 \qquad \text{for } 7 \leq X < 9$$
$$f[X, T_{end} - 2] = 0.9 \qquad \text{for } X = 6$$
$$f[X, T_{end} - 2] = 0 \qquad \text{for } X < 6$$

- Stepping further backwards, we find with similar reasoning that the optimal fitness values become:

$$f[X, T_{end} - 3] = 2.825 \qquad \text{for } X \geq 11$$
$$f[X, T_{end} - 3] = 1.95 \qquad \text{for } 9 \leq X < 11$$
$$f[X, T_{end} - 3] = 1.855 \qquad \text{for } X = 8$$
$$f[X, T_{end} - 3] = 1 \qquad \text{for } X = 7$$
$$f[X, T_{end} - 3] = 0.9 \qquad \text{for } X = 6$$
$$f[X, T_{end} - 3] = 0 \qquad \text{for } X = 5$$

All animals still choose to reproduce if their biomass is at least 7 at this step

- An interesting change occurs, however, at the next backward step in time, for animals of biomass value 7. If we examine V_1 and V_2 for X=7 at this time step, we find:

$$V_1(7, T_{end} - 4) = 0 + 0.9 \cdot f[8, T_{end} - 3] = 0.9 \cdot 1.855 = 1.6695$$
$$V_2(7, T_{end} - 4) = 1 + 0.95 \cdot f[5, T_{end} - 3] = 1 + 0 = 1$$

and, in contrast to the (later) steps discussed previously, animals of biomass 7 choose to feed and not to reproduce at this time step. This is also true at all prior time steps for animals of this biomass. The reproduction they miss in the actual time step is more than compensated for by the higher fitness associated with an increase in biomass.

Figure 10.2A illustrates the optimal choices as a function of biomass and time in the example discussed. Superimposed on the figure is a contour plot of fitness of the individuals.

It is interesting to compare this situation with the model output in Fig. 10.2B. This output is produced with exactly the same model and parameter settings, except that here we model an iteroparous population. At T_{end}, the fitness is proportional (in fact: equal in numbers) to their biomass. It can be seen that in this case animals will spend some time in reproducing in the first part of the simulation, but will then switch to feeding so as to secure their future fitness. The comparison of both cases demonstrates how important the quantification of fitness at the end of the simulation is for the result.

10.4 Case Study in R: The Patch-Selection Model

The previous example has shown that, for a small problem, it may be possible to perform the calculations by hand, but the clutter is increasing rapidly as the simulation proceeds. Much clarity is to be gained by writing a computer programme, which is what we will do next, based on a more complex example from the book of Clark and Mangel (2000).

The organism has a choice out of three patches of land to forage or reproduce. Each patch has a different predation risk (or probability of survival), and different energy reserve requirements. Two patches are used for foraging, with different probability

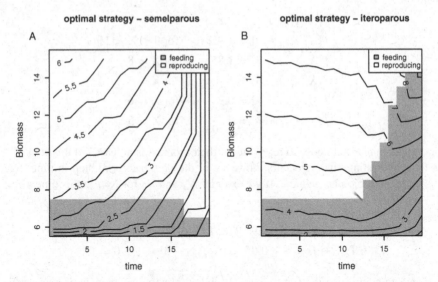

Fig. 10.2 Results of the simple patch selection model discussed in the text. **A.** Semelparous population in which all individuals have zero fitness at the end of the simulation. **B.** Iteroparous population where individuals have a fitness numerically equal to their biomass at the end. Shading indicates the optimal decision for either the feeding or reproduction patch, as a function of biomass and time. Superimposed contours indicate the fitness value of individuals. See text for parameter settings and other details of the model

of finding food, and the food is of different quality (i.e. the gain obtained through feeding differs). One other patch can be used for reproduction, but offers no food.

There is a minimal energy threshold (x_crit) below which the organism dies. There is also a maximal energy capacity (x_max), i.e. the organism can not become too fat. We consider all organisms between the minimal and maximal biomasses.

By choosing, at each time step, the best patch, organisms maximize their total reproductive success.

Dynamic programming is used to estimate optimal patch selection that maximises the organism's total reproductive success over 20 days and for the different size classes.

Clark and Mangel (2000) assume that the animal's biomass is restricted between 0 and 30 (units are not specified), and there are 31 biomass classes (0, 1, 2, 3, . . .30). Reproduction can only proceed above a critical biomass (x_rep=4).

At any time step t, animals have the choice of visiting 3 different patches, each with different probability of surviving and finding food and each with different costs, feeding and reproductive (fitness) gains. The parameter settings for each of these patches are given in the table below:

		Patch 1	Patch 2	Patch 3
Probability of survival	psurvive	0.99	0.95	0.98
Probability of feeding	pfood	0.2	0.5	0.0
Cost	cost	1	1	1
Feeding gain	Gain	2	4	0
Reproductive (fitness) gain	Repr	0	0	4

From these daily probabilities, costs and gains, it is clear that patches 1 and 2 are used to feed, but not to reproduce. The third patch is the reproductive patch: its probability to find food is 0, but it has nonzero probability to reproduce. Food patch 2 has higher probability of finding food than patch 1, but the mortality risk is also larger (or the probability of survival is smaller).

We now discuss how to solve this model together with its implementation into R. We first define the model parameters.

The model is expressed in terms of biomass classes (x_class) ranging between a minimal (x_crit) and maximal value (x_max). Reproduction starts at a certain biomass (x_rep). The model is run for 20 time steps (times) ; the last time step is the final condition.

```
x_crit  <- 0              # critical mass to survive
x_max   <- 30             # maximal mass
x_rep   <- 4              # critical mass for reproduction
x_class <- x_crit:x_max   # biomass classes
nmass   <- length(x_class) # number of mass classes

t_max   <- 20             # number of time steps
times   <- 1:(t_max-1)
```

Note the use of brackets surrounding t_max-1 when we define the time sequence. This is necessary, as the statement '1:t_max-1' is interpreted by R as (1:t_max)-1, or 0:(t_max-1), which is not the same.

Animals can visit 3 different patches (npatch), each with different probability of survival (psurvive), finding food (pfood), and with different costs (cost), feeding gains (gain), and reproduction gains (repr).

```
npatch    <- 3                    # number of patches

psurvive  <- c(0.99,0.95,0.98)    # probability of surviving
pfood     <- c(0.2 ,0.5 ,0 )      # probability of feeding
cost      <- c(1 ,1 ,1 )          # cost of a patch
feedgain  <- c(2 ,4 ,0 )          # gain of feeding
repr      <- c(0 ,0 ,4 )          # max reproduction
```

During the dynamic programming operation, two matrices will be filled containing the results. For each time step and biomass class they contain the optimal fitness values (f) and the optimal decision (bestpatch). These matrices are initialized as 0. One vector (V) records the fitness at the current time step and for each patch.

```
f         <- matrix(nrow=t_max,ncol=nmass ,0)    # optimal fitness
bestpatch <- matrix(nrow=t_max-1,ncol=nmass-1,0) # best patch
V         <- vector(length=npatch)               # current fitness
```

Biomasses range from 0 (x_crit) to 30 (x_max), and increase with one for each class. In R, matrices cannot start from an offset other than 1. Thus, f[1,t] will contain the fitness for individuals with biomass 0 at time t, whilst f[31,t] will contain the fitness of the last biomass class. As we may also need to calculate fitness of animals with biomasses beyond 0 and 30, we make a function that performs the correct cut-offs and retrieves the correct fitness value, for a biomass x and time t, from f:

```
fitness <- function(x,t)

{
xx <- pmin(x ,x_max)
xx <- pmax(xx,x_crit)
fitness <- f[t,xx-x_crit+1]
}
```

Note the use of pmin and pmax rather than min and max. In contrast to min and max which return the maximum or minimum of all the values present in their arguments, pmin and pmax can take several vectors (or matrices) as arguments. Shorter arguments are recycled until they reach the same length as the longer arguments. The functions return a single vector containing the 'parallel' maxima (or minima) of the vectors. Thus, in the statement xx <- pmin(x ,x_max) every value of the vector x is compared with x_max, and the vector xx will be equal to the vector x, except that all values of x which exceed x_max will have been replaced by x_max.

As dynamic programming models run backwards, the 'initial condition' is the final value, here the fitness, at t_max.

For this model, Clark and Mangel (2000) assume that the final fitness increases with biomass, according to a Monod-like function (Section 2.5.3) with half-saturation coefficient kx and maximal fitness 60. The statement ' f[,t_max] <-' changes all elements in the last column of f at once.

```
# final fitness, at t_max
fend       <- 60
kx         <- 0.25*x_max

f[t_max,]  <- fend*(x_class-x_crit)/(x_class-x_crit+kx)
```

We finally perform the main optimization loop, running backward in time (R's statement 'rev' reverses a sequence), estimating the new fitness associated to each patch and for each biomass class, and selecting the optimal decision.

The fitness value for the three patches, V, is calculated as the sum of the immediate gain and the expected fitness of the next state.

- If the animal reproduces, there is an immediate gain in fitness (dfit). An animal can only reproduce in those patches where repr is > 0, and when its biomass is higher than the critical biomass x_rep.

  The R-statement 'dfit<-pmax(0, pmin(x-x_rep,repr))' estimates the immediate fitness gain by reproduction in the three patches. It will return zero for animals whose biomass x is lower than the critical reproduction weight, x_rep; for animals that are heavier than x_rep, the fitness gain will be proportional to the excess biomass (x-x_rep), but not exceeding a maximal gain, repr.

- The expected fitness of the states (biomasses) at the next time step was calculated before (remember: we run backwards). We still need to calculate which biomass the animal will have at the next time step, given the decision that it takes. The different decisions have different losses and gains:

- There is a cost incurred in visiting all patches (cost).

- Reproduction consumes energy and leads to a loss of biomass (dfit, calculated above).

- When an animal moves to a patch, it may find food (at least, when pfood for that patch is >0), but it is never certain that this will also happen. However, it is not the animal's decision whether it will actually find food or not, but a chance process. In case it finds food (with probability pfood), its biomass in the next time step will equal x-cost-dfit+gain. In case it does not find food (this is also possible when it visits a food patch, namely with probability 1-pfood), its next biomass will be x-cost-dfit. To obtain the fitness consequences of the new biomass at time t, we make use of the previously calculated and already stored data for fitness at t+1. The *expected* fitness of an animal visiting a patch is then given by pfood*fitness(animal which has fed,t+1) + (1-pfood)*fitness(animal which has not fed, t+1).

- An animal is never certain it will survive. When it dies during a time step, its future fitness will not add to the total fitness of its actual step. As dying is not a decision of the animal either, we calculate again the *expected* fitness of an animal when entering one of the three patches. In principle, this is done in the same way as we did before for the probability of finding food. However, future fitness is zero when the animal does not survive, and so we can drop the term (1-psurvive)*0 After calculating the fitness for each patch (V), the highest fitness (f[]) and optimal patch (bestpatch) are saved. All these calculations require only little amount of code in R.

```
for (t in rev(times))                    # backward in time
{
 for (x in x_class[-1])                   # for each class, except x-crit
 {
  dfit <- pmax(0,pmin(x-x_rep,repr))      # reproduction
  expectgain <- psurvive*( pfood   *fitness(x-cost+feedgain-dfit,t+1) +
                          (1-pfood)*fitness(x-cost-dfit ,t+1) )
  V     <- dfit + expectgain
  V[expectgain == 0] <- 0                 # dead
  f[t,x-x_crit+1]           <- max(V)      # optimal fitness
  bestpatch[t,x-x_crit] <- which.max(V)    # best patch

 }                                        # next biomass class x

}                                         # next time t
```

optimal patch

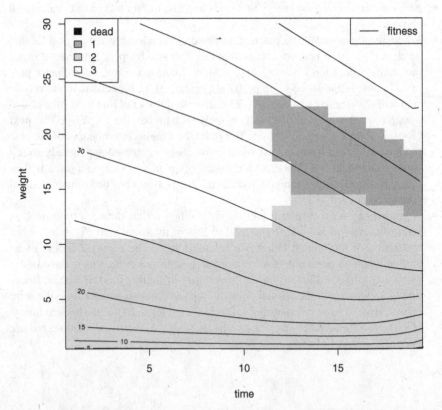

Fig. 10.3 Optimal patch selection for animals of different weight, and offered the choice between three different patches. The animals try to optimize the ultimate reproduction after 20 steps. *Grey shading* indicates the optimal patch to choose for an animal of a certain biomass (*on y axis*) at a certain time (*on x axis*). Line contours give the optimal fitness for these animals. See text for parameter settings and for the R-code to generate this output

The optimal patch selection (in matrix `bestpatch`) is finally depicted as a 2-D image, and a legend added. It is simple to also add optimal fitness values, for each time step and biomass class, as a contour plot.

```
par(mfrow=c(1,1))

image(x=times,y=x_class[-
1],z=bestpatch,ylab="weight",xlab="time",zlim=c(0,3),
      main="optimal patch",col=c("black","darkgrey","lightgrey","white"))
box()
legend("topleft",fill=c("black","darkgrey","lightgrey","white"),
legend=c("de ad","1","2","3"))
contour(x=1:t_max,y=x_class,z=f,add=TRUE)
legend("topright",legend="fitness",lty=1)
```

The results are depicted in Fig. 10.3.

When biomass is sufficiently low, the optimal strategy is to feed in food patch 2. Although the predatory mortality is highest here (the probability of survival is the lowest), this patch has the highest feeding gain. As the biomass increases, either patch 1 or patch 3, the reproductive patch, is visited.

The careful balancing of risk and reward is a characteristic of sequential decision processes.

Chapter 11
Testing and Validating the Model

By now we have the necessary background to transform an ecological problem in mathematical form, to solve the resulting equations and to investigate their stability properties. How can we be sure that our creation is meaningful?

There is no easy way to be sure. Models that are conceptually wrong or that are not carefully solved will produce results, albeit bad ones. This is not different from other scientific work. A badly designed or badly performed experiment will also yield results, but one will not learn anything useful from it.

Very early in the book, we already introduced the technique of checking whether the model equations make sense, i.e. whether they are dimensionally correct and whether they conserve energy and mass (Section 2.1.3).

Here we outline a number of other principles and tools that can be used to check the correctness of the model solution, the model logic and its realism. We also discuss methods to explore model behaviour, other than exploring the stability properties of previous chapters, in order to obtain a better insight into the dynamics of the model.

11.1 Coupled BOD-O2 Model Revisited

To illustrate many of the tools in this chapter, we reconsider the model of coupled oxygen (O2) and BOD (biochemical oxygen demand) dynamics, in an open sewage system, where sewage is expressed as BOD (see Fig. 11.1).

There are two coupled differential equations, expressing the rate of change for BOD and oxygen:

$$\frac{d\text{BOD}}{dt} = -r \cdot \text{BOD}$$

$$\frac{d\text{O2}}{dt} = -r \cdot \text{BOD} + k \cdot (O2^* - \text{O2}) \tag{11.1}$$

The first term denotes the oxidation of BOD and consumption of oxygen, the second term in the oxygen equation is the reaeration of oxygen; r is the first-order

K. Soetaert, P.M.J. Herman, *A Practical Guide to Ecological Modelling*,
© Springer Science+Business Media B.V. 2009

Fig. 11.1 Schematic
representation of the coupled
BOD-O$_2$ model

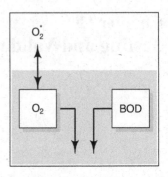

decay rate of BOD, $O2^*$ is the saturated oxygen concentration, and k is the reaeration coefficient.

At t=0, both the initial BOD and oxygen concentration are known:

$$BOD = BOD_0, O2 = O2_0.$$

11.2 Testing the Correctness of the Model Solution

After we have solved the model, first thing to test is the *correctness of the mathematical solution*. It is fairly common to make mistakes, so a thorough analysis of
model output should give confidence that this has not happened. This step is also
called model verification.

If possible we should compare our model output with published model output.
This may give relief when the published output is faithfully reproduced, but may
lead to confusion (who is in error?) when this is not the case.

For numerical models, it may be possible to *find an analytical solution*, under
certain simplified conditions. If this is the case, we should perform the test whether
our numerical solution compares favourably with the true solution.

This may sound confusing as we generally turn to numerical solution methods
when the model has become too complex to allow solving it analytically. However,
it is generally possible to run a numerical model in a simple setting that *can* be
solved analytically.

For the example BOD-O2 model, an analytical solution exists for constant r, k,
and $O2^*$ (Section 5.3.4).

$$BOD_t = BOD_0 \cdot e^{-r \cdot t}$$

$$O2_t = BOD_0 \cdot r \frac{e^{-k \cdot t} - e^{-r \cdot t}}{k - r} + O2_0 \cdot e^{-k \cdot t} + O2^* \cdot (1 - e^{-k \cdot t}) \qquad (11.2)$$

Assume that we want to make the saturated oxygen concentration and the decay rate a variable function of time, for instance because the temperature in the water is changing. Thus, rather than using this analytical solution, we need to solve this model numerically. However, it is a good idea to run the model numerically, and with constant parameters and compare the solution with the known analytical solution. The results are depicted in Fig. 11.2 A. As we solved the numerical model using R's integration function `ode`, there is a quasi exact match between both model runs. This gives confidence that at least the model implementation and numerical integration is accurate.

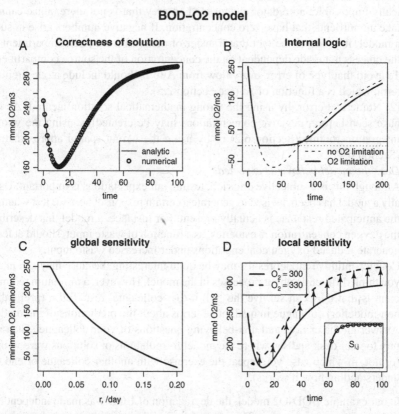

Fig. 11.2 Testing the coupled BOD-O_2 model; parameter values: $k = 0.1\,d^{-1}$, $O_2^* = 300$ mmol m^{-3}, r=0.05 d^{-1}, $O2_0 = 250$ mmol m^{-3}, $BOD_0 = 500$ mmol $O_2\,m^{-3}$. **A**. The numerical integration output (using R-function ode) cannot be distinguished from the analytical solution. **B**. If the reaeration is decreased ($k = 0.01\,d^{-1}$), negative O2 concentrations are generated. This necessitates reformulating the model (*solid line*). **C**. Effect of the oxygen consumption rate parameter r on the minimum oxygen concentration during a 100 days period. **D**. The effect of a small change in parameter O_2^* on the oxygen concentration (main figure) and the corresponding sensitivity functions S_{ij} (inset) versus time; dots represent S_{ij} corresponding to the arrows in main figure

11.3 Testing the Internal Logic of the Model

To test the internal logic of the model, we can perform several checks:

- *Do state variables remain positive?*
  Most of the ecological quantities (density, biomass, concentration, ..) are positive numbers and this positivity can not be violated. Negative concentrations have NO biological meaning and cannot be tolerated!
  There are two reasons why models can generate negative quantities. (1) There is an error in the model equations. Usually, negative numbers arise when material leaves a state variable whose value was zero (or already negative). This is biologically impossible: a predator cannot feed on prey that is not there, algae cannot take up nutrients that have zero concentration. If negative numbers arise in such a model, it is due to the fact that transfer of material from one compartment to the other is not made dependent on the concentration of the source compartment. To avoid that type of error, every flow from A to B should include a rate limiting term, which is a function of A. (see Section 2.5).
  (2) Numerical error, by using the wrong mathematical solution method. As we have seen before, negative concentrations may be created by using the wrong integration routine (Section 6.2) or by using the wrong spatial approximation (Section 6.4).
- *Does the model behave as expected?*
  Although this is a subjective criterion, testing our expectations is important. Usually a model has been devised to generate a certain response; here we test whether the anticipated response is actually present. For instance, a model that describes the oxygen concentration in estuaries, as a function of waste input, should at least generate reduced oxygen concentrations under increased waste input.
  Counter-intuitive model results may be quite interesting, because they can make you think differently about processes in the model. However, a minimum requirement is that they can survive the 'tell-it-to-a-colleague' test. Tell a (preferably non-modeller) colleague in qualitative terms about the predictions of the model. Answer all the critical and non-believing questions of your colleague. Then return to your desk and see where the model formulations or solutions were wrong (or not, in which case you repeat the exercise with another colleague or end up understanding how it works).

In our example BOD-O2 model, the degradation of BOD was made independent from the oxygen concentration. This was a mistake! (but a mistake made for didactic reasons: a more realistic model cannot be solved analytically).

If we cover the water surface with a semi-permeable sheet, the exchange with the air is reduced, and this model will predict negative oxygen concentration. This is due to the fact that BOD continues to degrade, even in the absence of oxygen. To make things worse, the negative oxygen continues to be exchanged with the air. In order to make the model more realistic, we add a rate-limiting term that reduces the BOD-oxidation rate at low oxygen concentrations and that eventually stops oxidation when oxygen concentration is 0. Thus, the improved model is:

$$\frac{d\text{BOD}}{dt} = -r \cdot \text{BOD} \cdot \frac{O2}{O2 + k_s}$$

$$\frac{dO2}{dt} = -r \cdot \text{BOD} \cdot \frac{O2}{O2 + k_s} + k \cdot (O2^* - O2) \tag{11.3}$$

The results of this new model can be admired in Fig. 11.2B.

11.4 Model Verification and Validity

During model verification we test whether the model is able to reproduce qualitatively and quantitatively the data.

This is a critical step in model building because here we are confronted with the consequences of our model. If observations agree with the outcome predicted by the model, it can be considered verified *for the time being*. This does not mean that the model is right, only that we were not able to prove it false. If the model outcome is incompatible with the data, then we need to refine the model such that it generates more faithful predictions.

Three courses of action can be taken when confronted with a mismatch between data and models: we may question the data, the model or the parameters.

- First of all we may take into account the precision of the *data*. Error is a fundamental characteristic of most ecological measurements. Mismatch between model outcome and data is less critical for uncertain data, than for data that are known with high precision. If we are lucky, it may be possible to estimate the probability distribution of observations or their variance. Often though, for 'expensive' data sets this is not possible, and every data point has to be considered equally likely (or unlikely).
- Secondly, we may consider revising the *model* assumptions, relationships, and model complexity. If one critical process has been omitted, then the model will not be able to reproduce faithfully even the qualitative aspects of the observations. The mismatch between model and data then will pinpoint gaps in our understanding of the system.
- Finally, it may be that the *parameter* values need fine-tuning. This is done in a calibration analysis as described in Section 4.3.

Although the model may have been verified, we must be wary of applications in which the model is used beyond the ranges of validity. Mathematical models have built-in assumptions, and therefore have restricted ranges of validity.

Often these ranges are broad enough to cover the problem for which the model was designed, but not for other applications. For instance, if you model primary production in an estuary with very high suspended load and high nutrient concentrations, you can have good correspondence between model and data by just specifying photosynthesis-irradiation (P-I) relations correctly. However, your formulations or parameter values on nutrient limitation may be far off, since the occurrence of nutrient limitation in the modelled estuary is too rare to test this aspect of the model. Applying this model to a clear-water system with very low nutrient concentrations may produce strange results. Similarly, the BOD-O2 model can only be meaningfully applied in those situations where reaeration is large enough to compensate for to oxygen demand by the BOD (see previous section).

11.5 Model Sensitivity

During sensitivity analysis, the effect of parameter values on the model outcome is estimated.

In practice, this analysis is carried out by changing one of the parameter values and inspecting the response of the model output.

Roughly speaking, sensitivity analysis can be subdivided in two types:

- During **global** sensitivity analysis parameter values are drawn within a relatively large plausible range and the effect on a (selection of) model output variable(s) is recorded. This allows establishing in a simple, yet efficient, way a cause-effect relationship. Several examples of global sensitivity analysis have already been given. For instance, in a bifurcation graph, the extreme values or steady-state conditions (y-axis) are depicted as a function of one parameter (x-axis). We have applied bifurcation analyses when we discussed the stability properties of models (Section 7.2.3) and discrete time models (Chapter 9). Another example of global sensitivity analysis was given in Section 5.4.3, when the critical size of a flat, cylindrical and spherical organism (the model variable) was estimated as a function of surrounding oxygen concentration (the model parameter). In Fig. 11.2C, the effect of the oxygen consumption rate parameter r on the minimum oxygen concentration during a 100 days period is depicted. The larger the oxygen consumption rate, the more oxygen is depleted.
- In a **local** sensitivity analysis, the effect of a change in parameter(s) in an infinitesimally small region is estimated. This can be done by means of a sensitivity function. These functions represent the sensitivity of model variables y_i to single parameters θ_j, weighted for the scaling of variables Δy_i (necessary if they have different units) and for the parameter uncertainty $\Delta \theta_j$:

$$S_{ij} = \frac{\Delta \theta_j}{\Delta y_i} \frac{\partial y_i}{\partial \theta_j} \tag{11.4}$$

Figure 11.2 D shows the local sensitivity analysis for the BOD-O2 model. If the oxygen saturation concentration is increased from 300 (the default, solid line) to 330 mmol m^{-3} (dotted line), the modelled oxygen concentration is higher. The difference between these model outputs, as denoted by the arrows in the figure are examples of sensitivity functions. Their values are depicted in the inset.

The mean sensitivity of all model variables with respect to each parameter can then be calculated and this information used to rank the model parameters according to the effect they have on the model output. Thus, this type of analysis is used to determine the most influential parameters from a model. It does not make sense to invest large amounts of time trying to find good values for parameters that have only small effect on model outcome.

Often many parameters affect model outcome, so rather than looking at one parameter at a time (so-called **univariate** sensitivity), it is also instructive to look at the interaction between parameters. This is the subject of **bivariate** or **multivariate** sensitivity analysis.

11.6 Case Studies in R

11.6.1 Time-Varying Oxygen Consumption in a Small Cylindrical Organism

We will develop and test a model that describes the impact of time-varying external oxygen concentrations on the oxygen levels in the body of a small cylindrical organism (Fig. 11.3). As we want to run the model dynamically, using a temporally changing external oxygen concentration, we certainly need to solve it numerically. However, finding a good numerical approximation for the spatial derivative is not that simple – it is very easy to make mistakes. To avoid errors, it is best to compare the numerical approximation with a known analytical solution. An analytical solution exists for constant external concentrations, and thus we will apply the numerical model under this condition to test whether our spatial numerical approximation is accurate enough.

For the numerical approximation of the time-varying model, it is simplest to start with the general 1-dimensional *transport-reaction* equation, describing the oxygen concentration as a function of body radius r for a cylindrical organism:

$$\frac{\partial O_2}{\partial t} = -\frac{1}{A} \cdot \frac{\partial A \cdot J}{\partial r} - Q \tag{11.5}$$

where the surface of a cylinder with radius r is $A = 2\pi\, r\, L$, with L the organism length, Q is the zero-order oxygen consumption rate, and J is the diffusive flux, given by:

$$J = -D_a \frac{\partial O_2}{\partial r} \tag{11.6}$$

(D_a is the diffusion coefficient). The boundary conditions at the centre of the cylinder (r=0) and at the outer surface (r=R) read:

$$\left.\frac{\partial O_2}{\partial r}\right|_{r=0} = 0; \; O_2|_{r=R} = BW O_2(t) \tag{11.7}$$

Fig. 11.3 Schematic representation of the model describing the oxygen budget in a cylindrical body with varying external O_2 concentration

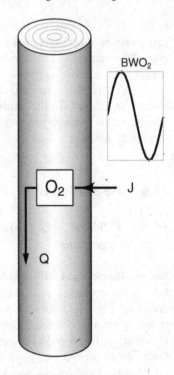

The steady-state *analytical* solution of this model and for constant external oxygen concentrations was discussed in Section 5.4.3:

$$O2(r) = \frac{Q}{4D_a}(r^2 - R^2) + BW \tag{11.8}$$

For the numerical implementation in R, we start by defining the parameters (see Section 5.4.3).

```
BW        <- 2             # mmol/m3
Da        <- 0.5           # cm2/d
R         <- 0.0025        # cm
Q         <- 250000        # nM/cm3/d
L         <- 0.1           # cm
```

The organism is subdivided into 100 (N) telescoping cylinders, which have equal thickness (dx). The distance, from the body centre to the layer interfaces and to the centre of each layer is given by xi and x respectively. These values are used to estimate the surface area at the layer interface and at the centre of layers (Ai and A). Finally, we also need dispersion distances, dxi. These are the distances from centre to centre of each layer, except for the dispersion distance at the outer interface: this only bridges the distance from the middle of the last layer to the outer surface, and is thus half the size of the other layers.

```
N   <- 100
dx  <- R/N
x   <- seq(dx/2,by=dx,length.out=N)
xi  <- seq(0 ,by=dx,length.out=N+1)
A   <- 2*pi*x *L
Ai  <- 2*pi*xi*L
dxi <- c(rep(dx,times=N),dx/2)
```

The model subroutine (`oxygen`) estimates the rate of change of the oxygen concentration (`dO2`) as the negative of the flux gradient (`-diff(Flux)`) minus the oxygen consumption (`Q`). In order to estimate the fluxes, the outer concentration is imposed (`BWO2`), while the internal boundary is a zero gradient boundary (i.e. the boundary concentration is simply the concentration of the first box, `O2[1]`).

The bottom water oxygen concentration is made a sinusoidal varying function of time, with a period of one hour. It varies between 0.4 and 3.6 mmol m^{-3} (see Fig. 11.4 C).

```
oxygen <- function (time, O2, pars)
{
   BWO2 <- BW* (1-0.8*sin(2*pi*time*24))
   Flux <- -Da * diff(c(O2[1], O2, BWO2))/dxi
   dO2  <- - diff(Ai*Flux)/A/dx -Q
   return  (list(dO2=dO2,c(Flux=Flux,BWO2=BWO2)))
}
```

We use the steady-state solver 'steady.1D' from R-package `rootSolve` to retrieve the oxygen concentration. This package has to be loaded first. Only about 4 iterations are necessary to find the steady-state condition.

```
require(rootSolve)

CONC  <- steady.1D(y=runif(N),func=oxygen,nspec=1,atol=1e-10)
O2    <- CONC$y
```

Then the steady-state oxygen concentration is plotted and compared with the analytical steady-state solution:

```
plot(x,O2,xlab="distance, cm",ylab="oxygen, μ mol/l")
lines(x, BW+Q/(4*Da)*(x^2-R^2))

legend ("topleft",lty=c(1,NA),pch=c(NA,1),
    c("analytical solution","numerical approximation"))
```

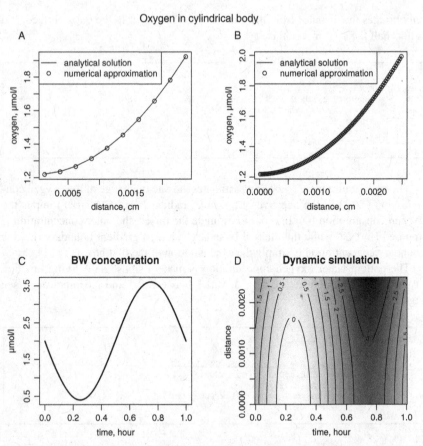

Fig. 11.4 For this model that describes the oxygen budget in a cylindrical organism, the correctness of the numerical approximation is first tested. To do that, its steady-state solution is compared with the known analytical steady-state solution, for varying number of discrete layers (**A**: 10 layers, **B**: 100 layers). As this comparison is favourable, we can then confidently apply the model in a dynamic simulation (**D**) where the bottom water oxygen concentration varies in time (**C**)

The numerical approximation almost exactly matches the analytical solution (Fig. 11.4 A, B), so we can now confidently run the model dynamically (for one hour) and plot the dynamic results (depicted in Fig. 11.4 D):

```
require (deSolve)

times<- seq(0,1/24,length.out=100)
out   <-as.data.frame(ode.1D(O2,times,oxygen,parms=0,nspec=1))
oxy   <- out[,2:101]
image(times*24,y=x,z=as.matrix(oxy),
      xlab="time, hour",ylab="distance ",
      col=grey(seq(1, 0.1,len=100)))
contour(times*24,y=x,z=as.matrix(oxy),add=TRUE)
```

11.6.2 R for Validation and Verification

The more complex a model becomes, the more cumbersome to inspect the results. R is especially strong for automating graphics.

Here is a script that plots all output variables, kept in a table called output, 4 on a page. Just click <ESC> to stop the display.

First it is specified that the display of the figure should be outlined in 2 rows and 2 columns. R-statement ask=TRUE ensures that a new figure will be plotted only when a key is struck

```
par(ask=TRUE , mfrow=c(2,2))
for (i in 2:ncol(table))
plot (x = output[,1],y = output[,i],
      sub="hit ESC to quit" )
```

We may also store the model output in a binary file, so that we can later have another look at the output, without the need to re-run the model. Binary files are much smaller than plain ASCII files and faster read by R.

Thus, to write table output to file out, we write:

```
save(file="out",output)
```

And in later sessions we can just recover the previously saved file:

```
load(file="out")
```

11.6.3 Univariate Local Sensitivity Analysis

We finally demonstrate how to perform local sensitivity analysis in R, using a simple numerical model of bacterial kinetics in a well-stirred, closed system (Fig. 11.5 A).

11.6.3.1 The model

The model describes bacterial biomass (B, molC m^{-3}) and glucose (S, mol C m^{-3}), the substrate on which they grow. The governing equations are:

$$\frac{dB}{dt} = g \cdot \gamma \cdot \frac{S}{S + k_S} \cdot B - d \cdot B - r \cdot B$$

$$\frac{dS}{dt} = -g \cdot \frac{S}{S + k_S} \cdot B + d \cdot B \qquad (11.9)$$

where g $(0.2\,hr^{-1})$ is the maximal uptake rate when substrate is abundant, k_s $(10\,mol\,C\,m^{-3})$ is the half-saturation constant, γ (0.5) is the bacterial growth efficiency, r $(0.01\,hr^{-1})$ is basal respiration and d $(0.01\,hr^{-1})$ is the bacterial mortality rate.

The initial conditions of bacteria and glucose are $0.1\,mol\,C\,m^{-3}$ and $100\,molC\,m^{-3}$ respectively.

Implementation of this model in R is straightforward: we start by defining the parameter values, and a function (model) that takes as input the simulation time and the values of the state variables and parameters, and that returns the rates of changes, as a list. We then define the times at which we want to obtain model output, set the initial conditions of the state variables and run the model using R's integration routine ode (from package deSolve). Finally the evolution of bacterial carbon, glucose and total organic carbon (the sum of both) is plotted as a function of time and a legend added (Fig. 11.5 B).

```
pars <- list(Bini=0.1,Sini=100,gmax =0.5,eff = 0.5,
             ks =0.5, rB =0.01, dB =0.01)

model <- function(t,state,pars)
{
with (as.list(c(state,pars)), {

dBact = gmax*eff*Sub/(Sub+ks)*Bact - dB*Bact - rB*Bact
dSub  =-gmax *Sub/(Sub+ks)*Bact + dB*Bact

return(list(c(dBact,dSub)))
                               })
}

tout     <- seq(0,50,by=0.5)
state    <- c(Bact=pars$Bini,Sub =pars$Sini)
require(deSolve)
out      <- as.data.frame(ode(state,tout,model,pars))

plot(out$time,out$Bact,ylim=range(c(out$Bact,out$Sub)),
     xlab="time, hour",ylab="molC/m3",type="l",lwd=2)
lines(out$time,out$Sub,lty=2,lwd=2)
lines(out$time,out$Sub+out$Bact)

legend("topright",c("Bacteria","Glucose","TOC"),
       lty=c(1,2,1),lwd=c(2,2,1))
```

11.6.3.2 Graphical analysis

What is the effect on model output if we perturb one parameter slightly about the mean? The simplest way to investigate that is graphically: we slightly increase each parameter with a small factor, say 10%, one by one, run the model with these altered values and compare the output with the original output.

Here is how to do that in R: first take a copy of the original output (the Reference), and of the parameter values (pp). We use R-function unlist, as we want to treat the parameters as one vector, not as a list). We specify that we want to visualize the graphs in 3 rows and columns (mfrow). Then each parameter value

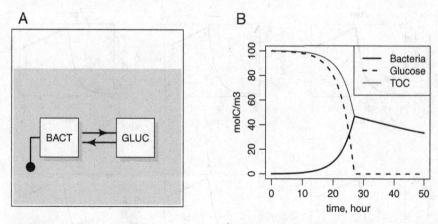

Fig. 11.5 Bacterial C-dynamics in a well-mixed tank. **A.** Bacteria take up glucose, and respire. When bacteria die, they add to the glucose pool. **B.** Results of the model simulation- see text for explanation

is increased with 10% (`tiny`), the model initialized (`state`) and run and plotted, comparing the new with the reference output. After resetting the original parameter value the procedure is repeated.

```
Reference  <- out
pp         <- unlist(pars)
tiny       <- 0.1
par(mfrow=c(3,3))
for (i in 1:length(pars))
{
  pars[i]  <- pp[i]*(1+tiny)
  state    <- c(Bact=pars$Bini,Sub =pars$Sini)
  out      <- as.data.frame(ode(state,tout,model,pars))

  plot(out$time,out$Bact,xlab="hour",ylab="molC/m3",
       type="l",lwd=2,main=names(pars)[i])
  lines(Reference$time,Reference$Bact,lty=2)
  pars[i] <- pp[i]
}

plot(0,axes=FALSE,xlab="",ylab="",type="n")
legend("center",c("perturbed","reference"),lwd=c(2,1),
       lty=c(1,2))
```

The last two lines first open a new plot window, without labels or lines (`type="n"`), and then add a legend.

Clearly, the maximal uptake rate, *gmax*, and the growth efficiency, *eff*, affect the bacterial biomass over the entire period, whilst the effect of initial substrate concentration, *Sini*, is discernible only when the glucose concentration has dropped to very low values. Increasing the four other parameters with 10% has very little impact on bacterial biomass (Fig. 11.6).

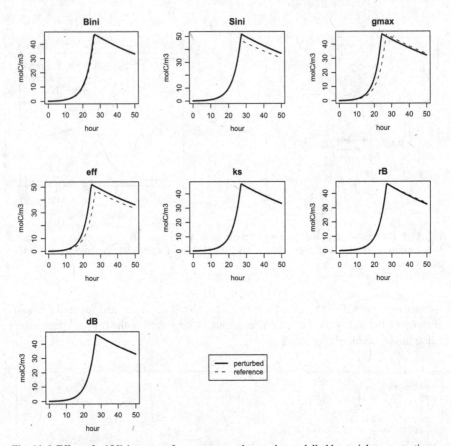

Fig. 11.6 Effect of a 10% increase of a parameter value on the modelled bacterial concentration

11.6.3.3 Sensitivity functions

We now condense this graphical information into a series of numbers, the so-called sensitivity functions. These functions represent the sensitivity of model variables y_i to single parameters θ_j, weighted for the scaling of variables Δy_i and for the parameter uncertainty $\Delta \theta_j$.

$$S_{ij} = \frac{\Delta \theta_j}{\Delta y_i} \frac{\partial y_i}{\partial \theta_j} \qquad (11.10)$$

This weighing of variables is necessary only if they have different units, while the weighing of parameters allows to make a distinction between parameters that are known with high precision, and those that are much more uncertain. For a model that is solved numerically, it is simplest to approximate the derivate of the output with respect to the parameters numerically:

$$\frac{\partial y_i}{\partial \theta_j} \approx \frac{y_i|_{\theta_j*} - y_i|_{\theta_j}}{\theta_j* - \theta_j} \tag{11.11}$$

where θ_j* is the slightly perturbed parameter value and $y_i|_{\theta_j*}$ is the model output using this value.

The implementation in R is similar to the graphical procedure, except that the parameters are much less perturbed (value `tiny` is now much smaller), and thus stay more closely in the vicinity of their nominal value. We estimate the sensitivity only for the bacteria, and at the selected time intervals (`yRef`). The parameters are scaled with their nominal value (`pp`); the output variables are not scaled.

```
State    <- c(Bact=pars$Bini, Sub =pars$Sini)
yRef     <- as.data.frame(ode(state,tout,model,pars))$Bact
pp       <- unlist(pars)
nout     <- length(yRef)
npar     <- length(pars)

tiny     <- 1e-8
dp       <- pp*tiny

Sens     <- matrix(nrow=nout,ncol=npar,NA)

for (i in 1:npar)
{
   dval     <- pp[i]+dp[i]
   pars[i] <- dval
   state    <- c(Bact=pars$Bini,Sub =pars$Sini)
   yPert    <- as.data.frame(ode(state,tout,model,pars))$Bact
   Sens[,i]<- (yPert-yRef)/dp[i]*pp[i]
   pars[i] <- pp[i]
}
colnames(Sens) <- names(pars)
rownames(Sens) <- tout
format(as.data.frame(Sens[1:5,]),digits=2)
```

The last line prints the first 5 rows to the screen, with not too many significant digits:

	Bini	Sini	gmax	eff	ks	rB	dB
0	0.10	0.00000	0.000	0.000	0.0e+00	0.00000	0.00000
0.5	0.11	0.00007	0.014	0.014	-6.8e-05	-0.00056	-0.00056
1	0.13	0.00016	0.031	0.031	-1.5e-04	-0.00126	-0.00126
1.5	0.14	0.00026	0.053	0.053	-2.6e-04	-0.00211	-0.00211
2	0.16	0.00039	0.079	0.079	-3.9e-04	-0.00316	-0.00316

The graph of these sensitivity functions show how they vary with time; some parameters have very similar sensitivity functions (Fig. 11.7).

```
matplot(tout,Sens,type="l",lty=1~:10,col=1)
legend("topright",names(pars),lty=1:10,col=1)
```

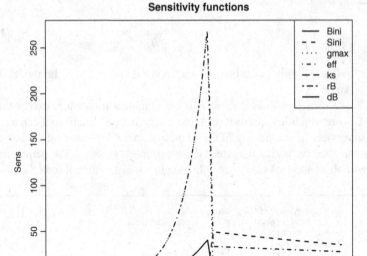

Fig. 11.7 Sensitivity functions (see text for definition) of the different parameters in the bacterial growth model

The overall sensitivity of the output with respect to each parameter can now be calculated by a number of summarizing measures, e.g.:

$$\delta_i^{Sqr} = \sqrt{\frac{1}{n} \cdot \sum_{j=1}^{n} S_{ij}^2} \quad \delta_i^{Abs} = \frac{1}{n} \cdot \sum_{j=1}^{n} |S_{ij}| \tag{11.12}$$

This allows ranking of the importance of the different parameters, according to decreasing sensitivity.

```
mabs <- colMeans(abs(Sens))
msqr <- sqrt(colSums(Sens*Sens)/nout)
format(data.frame(msqr,mabs),digits=2)
```

```
        msqr   mabs
Bini    9.59   5.21
Sini   28.62  19.57
gmax   58.88  29.51
eff    62.43  37.12
```

```
ks    0.37  0.17
rB    4.65  3.47
dB    2.98  2.06
```

These values for the mean sensitivity $\mathtt{msqr}$ show that the model is relatively insensitive to changes in the half-saturation coefficient ($\mathrm{msqr}_{ks} = 0.4$). The growth efficiency and maximal growth rate are the most sensitive parameters ($\mathrm{msqr}_{gmax} = 58.9$, $\mathrm{msqr}_{eff} = 62.4$), followed by the initial conditions of substrate and bacteria and finally the basal respiration and mortality rate of bacteria.
This completes the univariate sensitivity study.

11.6.4 Bivariate Local Sensitivity Analysis

We can now go on and test whether two parameters have a similar effect on model output (so-called bivariate sensitivity), or whether the effect of parameters can be compensated by the effect of a *combination* of other parameters (multivariate sensitivity). A nice description of these various types of sensitivity analysis is given in Brun et al. (2001). Here we will restrict ourselves to pairwise (bivariate) analysis, since for multivariate analysis a rather advanced knowledge of multivariate statistics is needed. Interested readers are referred to Brun et al. (2001).

Before proceeding, we remove the insensitive parameter *ks* from the subsequent analysis. This reduces the number of parameters (npar)

```
Sens <- Sens[ ,colnames(Sens) !="ks"]
npar <- ncol(Sens)
```

To check whether two parameters have a similar sensitivity on model output, their sensitivity functions can be compared graphically and their pairwise correlation estimated. R's graphical function $\mathtt{pairs}$ produces pair-wise scatterplots, whilst the statistical function $\mathtt{cor}$ estimates pair-wise correlation coefficients.

In the script below, we combine both functionalities: the lower panel produced by the $\mathtt{pairs}$ function plots the values of the sensitivity functions versus one another (the default), whilst the upper panel prints the correlation coefficients between x and y (with 2 significant digits), in the centre of both ranges. The function `panel.cor` does that (note: a slightly more complex $\mathtt{panel.cor}$ function is provided in the help screen of $\mathtt{pairs}$ – type `?pairs` to see it).

```
panel.cor <- function(x, y)
              text(x=mean(range(x)),y=mean(range(y)),
              labels=format(cor(x,y),digits=2))

pairs(Sens,upper.panel=panel.cor)
mtext(outer=TRUE,side=3,line=-2,
      "Sensitivity functions",cex=1.5)
```

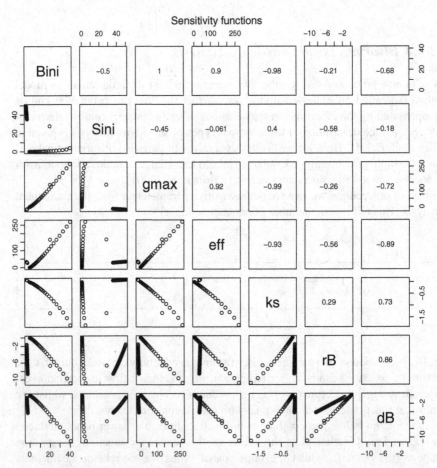

Fig. 11.8 Bivariate correlation graphs between the different pairs of parameters in the bacterial growth model. The upper half of the graph gives the linear correlation coefficients, the lower half plots the scaled sensitivity functions pair wise. See text for details

The resulting graph (Fig. 11.8) shows the nearly perfect linear dependence of the sensitivity functions of several parameter pairs, e.g *gmax* and *Bini*, *eff* and *gmax*, *rB* and *dB* (pos correlation), *dB* and *eff* (negative correlation) and many others.

This means that, for instance, changes in the model output caused by increasing the maximal growth rate, *gmax*, can also be achieved by increasing the initial concentration of bacteria, *Bini*.

Chapter 12
Further Reading and References

In the main text we have deliberately omitted reference to the literature, except for some rare cases, mainly when we dealt with specific examples. Here we make amends for this default.

As for most books, there is skewness in the citations. Where several references or examples are available, we tend to select those of our nearest colleagues, from the Netherlands Institute of Ecology (NIOO). We have good reasons to do so: not only have they provided multiple feedback to the book, there is a particularly high 'bio'diversity of modellers at NIOO, covering almost all facets of modelling, bridging theory and practice, micro- and macroscopic descriptions, simple to complex, deterministic and stochastic modelling approaches. In addition, as we have more affinity with marine and estuarine research, there is also skewness with respect to marine model applications.

12.1 Model Formulations

Several general textbooks (Yodzis, 1989; Renshaw, 1991; Jørgensen, S.E., 1994, Haefner, 1996, Hannon and Ruth, 1997; Gilman and Hails, 1997; Gurney and Nisbet, 1998; Case 2000; Edelstein-Keshet, 2005) provide introductory explanations for several aspects of modelling. Of these, Haefner (1996), Gurney and Nisbet (1998) and Edelstein-Keshet (2005) are closest to the current book.

Simple mathematical models for the marine phytoplankton were introduced in the forties (Riley 1946). Steele (1958, in Gentleman, 2002) developed a nutrient-phytoplankton-zooplankton (NPZ) model in a two-layered sea. The NPZD model discussed in Section (2.9.1) extends this model with an explicit description of detritus. A classical paper (Fasham et al., 1990) includes more detail and has been widely used. Steele and Henderson (1992) discuss the importance of the closure term in plankton models. Gentleman (2002) reviews the history of plankton modelling in detail. The book by Fennel and Neumann (2004) exclusively deals with marine ecosystem modelling.

The model of Shuter (1979) was one of the first dealing with the physiological adaptation of algae to external conditions; the aquaphy model (Lancelot et al., 1991), discussed in Section 2.9.2 is a somewhat more complex version.

Kooijman (2000) extensively discusses physiology-based models for uptake and use of food in individuals and populations, called Dynamic Energy Budget models. This book is quite difficult to grasp! Van der Meer (2006) gives a more accessible account and reviews the use of these models in marine populations and the problem of parameter estimation.

12.2 Spatial Pattern

Czárán (1997) deals with both discrete and continuous spatial models; it makes a good introduction to discrete spatial modelling. Wilson (2000) also describes some discrete spatial models and their implementation in the programming language C; the emphasis is on individual-based (microscopic) models. Forest gap models are an important class of individual-based models, which started off by a pioneering model by Botkin et al. (1972). Bugmann (2001) reviews many applications. Grimm and Railsback (2005) give a recent overview of individual-based and agent-based modelling approaches, with due attention to software problems. Mooij and Osiris (1996) provide a modelling framework for individual-based models; Vos et al. (2002) provide an aquatic example; SWARM (http://www.swarm.org/) is a framework for agent-based simulations. R-package simecol (Petzoldt and Rinke, 2007) provides some utility functions to speed up running cellular automata models in R.

The classic books on continuous descriptions of diffusion are Crank (1975) which treats the subject from a mathematical perspective and Okubo (1980) or the updated version (Okubo and Levin, 2001), which provide numerous ecological examples.

Modelling approaches that consider spatial interaction between sedentary organisms (plants) are neighbourhood models (Pacala, 1986), zone-of-influence models (Wyszomirski, 1986; Schwinning and Weiner, 1998), ecological field models (Wu et al., 1985) and field-of-neighbourhood models (Berger and Hildenbrandt, 2000). Cellular automata originated with the paper from Wolfram (1983); Ermentrout and Edelstein-Keshet (1993) discuss several biological applications.

Jamart and others (1977) were amongst the first to simulate the dynamic response of a marine phytoplankton-nutrient systems in a 1-dimensional water column and with time-variable forcing functions. The model of Wroblewski (1977) was one of the first attempts to include marine food web descriptions in a circulation model.

Berner (1980), and Boudreau (1997) are the major reference works on sediment biogeochemical models, also referred to as diagenetic models. A number of models that described oxygen and/or nutrient cycles in coupled discrete layers were published in the late seventies and eighties (e.g. Vanderborght and Billen, 1975; Goloway and Bender, 1982). The development of complex continuous numerical diagenetic models is a more recent phenomenon (Soetaert et al., 1996b; Wang and Van Cappellen, 1996). A particularly nice account on sediment bioturbation can be found in Meysman et al. (2006).

12.3 Parameterization

A useful point of entry into the vast subject of mathematical minimization is Press et al. (1992). The Levenberg-Marquardt algorithm, which is discussed in their chapter on 'modelling of data' is, in our experience, generally the most efficient mathematical routine. Introductory books on genetic algorithms are Goldberg (1989) and Coley (1999). Kruger (1993) gives a comprehensive discussion of the simulated annealing algorithm; Price (1977) introduced the (wonderfully simple) controlled random search method that was discussed in Section 4.4.3. The use of observations to infer the parameter values of a model is called an 'inverse problem'. For the statistically and mathematically inclined reader, the book of Tarantola (2005) takes a probabilistic approach to such problems. For the less affluent, this book is downloadable from the web.

McCallum (2000) discusses many simple models, and how to parameterise them. With respect to linear and nonlinear fitting, several R introductory texts are relevant (see the R-website www.r-project.org).

12.4 Model Solution

The mathematical reference book for numerical analysis, written for the non-mathematician is Press et al. (1992), which also discusses implementations in several computer languages, Pascal, Fortran, C, all in different books. Note though that these authors have quite strict rules about the use of their code: although code is published in their books, one is allowed only to use ideas, not the actual routines.

Abbott and Basco (1989) is a reference book for the mathematically skilled (!) reader, interested in numerical transport schemes and how to keep numerical problems in check. Some books that deal with ecological or biochemical models include very comprehensive accounts on numerical solutions and their problems, e.g. Thomann and Mueller (1987), Haefner (1996) and Chapra (1997). Various software repositories on the internet provide access to freely usable numerical solution methods. An invaluable source for the Fortran programmer are the Linpack routines (http://www.netlib.org/linpack/).

Best source of analytical model solutions are software packages that can solve them in symbolic form, e.g. MAPLE, Mathematica,... The book of King et al., (2003), although readable and understandable, is written for the mathematically-proficient.

Using R to solve dynamic simulation models became possible with the introduction of an R-package, odesolve, containing integration routines (Setzer, 2001) and a description of how to do this in the R-newsletter (Petzoldt, 2003). Package deSolve (Soetaert et al., 2008a), the successor of odesolve contains a more complete set of integration routines, including functions to solve 1-dimensional and 2-dimensional problems.

12.5 Stability and Equilibrium Analysis

Ludwig et al. (1978) wrote the first ecological analysis of bifurcation and multiple states (the spruce budworm system), whilst May (1977) reviewed the conditions under which bifurcations and multiple states may arise. A good account of the subject is the book of Clark (1990), the classic reference for optimal management of renewable resources, mainly popular by economists. An invaluable source of inspiration for Chapter 7 was Ludwig et al. (2002). The book by Kot (2001) deals in extenso with stability, bifurcations in population ecology; the models are simple, the mathematics rather complex. Also recommended is Wiggins (1990). Since its first description, multiple stable states have been documented in many other areas, e.g. lakes (Scheffer, 2001) or tidal flats (van de Koppel et al., 2001).

Many ecological modelling books also deal with equilibrium formulations (e.g. Edelstein-Keshet, 2005). Auger et al. (2008) apply the technique to population dynamics models. Hofmann et al. (2008) give a step-by-step account of how to derive the equilibrium formulations in aquatic pH models.

Package rootSolve (Soetaert, 2008) was devised in order to facilitate stability and steady-state analysis of ecological models in R. It comprises an implementation of the Newton-Raphson method and includes special-purpose solvers for 1- and 2-dimensional models.

12.6 Discrete Time and Dynamic Programming Models

A discussion of insect predator-prey or parasitoid-host interactions generally starts with the models of Nicholson (1933) and Nicholson and Bailey (1935). Rogers (1972) derives more complex models. The paper by Ricker (1954) is a pivotal classic in the field of population biology and of discrete time models.

In many books, discrete time models receive equal or even more attention than continuous time models (you may have noticed that we do not share this view). Kot, 2001, Gilman and Hails, 1997; Gurney and Nisbet (1998) deal, amongst other things, with the subject.

Matrix models were introduced by Leslie in the 1940s. (Leslie, 1945), and rediscovered by ecologists in the sixties (Lefkovitz, 1965) and seventies (Sarukhán and Gadgil, 1974; Werner and Caswell, 1977). The monograph by Caswell (2001) deals with every possible facet of matrix models, and also covers topics such as parameter estimation, sensitivity analysis, stochasticity. It includes many references to the literature, but is not easy to digest. Gotelli, 2001 is an easy-reading book that also deals, amongst others, with matrix models. In addition, the book by Ebert (1999), with its emphasis on methodological aspects, provides a gentle learning curve into the field of demographic models.

demogR (Jones, 2007) and popbio (Stubben and Milligan, 2007) are two R-packages that deal with the implementation and analysis of demographic models in R.

Notwithstanding its title ('Dynamic state variable models in ecology') the book by Clark and Mangel (2000) deals with dynamic programming applications only! There are plenty of references in this book. The book by Houston and McNamara (1999) discusses other ecological problems that can be tackled using dynamic programming. Klaassen et al. (2006) provide a nice example applied to migration of geese. The technique can also be applied in the field of nature conservation (McCarthy and Possingham, 2007) or include social interactions, which leads to state-dependent dynamic games (McNamara et al., 1997). Denardo (2003) takes a more theoretical and technical approach to sequential decision or dynamical programming problems but does not deal with ecological applications.

Appendix A
About R

R (R Development Core Team, 2007) is the open-source (read: free-of-charge) version of the language S. It is best known as a package that performs statistical analysis and graphics. However, R is much more: it is a high-level language in which one can implement new methods, and make high-quality graphs. R has high-level commands to operate on matrices, perform numerical integration, advanced statistics, . . . which are easily triggered and which make it ideally suited for modelling.

How does one learn to work with R? Not easily . . . if one wants to take advantage of the high-level commands. If you start with R, use it for a significant project, or you may be browsing the 'R for beginners' guide for months or years. The slow start is the price to be paid for the absence of point-and-click technology. However, flexibility is gained: menu-driven programmes are convenient only for routine tasks. Many interesting applications are not routine and therefore cannot be accomplished from the icons provided by program developers.

There are many excellent sources for learning the R (or S) language. R comes with several manuals that can be consulted from the Rgui window (Help/Manuals). R-intro.pdf is a good start. Many other good introductions to R are available, some freely on the web, and accessible via the R web site (www.r-project.org). You may find a more extensive introduction to R, which includes exercises, on the books webiste. Sometimes the best help is provided by the very active mailing list. If you have a specific problem, just type R: <problem> on your search engine. Chances are that someone encountered the problem and it was already solved.

A.1 Installing R

R is downloadable from the following web site: http://www.r-project.org/
Choose the precompiled binary distributions.
On this website, you will also find useful documentation.
To use R as a tool for developing/running the ecological models for this book, several packages also need to be downloaded.

- deSolve. Performs model integration. (Soetaert et al., 2008).
- rootSolve. Steady-state analysis (Soetaert, 2008).
- seacarb. Aquatic chemistry (Proye et al., 2007).

Downloading specific packages can best be done within the R program itself. Select menu item 'packages/install packages', choose a nearby mirror site and select the package you want. You are now ready to run the ecological models.

We prefer to run R from within the Tinn-R editor, which can be downloaded from URL http://www.sciviews.org/Tinn-R/ and http://sourceforge.net/projects/tinn-r. This editor provides R-sensitive syntax and help.

From within the Tinn-R program, you launch R via the menu (R/start preferred Rgui).

A.2 A Very Short Introduction

R-code is highly readable, once you realise that:

- '<-' is the assignment operator.
- everything starting with '#' is considered a comment.
- R is case-sensitive: 'a' and 'A' are two different objects.

A.2.1 Console Versus Scripts

There are two ways in which to work with R.

1. We can type commands into the R **console** window at the command prompt (>) and use R as a powerful scientific calculator.
   For instance, typing in the console window:

```
> pi*0.795^2 ; 25*6/sqrt(67) ; log(25)
```

will give as answer:

```
[1] 1.985565
[1] 18.32542
[1] 3.218876
```

Here sqrt and log are built-in functions in R; pi is a built-in constant.
2. Alternatively, we can create **R-scripts** that are sequences of R-commands and functions, and save them in a file ('filename.R') for later re-use. These scripts should be submitted to R before they are executed. This can be done either by typing, in the R-console window: source ('filename.R') or by opening the file, selecting the R-script and submitting the text to R.

Within the tinn-R editor, you can either submit the entire file at once to R, parts of the text, or line-by-line.

A.2.2 Getting Help

R has an extensive **help** facility. For instance, typing

```
>   ?log
```

will explain about this function, and the related functions log10, log2, and other.

```
>   ?Arithmetic
```

lists the arithmetic operations in R.

```
> help.search(''integration'')
```

will list occurrences of the word integration in R-commands.

A.2.3 Vectors and Matrices

R can take as arguments for its functions single **numbers**, **vectors**, or **matrices**.

Vectors can be created in many ways:

The operator c() combines numbers into a vector

A sequence of values, each 1 larger than the previous one can be created by the operator ' : '

A more general sequence can be generated by the command 'seq'

For instance:

```
> sin( c(0,pi, pi/2,pi,3*pi/2,2*pi) )
> sin( seq(from=0,to=2*pi, by=pi/2 ))
> sin( seq(0, 2*pi, pi/2 ))
```

will all calculate the sine of 0, π, ... $2*\pi$:

```
[1] 0.000000e+00 1.000000e+00 1.224606e-16 -1.000000e+00 -
2.449213e-16
```

Note that R-command 'seq' takes as input (amongst others) parameters from, to and by. If the order is kept, they need not be specified by name (3rd example).

```
>V <- 1:100
>sqrt(V)
```

will create a sequence of integers between 1 and 100 and take the square root of all of them. The operator ' <- ' assigns the sequence to vector V.

```
>exp(seq(0.1,1,by = 0.1))
```

will display the exponent of the sequence of values, 0.1,0.2,....1.0
    The command

```
> M<-matrix(nrow = 2,data = c(1,2,3,4))
```

will create a **matrix**, with two rows (and two columns). Note that in R, matrices are filled column-wise.
    The next two statements display the matrix and the square root of its elements:

```
>M
```

```
     [,1] [,2]
[1,]   1    3
[2,]   2    4
```

```
> sqrt(M)
```

```
        [,1]       [,2]
[1,] 1.000000  1.732051
[2,] 1.414214  2.000000
```

Matrix algebra is very simple. For instance:

```
> solve(M) %*% M
```

will invert matrix M (solve(M)), and multiply with M, giving the unity matrix:

```
       [,1] [,2]
[1,]    1    0
[2,]    0    1
```

A multi-dimensional array is created as follows:

```
>AR<-array(dim = c(2,3,2),data = 1)
```

In this case AR is a 2*3*2 array, and its elements are 1.

The elements of vectors, matrices and arrays can be indexed using the '[]' operator:

```
solve(M)[1,1]
M[,2]
M[1,]
V[1:10]
```

The above code takes the element on the first row, first column of the inverse of matrix M (1st line), next the 2nd column (2nd line) and first row of M is selected and displayed (3rd line). The last line then takes the first 10 elements of vector V.

A.2.4 More Complex Data Structures

R also allows creating more complex structures such as:

- data.frame
This is a combination of different data types (e.g. characters, integers, logicals, reals) in tabular format. Data frames have the advantage over matrices that their columns can be assessed by name rather than by number. For instance, if 'out' is a data.frame that contains 'time', 'DIN' and 'DET' as columns, then

```
out$time
```

will extract the column with time values.

- list
A list is a combination of several structures. Similarly as for data frames, these structures can also be accessed by their names. See next chapter for how lists are used.

A.2.5 User-defined Functions and Programming

One of the strengths of R is that we can make our own functions that add to R-s built-in functions.

For instance, after submitting to R:

```
> Circlesurface <- function (radius) pi*radius^2
```

We can use the function Circlesurface to calculate the surfaces of circles with given radius:

```
>Circlesurface(10)
```

or

```
>Circlesurface(1:100)
```

which will calculate the surface of circles with radiuses 1, 2, ...,100.

More complicated functions may return more than one element:

```
Sphere <- function(radius)

{
 volume  <- 4/3*pi*radius^3
 surface <- 4 *pi*radius^2
 return(list(volume=volume,surface=surface))
}

Sphere(2)
```

Which will output the volume and surface of a sphere with radius 2:

```
$volume

[1] 33.51032

$surface

[1] 50.26548
```

R has all the features of a high-level programming language:

- If, else, ifelse

```
if ( x<0 )
   string <- 'x<0'
else if ( x <2 )
   string <- '0>=x<2'
else
   string <- 'x>=0'
print(string)
```

- Loops allow a set of commands to be executed a certain number of times (for) or until a specified condition is met (while, repeat).

```
for (i in 1:10) print(c(i,2*i,3*i))
i<-1 ; while(i<11) {print(i); i<-i+1}
i<-1 ; repeat{print(i);i<-i+1;if(i>10) break}
```

The curly braces '{}' embrace multiple statements that are executed within each iteration.

Note: loops are implemented very inefficiently in R and should be avoided as much as possible.

A.2.6 R Packages

A package in R is a file that contains many objects that perform certain related tasks. Packages can be downloaded from the R website (see Section 1.1). Once installed, we generate a list of all available packages, we load a package and we obtain a list with its contents by the following commands:

```
library()
library(deSolve)
library(help=deSolve)
```

A.2.7 Graphics

R has extensive graphical capabilities.
   Try:

```
>demo(graphics)
>demo(image)
>demo(persp)
```

to obtain a display of R's simple, image-like(2-D) and perspective (3-D) capabilities.

Throughout this book, we mostly use R's graphical commands curve, plot and matplot:

```
> curve(0.5*x^3-2*x^+2)
```

A.2.8 Minor Things to Remember

1. Pathnames in R are written with forward slashes '/', although in windows, back-slashes are used. So to set a working directory or to specify a path:

```
path <- "C:/Documents and Settings/Karline Soetaert/My
Documents/R code/"
setwd(paste(path,''book'',sep=''''))
```

2. If, in a script, a sentence on one line is syntaxically correct, R will execute it, even if we intend it to proceed on the next line. For instance if we write:

```
dZoo <- ingestion - respiration
   - mortality
```

then dZoo will get the value (ingestion-respiration) and, if not in a function, R will print the value of (−mortality). In case the statement is in a function, R will even not print, such that these errors are very hard to trace (but a warned person is worth two).

A correct example is to include the entire statement between brackets:

```
dZoo <- (ingestion - respiration
                   - mortality)
```

or to write:

```
dZoo <- ingestion - respiration -
        mortality
```

A.3 Interfacing R with Low-Level Languages: The Competition in a Lattice Grid Model Revisited

We now re-implement the model of Silvertown et al. (1992), describing spatial competition amongst five species of grasses in a 40*40 lattice (see Section 3.6.4).

Due to the nesting of three loops, the R-code of this model was insupportably slow (in the order of a minute).

Fortunately, it is relatively simple to call a subroutine, written in another programming language from within R. Thus, R is used as a trigger for this subroutine, and for further statistical analysis and graphical output. Code generated by a compiled language may considerably speed up the calculations. This is because compiled languages transform programs directly into machine code, instructions that a CPU understands. In contrast, R is an interpreted language, and thus processes a program at runtime. Every line is read, analysed, and executed. Within a loop, each line has to be processed every time, which makes interpreted code significantly slower than compiled code. Thus, loops should be avoided as much as possible, or the looping should be performed in code generated with a compiled language.

To illustrate the interfacing concept we have re-implemented the lattice grid competition model as a subroutine in Fortran 77, which is then compiled as a dynamic link library (DLL, Microsoft windows® only) and invoked by R.

From the R-programme in Section 3.6.4, only the competition function is changed:

R-statement .Fortran('lattice', ...) invokes an external function called 'lattice', which was created with the Fortran 77 language. Of course, the function 'lattice', which is embedded in a DLL, has to be loaded first. R-function dyn.load does that (note that the file 'competition.dll' has to be in R's working directory for this to work). Note the coercion of all arguments to the correct type before calling the subroutine. Mistakes in matching types are difficult-to-trace – so it is better to avoid them.

```
dyn.load(``competition.dll'')

competition <- function(cells,nstep=100)

{
 nspec    <-  nrow(replacement)
 ncell    <-  nrow(cells)
 sumdens  <-  matrix(nrow=nstep,ncol=nspec,as.integer(0))
 seed     <-  runif(1)*-10
 res <- .Fortran("lattice",nspec=as.integer(nspec),ncell=as.integer(ncell),
                 nstep=as.integer(nstep), cells=as.integer(cells),
                 replacement=as.double(replacement),
                 sumdens=sumdens,seed=as.integer(seed))

 return(list (cells=matrix(nr=ncell,nc=ncell,res$cells),
              density=matrix(nr=nstep,nc=nspec,res$sumdens)))
}
```

The Fortran subroutine is given below. Note that the sequence of variables as they are called in R has to be the same in the Fortran source.

```
      SUBROUTINE lattice(nspec,ncell,nstep,cell,replace,sumdens,seed)
      IMPLICIT NONE
      INTEGER nspec,ncell,nstep,seed,ii
      INTEGER cell(ncell,ncell),newcell(ncell,ncell)
      INTEGER sumdens(nstep,nspec)
      DOUBLE PRECISION replace(nspec,nspec)

      INTEGER I,J,K,L
      INTEGER neighbour(4)
      DOUBLE PRECISION replacement(4),rep
      DOUBLE PRECISION rnd
      REAL RAND
C initialize random number generator
      call sRand(seed)
      DO I = 1,nstep
        DO J = 1, ncell
          DO K = 1, ncell

          rnd = DBLE(Rand())

          ii = cell(J,K)

          IF(J .GT. 1) THEN
             neighbour(1)= cell(J-1,K)
          ELSE
             neighbour(1)= cell(ncell,K)
          ENDIF

          IF(J .LT. ncell) THEN
             neighbour(2)= cell(J+1,K)
          ELSE
             neighbour(2)= cell(1,K)
          ENDIF
```

```
            IF(K .GT. 1) THEN
              neighbour(3)= cell(J,K-1)
            ELSE
              neighbour(3)= cell(J,ncell)
            ENDIF

            IF(K .LT. ncell) THEN
              neighbour(4)= cell(J,K+1)
            ELSE
              neighbour(4)= cell(J,1)
            ENDIF

            Rep         = 0
            newcell(J,K) = ii
C cumulative probabilities
            DO L = 1, 4
              replacement(L) = rep+replace(neighbour(L),ii)/4

              IF (rnd .GE. rep .AND. rnd.LT.replacement(L)) THEN
                newcell(J,K) = neighbour(L)
                exit
              ENDIF
              rep = replacement(L)

            ENDDO
          ENDDO
        ENDDO

        DO J = 1, nspec
          sumdens(I,J) = 0
        ENDDO
        DO J = 1, ncell
          DO K = 1, ncell
            ii        = newcell(J,K)
            cell(J,K) = ii
            sumdens(i,ii) = sumdens(i,ii)+1

          ENDDO
          ENDDO
        ENDDO

        END SUBROUTINE
```

To create the dynamic linked library, the Fortran code needs to be compiled. It is simple to do that directly from WITHIN the R-code!

Executing an external command in R is done by statement 'system'. R CMD SHLIB creates a dynamically linked library; it finds a suitable compiler to do so.

```
system("R CMD SHLIB competition.f")
```

Running this version of the model is significantly faster than the version that was entirely implemented in R: it takes less than a second for the output to appear.

Appendix B
Derivatives and Differential Equations

B.1 Derivatives

The concept of a derivative is at the core of a dynamic mathematical model, which is essentially specified as time derivatives of the state variables:

$$\frac{d\text{STATEVARIABLE}}{dt} = \text{Sources} - \text{Sinks}$$

The meaning of the derivative here is: the instantaneous rate (or 'speed') at which the value of the state variable changes in time. The word 'instantaneous' refers to the fact that the derivative, although defined for all time moments, obtains, in principle, different values at each instant t.

Its meaning can best be understood by considering a finite interval, Δt. At the start of this interval the state variable has a start value, say S_0, while it has the value S_1 at the end of the interval. Thus, over the interval the state variable has changed at the *average* rate of:

$$\frac{\Delta S}{\Delta t} = \frac{S_1 - S_0}{t_1 - t_0}$$

When we decrease the width of the interval Δt further and further, we get closer and closer to what is called the 'instantaneous rate'. In the limit we find the derivative, which is defined for any moment in time t, and expresses how the state variable will change (dS) over an infinitesimal time period (dt) starting at this time t.

$$\frac{dS}{dt} = \lim_{\Delta t \to 0} \frac{\Delta S}{\Delta t}$$

Note that there are also functions for which the derivative is not defined at all periods in time, e.g. because of sudden 'jumps' in the functions – we will not normally consider these situations but sometimes they can be relevant for ecological problems.

Equations involving derivatives are called 'differential equations'; they are dealt with next.

B.2 Taxonomy of Differential Equations

Differential equations express the *rate of change* of a constituent (C) along one or more dimensions, usually time and/or space.

For example, in the purely advective equation

$$\frac{\partial C}{\partial t} = -u \frac{\partial C}{\partial x}$$

$\partial C/\partial t$ is mathematical shorthand for the change of C with respect to time t (while holding space, x, constant); $\partial C/\partial x$ expresses the change along a spatial dimension while holding time t constant.

Constituents can also change along other dimensions, e.g. age, size. For instance, in the McKendrick-von Forster equation:

$$\frac{\partial N}{\partial t} + \frac{\partial N}{\partial a} = -m(a, t) \cdot N(a, t)$$

$\partial N/\partial a$ is the change in population density with age a.

Differential equations that contain only one independent variable (time, depth) are called *ordinary* differential equations. In contrast, in *partial* differential equations there are more than one independent variables e.g. time and depth, or two spatial dimensions. Both the advective and McKendrick-von Forster-equation are partial differential equations. Remark that in partial differential equations, the dC/dt sign is replaced by a $\partial C/\partial t$.

It is possible to take the derivative of a derivative. In this case, we obtain a *second-order* derivative. This notion is illustrated by means of a physical example. Consider the position of a moving particle in one dimension, x (m).

The velocity (m s^{-1}) of the particle at a particular point in time is given by the derivative of its position in time. In other words, it is the rate of change of the position with respect to time, a first-order derivative:

$$v = \frac{dx}{dt}$$

The acceleration, a (m s^{-2}) of the particle is nothing more than the rate of change of the velocity in time:

$$a = \frac{dv}{dt} = \frac{d}{dt}\left(\frac{dx}{dt}\right) = \frac{d^2x}{dt^2}$$

which is a first-order derivative in velocity and a second-order derivative in x.

The *order of a differential equation* equals the highest derivative in the equation. In the advection-diffusion-reaction equation for instance:

$$\frac{\partial C}{\partial t} = -u\frac{\partial C}{\partial x} + \frac{\partial}{\partial x}D\frac{\partial C}{\partial x} - kC$$

the highest derivative is second-order (the diffusive term), thus this is a second-order differential equation.

The *steady-state solution* of a differential equation is one where the derivative with respect to time (the rate of change) is 0 and the state variables do not change in time. Thus, for the advective-diffusive-reaction equation the steady-state solution is given by the solution of the following equation:

$$0 = \frac{\partial}{\partial x}D\frac{\partial C}{\partial x} - u\frac{\partial C}{\partial x} - kC$$

Solving a differential equation requires the introduction of *initial, and boundary* conditions.

The number of spatial boundary conditions in a differential equation is the same as the order of the highest derivative in space and the number of initial conditions is the same as the order of the highest derivative in time.

B.3 General Solutions of Often Used Differential Equations

B.3.1 Simple Time-Dependent Equations

Differential equation	General solution
$\dfrac{dN}{dt} = r \cdot N$	$N(t) = A \cdot \exp(r \cdot t)$
$\dfrac{dN}{dt} = r \cdot \left(1 - \dfrac{N}{K}\right) \cdot N + h \cdot N$	$N(t) = \dfrac{K \cdot (r+h)}{r + A \cdot K \cdot (r+h) \cdot \exp\left[-(r+h) \cdot t\right]}$
$\dfrac{dN}{dt} = r \cdot N \cdot \exp(-D \cdot t)$	$N(t) = A \cdot \exp\left[\dfrac{r \cdot \left[1 - \exp(-D \cdot t)\right]}{D}\right]$

B.3.2 Steady-State Transport – Reaction in 1-D, Constant Surface

Differential equation	General solution
$0 = D\dfrac{d^2 C}{d x^2} - u\dfrac{d C}{d x} - \gamma \cdot \exp(\delta \cdot x)$ $-k \cdot C + Q$	$C(x) = A \cdot \exp(\alpha_1 \cdot x) + B \cdot \exp(\alpha_2 \cdot x)$ $+ \dfrac{\gamma \exp(\delta \cdot x)}{D \cdot \delta^2 - u \cdot \delta - k} + \dfrac{Q}{k}$ $\alpha_1 = \dfrac{u - \sqrt{u^2 + 4D.k}}{2D}$ and $\alpha_2 = \dfrac{u + \sqrt{u^2 + 4D.k}}{2D}$
$0 = -u\dfrac{d C}{d x} - \gamma \exp(\delta \cdot x) - k \cdot C + Q$	$C(x) = A \cdot \exp\left(-\dfrac{k}{u} \cdot x\right) - \dfrac{\gamma \exp(\delta \cdot x)}{u \cdot \delta + k} + \dfrac{Q}{k}$
$0 = D\dfrac{d^2 C}{d x^2} - u\dfrac{dC}{dx} + Q$	$C(x) = A \cdot \dfrac{D \cdot \exp\left(\dfrac{u}{D}x\right)}{u} + \dfrac{Q \cdot x}{u} + B$
$0 = D\dfrac{d^2 C}{d x^2} - \gamma \cdot \exp(\delta \cdot x) + Q$	$C(x) = \dfrac{\gamma \cdot \exp(\delta \cdot x)}{D \cdot \delta^2} - \dfrac{1}{2} \cdot \dfrac{Q \cdot x^2}{D} + A \cdot x + B$

In all these equations, A and B are integration constants to be derived from the boundary conditions.

B.3.3 Steady-State Transport – Reaction in 1-D, Cylindrical Coordinates

Differential equation	General solution
$0 = \dfrac{D}{r}\dfrac{d}{dr}\left(r\dfrac{dC}{dr}\right) + Q$	$C(r) = -\dfrac{Q \cdot r^2}{4D} + A \cdot \log_e(r) + B$
$0 = \dfrac{D}{r}\dfrac{d}{dr}\left(r\dfrac{dC}{dr}\right) - k \cdot C + Q$	$C(r) = A \cdot I_0\left(\sqrt{\dfrac{k}{D}}r\right) + B \cdot Y_0\left(\sqrt{-\dfrac{k}{D}}r\right) + \dfrac{Q}{k}$

I0 and Y0 are the (modified) Bessel functions of the first and second kind, respectively, and of order 0.

$$\text{Note: } \frac{d\left(I_0\left(\sqrt{\dfrac{k}{D}}r\right)\right)}{dr} = \sqrt{\frac{k}{D}}I_1\left(\sqrt{\frac{k}{D}}r\right)$$

where I_1 is the Bessel function of the first kind and order 1.

B.3.4 Steady-State Transport – Reaction in 1-D, Spherical Coordinates

Differential equation	General solution
$0 = \dfrac{D}{r^2}\dfrac{d}{dr}\left(r^2\dfrac{dC}{dr}\right) - \gamma\exp(\delta \cdot x)$ $-k \cdot C$	$C(r) = \dfrac{A}{r}\cdot\sinh\left(\sqrt{\dfrac{k}{D}}r\right) + \dfrac{B}{r}\cdot\cosh\left(\sqrt{\dfrac{k}{D}}r\right)$ $-\dfrac{(-r\delta^2 D + 2\delta D + kr)\exp(\delta r)\,\gamma}{r(k - \delta^2 D)^2}$
$0 = \dfrac{D}{r^2}\dfrac{d}{dr}\left(r^2\dfrac{dC}{dr}\right) - \gamma\exp(\delta \cdot x)$ $-k \cdot C + Q$	$C(r) = \dfrac{A}{r}\cdot\sinh\left(\sqrt{\dfrac{k}{D}}r\right) + \dfrac{B}{r}\cdot\cosh\left(\sqrt{\dfrac{k}{D}}r\right)$ $-\dfrac{1}{kr(k - \delta^2 D)^2}\cdot\left\{\begin{array}{l}(-\gamma(kr - \delta D(\delta r - 2))k\cosh(\delta r)\\ -\gamma(kr - \delta D(\delta r - 2))k\sinh(\delta r)\\ +Qr(-\delta^2 D + k)^2)\end{array}\right\}$
$0 = \dfrac{D}{r^2}\dfrac{d}{dr}\left(r^2\dfrac{dC}{dr}\right) - k \cdot C + Q$	$C(r) = A\cdot\sinh\left(\sqrt{\dfrac{k}{D}}r\right) + B\cdot\cosh\left(\sqrt{\dfrac{k}{D}}r\right) + \dfrac{Q}{k}$ or: $C(r) = A\cdot r^a + B\cdot r^b + \dfrac{Q}{k}$ $a = -\dfrac{D - \sqrt{D^2 + 4kr^2 D}}{2D}$ $b = -\dfrac{D + \sqrt{D^2 + 4kr^2 D}}{2D}$
$0 = \dfrac{D}{r^2}\dfrac{d}{dr}\left(r^2\dfrac{dC}{dr}\right) + Q$	$C(r) = -\dfrac{Q\cdot r^2}{6D} - \dfrac{A}{r} + B$

B.4 Particular Solutions of Dynamic Diffusion-Reaction Equation

Differential equation	Initial, boundary conditions	Particular solution
$\dfrac{\partial C}{\partial t} = D\dfrac{\partial^2 C}{\partial x^2} + k \cdot C$	$C(t=0,x) = C_0$ for $x = 0$, $C(t=0,x) = 0$ for $x \neq 0$; $C(t,x=\pm\infty) = 0$	$C(x,t) = \dfrac{C_0\cdot\exp(k\cdot t)}{2\sqrt{\pi Dt}}\cdot\exp\left(-\dfrac{x^2}{4Dt}\right)$
$\dfrac{\partial C}{\partial t} = D\dfrac{\partial^2 C}{\partial x^2} - u\dfrac{\partial C}{\partial x} + k \cdot C$	$C(t=0,x) = C_0$ for $x = 0$, $C(t=0,x) = 0$ for $x \neq 0$; $C(t,x=\pm\infty) = 0$	$C(x,t) = \dfrac{C_0\cdot\exp(k\cdot t)}{2\sqrt{\pi Dt}}\cdot\exp\left(-\dfrac{(x - u\cdot t)^2}{4Dt}\right)$
$\dfrac{\partial C}{\partial t} = \dfrac{1}{r}D\dfrac{\partial}{\partial r}r\dfrac{\partial C}{\partial r} + k \cdot C$	$C(t=0,r)=C_0$ for $r = 0$, $C(t=0,r) = 0$ for $r \neq 0$; $C(t,r = \pm\infty) = 0$	$C(r,t) = \dfrac{C_0}{4\pi Dt}\cdot\exp\left(k\cdot t - \dfrac{r^2}{4Dt}\right)$

B.5 Derivatives and Integrals

$$\frac{d\left(a \cdot x^b\right)}{dx} = a \cdot b \cdot x^{b-1}$$

$$\frac{d\left(a \cdot e^{b \cdot x}\right)}{dx} = a \cdot b \cdot e^{b \cdot x}$$

$$\int a \cdot e^{b \cdot x} dx = \frac{a \cdot e^{b \cdot x}}{b}$$

$$\int_{x1}^{x2} a \cdot e^{b \cdot x} dx = \frac{a}{b} \cdot \left[e^{b \cdot x2} - e^{b \cdot x1}\right]$$

Appendix C
Matrix Algebra

C.1 Matrices

A *matrix* is a rectangular array of numbers. A matrix with m rows and n columns is said to be of order m × n; (m,n) is also referred to as the dimension of the matrix. It may be written as $\mathbf{A}_{(m \times n)}$. In expanded form, a matrix is written with coefficients a_{ij} for each entry; the first subscript refers to the rows, the second to the columns. A matrix is often symbolized with an uppercase, bold-faced letter.

$$\mathbf{A}_{(2 \times 2)} = \begin{bmatrix} a_{11} & a_{12} \\ a_{21} & a_{22} \end{bmatrix}$$

A matrix with column order one is called a column *vector*.

$$\mathbf{a} = \begin{bmatrix} a_1 \\ a_2 \end{bmatrix}$$

A row vector is a matrix with one row and several columns.

The *product* between matrices $\mathbf{A}$ and $\mathbf{B}$, written as $\mathbf{AB}$, is only defined if the number of columns of $\mathbf{A}$ equals the number of rows of $\mathbf{B}$. The product of a matrix with order m × n with a matrix with order n × p gives a new matrix with order m × p and whose ij-th element equals: $\sum_{k=1}^{n} a_{ik} b_{kj}$.

Example:

$$\begin{bmatrix} a_{11} & a_{12} \\ a_{21} & a_{22} \end{bmatrix} \cdot \begin{bmatrix} x_1 \\ x_2 \end{bmatrix} = \begin{bmatrix} a_{11} \cdot x_1 + a_{12} \cdot x_2 \\ a_{21} \cdot x_1 + a_{22} \cdot x_2 \end{bmatrix}$$

A *diagonal* matrix is one for which the values off diagonal are 0. The *identity* matrix $\mathbf{I}$ is a special form of a diagonal matrix where the diagonal elements are 1.

$$I = \begin{bmatrix} 1 & 0 & \ldots & 0 \\ 0 & 1 & \ldots & 0 \\ \ldots\ldots & & \ddots & \ldots \\ 0 & 0 & \ldots & 1 \end{bmatrix}$$

The identity matrix plays the same role as the number 1 in scalar notation, this is:
$AI = A = IA$

The *transpose* of an mxn matrix A, is a nxm matrix denoted by A^T or A' and having rows identical to the columns of A and vice-versa:

$$A^T = \begin{bmatrix} a_{11} & a_{21} \\ a_{12} & a_{22} \end{bmatrix}$$

Note: $(A^T)^T = A$; $(A+B)^T = A^T + B^T$; $(AB)^T = B^T A^T$

The *inverse* of a matrix A is the unique matrix A^{-1} which, when multiplied with A produces the identity matrix I:

$$A\ A^{-1} = I$$

The inverse can only be taken from square matrices; not all square matrices have an inverse. Note that $(cA)^{-1} = c^{-1}A^{-1}$; $(AB)^{-1} = B^{-1}\ A^{-1}$

C.2 Linear Equations

The m linear equations in the n*m unknowns x

$$a_{11}x_1 + a_{12}x_2 + \ldots a_{1n}x_n = b_1$$
$$\ldots$$
$$a_{m1}x_1 + a_{m2}x_2 + \ldots a_{mn}x_n = b_m$$

can be written in matrix form:

$$Ax = b$$

Where $A_{(mxn)}$ is called the coefficient matrix, vector x contains the unknowns and b is called the right hand side.

The solution of the linear system of **m** linearly independent equations in **m** unknowns

$$Ax = b \text{ is given by :}$$
$$x = A^{-1}b$$

where A^{-1} is the inverse of square matrix $A_{(m \times m)}$. (but not all square matrices have an inverse).

C.3 Eigenvalues and Eigenvectors, Determinants

Given a square (m x m) real matrix A, an *eigenvalue* λ is a scalar for which:

$$Ax = \lambda x \text{ for } x \neq 0 :$$

$$a_{11}x_1 + a_{12}x_2 + \ldots a_{1n}x_n = \lambda x_1$$
$$a_{21}x_1 + a_{22}x_2 + \ldots a_{2n}x_n = \lambda x_2$$
$$\ldots$$
$$a_{n1}x_1 + a_{n2}x_2 + \ldots A_{nn}x_n = \lambda x_n$$

the column vector x is called the **right** *eigenvector* of A for the eigenvalue λ; the eigenvector is defined only up to a scaling factor. For symmetric A the eigenvalues are real (otherwise they might be complex numbers).

The left eigenvector x_l is the row vector satisfying

$$x_l A = \lambda x \text{ for } x \neq 0 :$$

The left eigenvector of A equals the right eigenvector of the transpose of A.

C.4 R-Examples

In the following R-script, a vector and a matrix are first created (line $1 - 2$), and then the matrix is multiplied with scalar 2 and added to itself (line 3); element-wise multiplication is performed on line 4

The statement on the last line performs matrix multiplication (in R, matrix multiplication is done using %*%).

```
v <- c(1,2,3)
A <- matrix(nrow=2,ncol=2,data=2)
2*A+A
A*A
A%*%A
```

The next R-script first creates and shows a unity matrix with 10 rows and columns (the ' ; ' separates two statements).

Next a square matrix A is created and shown (embracing a statement between brackets executes the command and prints to the screen).

The statements t(A), and solve(A) take the transpose and the inverse respectively.

The last line multiplies A with its inverse, which outputs the unity matrix.

```
I <- diag(1,nrow=10) ; I

(A <- matrix(nrow=2,data=1:4))
t(A)
solve(A)
solve(A)%*%A
```

Solving a system of equations is also done using R's function solve:

```
A <- matrix(nrow=2,data=1:4)
B <- c(5,6)
X <- solve(A,B)

A%*%X-B
```

In R the eigenvalue and (right) eigenvectors of a matrix are calculated by 'eigen' (first line). This function returns a list that contains both the eigenvalues ($values) and the eigenvectors ($vectors). The latter are represented by the columns of the matrix.

Lines 2 and 3 calculate $\mathbf{Ax} - \lambda\mathbf{x}$ for both eigenvectors, which equals 0.

```
(er <- eigen(A))
A%*%er$vectors[,1]-er$values[1]*er$vectors[,1]
A%*%er$vectors[,2]-er$values[2]*er$vectors[,2]
```

The next 3 lines contain the same calculations, now for the left eigen vectors of **A**.

```
(el <- eigen(t(A)))
el$vectors[,1]%*%A-el$values[1]*el$vectors[,1]
el$vectors[,2]%*%A-el$values[2]*el$vectors[,2]
```

C.5 The Jacobian Matrix

One matrix that plays a prominent role in this book is the so-called Jacobian matrix. It is used in Chapter 7.6 to estimate the stability type of a steady-state solution, but it is also at the heart of numerical integration (Section 6.3.5) and nonlinear root solving (the Newton-Raphson method, Section 7.5.2. Here is a definition:

Consider n functions f_i in n elements x_i, i=1,... n.

The Jacobian matrix J of this function evaluated at a point $\hat{x} = (\hat{x}_1, \hat{x}_2, \ldots \hat{x}_n)$ is an n*n square matrix with on its i-th row and j-th column the element $\partial f_i / \partial x_j(\hat{\mathbf{x}})$, i.e. the partial derivate of f_i with respect to x_j evaluated at point $(\hat{\mathbf{x}})$.

$$J = \begin{bmatrix} \dfrac{\partial f_1}{\partial x_1}(\hat{\mathbf{x}}) \cdots \dfrac{\partial f_1}{\partial x_j}(\hat{\mathbf{x}}) \cdots \dfrac{\partial f_1}{\partial x_n}(\hat{\mathbf{x}}) \\ \cdots \quad \cdots \quad \cdots \quad \cdots \\ \dfrac{\partial f_i}{\partial x_1}(\hat{\mathbf{x}}) \cdots \dfrac{\partial f_i}{\partial x_j}(\hat{\mathbf{x}}) \cdots \dfrac{\partial f_i}{\partial x_n}(\hat{\mathbf{x}}) \\ \cdots \quad \cdots \quad \cdots \quad \cdots \\ \dfrac{\partial f_n}{\partial x_1}(\hat{\mathbf{x}}) \cdots \dfrac{\partial f_n}{\partial x_j}(\hat{\mathbf{x}}) \cdots \dfrac{\partial f_n}{\partial x_n}(\hat{\mathbf{x}}) \end{bmatrix}$$

If $f_i(x)$ is the differential equation of state variable x_i ($f_i = \dfrac{dx_i}{dt}$) then the Jacobian becomes:

$$J = \begin{bmatrix} \partial \dfrac{\dfrac{dx_1}{dt}}{\partial x_1} \cdots \partial \dfrac{\dfrac{dx_1}{dt}}{\partial x_j} \cdots \partial \dfrac{\dfrac{dx_1}{dt}}{\partial x_n} \\ \cdots \quad \cdots \quad \cdots \quad \cdots \\ \partial \dfrac{\dfrac{dx_i}{dt}}{\partial x_1} \cdots \partial \dfrac{\dfrac{dx_i}{dt}}{\partial x_j} \cdots \partial \dfrac{\dfrac{dx_i}{dt}}{\partial x_n} \\ \cdots \quad \cdots \quad \cdots \quad \cdots \\ \partial \dfrac{\dfrac{dx_n}{dt}}{\partial x_1} \cdots \partial \dfrac{\dfrac{dx_n}{dt}}{\partial x_j} \cdots \partial \dfrac{\dfrac{dx_n}{dt}}{\partial x_n} \end{bmatrix}$$

It contains, in its rows, elements which determine how the rate of change of one state variable is affected by a change in all the state variable values (the columns). This is the form of the Jacobian that is used in this book.

Appendix D
Statistical Distributions

D.1 Probability Distribution

For every value x and for an infinitesimal interval dx starting from x, the probability density function gives the probability of occurrence in the interval [x,x+dx].

Informally, a probability density function can be considered a smooth, continuous version of a histogram.

The integral of the probability density function between minus infinity and plus infinity equals 1.

D.2 Normal Distribution

The probability density function of the (univariate) normal distribution with mean μ and standard deviation σ is:

$$f(x;\mu,\sigma) = \frac{1}{\sqrt{2\pi\sigma^2}} \cdot \exp^{-\frac{(x-\mu)^2}{2\sigma^2}}$$

It is the familiar bell-shaped function. In R, the normal density function is generated with dnorm, the cumulative density function with pnorm, whilst random normally distributed numbers are generated with rnorm (see Fig. D.1):

```
par(oma=c(0,0,0,2))
plot(seq(-4,4,length.out=5000),rnorm(5000,0,1),
     xlab='''',ylab=''N(0,1)'',pch=''+'',col=''grey'',
     main=''Normal distribution'')
par(new=TRUE)
curve(pnorm(x,0,1),lwd=2,axes=FALSE,xlab='''',ylab='''')
curve(dnorm(x,0,1),lwd=2,add=TRUE,lty=2)
axis(side=4)
legend(''topleft'',c(''pnorm'',''dnorm'',''rnorm''),lty=c(1,2,N
A),pch=c(NA,NA,''+''),lwd=c(2,2,NA))
mtext(side=4,''p'',outer=TRUE)
```

Normal distribution

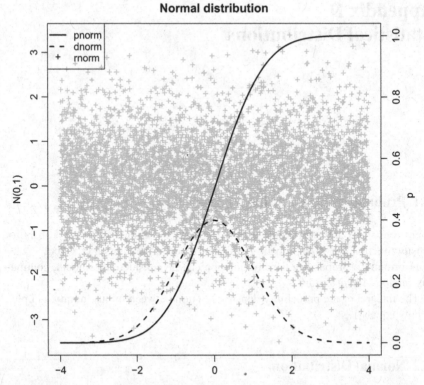

Fig. D.1 The normal probability density function (zero mean, unit variance) (dashed line, right axis); Cumulative normal probability density function (full line, right axis); Normally distributed random numbers (crosses, left axis).

D.3 The Poisson Distribution

This is a probability distribution that describes the occurrence of rare discrete random events. It was used in chapter 9.3 when we modelled the encounters between a parasitoid and its host. The probability that n events will occur in a time interval is given by: $p(n) = e^{-\mu} \mu^n / n!$ where μ is the average number of events in the time interval, and n! equals the n factorial $1*2*\ldots n - 1*n$.

References

Abbott, M.B., Basco, D.R., 1989. Computational fluid dynamics. An introduction for engineers. Longman Scientific & Technical, John Wiley & Sons, Inc, New York.

Allee, W.C., 1931. Animal aggressions. University of Chicago Press, Chicago.

Andersen, T., 1997. Pelagic nutrient cycles. Herbivores as sources and sinks. Springer. Ecological studies 129.

Andersson, H., Wijsman, J., Herman, P., Middelburg, J., Soetaert, K., Heip, C., 2004. Respiration patterns in the deep ocean. Geophysical Research Letters 31: LO3304.

Auger, P., Bravo de la Parra, R., Poggiale, J.-C., Sánchez, E., Nguyen-Huu, T., 2008. Aggregation of Variables and Applications to Population Dynamics. In: Structured Population Models in Biology and Epidemiology, Lecture Notes in Mathematics. Springer, Berlin, pp. 209–263.

Berger U., Hildenbrandt H., 2000. A new approach to spatially explicit modelling of forest dynamics: Spacing, ageing and neighbourhood competition of mangrove trees. Ecological Modelling 132: 287–302.

Berner R.A., 1980. Early Diagenesis- A Theoretical Approach. Princeton Univ. Press. Princeton.

Botkin, D.B., Janak, J.F., Wallis, J.R., 1972. Some Ecological Consequences of a Computer Model of Forest Growth, Journal of Ecology 60: 849–872.

Boudreau, B.P., 1997. Diagenetic Models and their Implementation. Modelling transport and Reactions in Aquatic Sediments. Springer, Berlin.

Brun, R., Reichert, P., Kunsch, H.R., 2001. Practical identifiability analysis of large environmental simulation models. Water Resources Research 37(4): 1015–1030.

Bugmann H., 2001. A Review of Forest Gap Models. Climatic Change, 51: 259–305.

Caswell, H., 2001. Matrix population models: construction, analysis, and interpretation. Second edition. Sinauer, Sunderland.

Chapra, S.C., 1997. Surface water-quality modelling. Mc Graw-Hill, Boston.

Clark, C.W., 1990. Mathematical bioeconomics: optimal management of renewable resources. John Wiley & Sons, New York.

Clark, C.W., Mangel, M., 2000. Dynamic state variable models in ecology. Methods and Applications. Oxford University Press, New York.

Coley, D.A., 1999. An introduction to genetic algorithms for scientists and engineers. World Scientific Publishing Co. Singapore.

Crank, J., 1975. The mathematics of diffusion. Oxford University Press, Oxford.

Czárán T., 1997. Spatiotemporal models of population and community dynamics. Chapman & Hall, London.

Denardo, E.V., 2003. Dynamic programming. Models and applications. Dover Publications Inc, Mineola New York.

Ebert, T.E., 1999. Plant and animal populations, methods in demography. Academic Press, London, 312p.

Edelstein-Keshet, L., 2005. Mathematical models in biology. Classics in applied mathematics. SIAM 46, New York.

Ermentrout G.B., Edelstein-Keshet, L., 1993. Cellular automata approaches to biological modelling. Journal of Theoretical Biology 160: 97–133.

Fasham, M.J.R., Ducklow, H.W., McKelvie, S.M., 1990. A nitrogen-based model of plankton dynamics in the oceanic mixed layer. Journal of Marine Research 48: 591–639.

Fennel, W., Neumann, T., 2004. Introduction to the modelling of marine ecosystems. Elsevier Oceanography Series 72. Elsevier, Amsterdam.

Fiadeiro Me, Veronis G., 1977. Weighted-mean schemes for finite-difference approximation to advection-diffusion equation. Tellus v 29: 512–522.

Gentleman, W., 2002. A chronology of plankton dynamics in silico: how computer models have been used to study marine ecosystems. Hydrobiologia 480: 69–85.

Gielen, J.L.W., Kranenbarg, S., 2002. Oxygen balance for small organisms: an analytical model. Bulletin of Mathematical Biology 64: 175–207.

Gilman, M., Hails, R., 1997. An introduction to ecological modelling. Putting practice into theory. In: Methods in ecology. Blackwell science, Oxford.

Goloway, F., Bender, M., 1982. Diagenetic models of interstitial nitrate profiles in deep sea suboxic sediments. Limnology and Oceanography 27(4): 624–638.

Goldberg, D.E., 1989. Genetic algorithms in search, optimization & machine learning. Addison-Wesley publishing company, inc, Reading.

Gotelli, N.J., 2001. A primer of ecology. Sinauer Associates inc., Sunderland, 265p.

Grimm, V., Railsback, S.F., 2005. Individual-based modelling and ecology. Princeton series in theoretical and computational biology. Princeton University Press, Princeton.

Gurney, W.S.C., Nisbet, R.M., 1998. Ecological dynamics. Oxford University Press, New York.

Haefner, J.W., 1996. Modelling biological systems. Principles and applications. Chapman and Hall, New York.

Hannon, B., Ruth, M., 1997. Modelling dynamic biological systems. Springer, New York.

Hansen, P.J., Bjornsen, P.K., Hansen, B.W., 1997. Zooplankton grazing and growth: Scaling within the 2–2000-μm body size range. Limnology and Oceanography 42: 687–704.

Hofmann, A.F., Meysman, F.J.R., Soetaert, K., Middelburg, J.J., 2008. A step-by-step procedure for pH model construction in aquatic systems. Biogeosciences 5: 227–251.

Holling, C.S., 1959. Some characteristics of simple types of predation and parasitism. Canadian Entomology 91: 824–839.

Houston, A.I., McNamara, J.M., 1999. Models of adaptive behaviour: An approach based on state (Hardcover). Cambridge University Press, New York.

Jamart, B.M., Winter, D.F., Banse, K., Anderson, G.C., Lam, R.K., 1977. A theoretical study of phytoplankton growth and nutrient distribution in the Pacific Ocean off the northwestern US coast. Deep-Sea Research 24: 753–773.

Jones, J.H., 2007. demogR: A package for evolutionary demographic analysis in R. Journal of Statistical Software 22(10).

Jørgensen, S.E., 1994. Fundamentals of ecological modelling (2nd edition). Developments in environmental modelling 19. Elsevier, Amsterdam.

Kaitala, V., Ylikarjula, J., Heino, M., 2000. Non-unique population dynamics: basic patterns. Ecological Modelling 135: 127–134.

Keyfitz, N., Flieger, W., 1971. Population: facts and methods of demography. W.H. Freeman, Sna Francisco, California, USA.

King, A.C., Billingham, J., Otto, S.R., 2003. Differential equations, linear, nonlinear, ordinary, partial. Cambridge University Press, Cambridge.

Klaassen, M., Bauer, S., Madsen, J., Ingunn, T., 2006. Modelling behavioural and fitness consequences of disturbance for geese along their spring flyway. Journal of Applied Ecology 43: 92–100.

Kooijman, S.A.L.M., 2000, Dynamic energy and mass budgets in biological systems, Cambridge University Press, Cambridge.

Kot, M., 2001. Elements of mathematical ecology. Cambridge University Press, Cambridge.

Kruger, J., 1993. Simulated annealing: a tool for data assimilation into an almost steady model state. Journal of Physical Oceanography 23: 679–688.

Lancelot, C., Veth, C., Mathot, S. 1991. Modelling ice-edge phytoplankton bloom in the Scotia-Weddel sea sector of the Southern Ocean during spring 1988. Journal of Marine Systems 2: 333–346.

Lefkovitch, L.P., 1965. The study of population growth in organisms grouped by stages. Biometrics 21: 1–18.

Leslie, P.H., 1945. On the use of matrices in certain population mathematics. Biometrika 33: 183–212.

Lotka, A.J., 1925. Elements of physical biology. Williams & Wilkins Co., Baltimore.

Lorenz, E.N., 1963. Deterministic non-periodic flows. Journal of Atmospheric Sciences 20: 130–141.

Ludwig, D., Jones, D.D., Holling, C.S., 1978. Qualitative analysis of insect outbreak systems: the spruce budworm and forest. Journal of Animal Ecology, 47: 315–322.

Ludwig, D., Walker, B.H., Holling, C.S., 2002. Models and Metaphors of Sustainability, Stability and Resilience. In: Gunderson, L.H., and Pritchard Jr., L. (eds). Resilience and the Behavior of Large-Scale Systems. SCOPE 60. Washington, Island Press, 21–48.

May, R.M., 1977. Thresholds and breakpoints in ecosystems with a multiplicity of stable states. Nature 269: 471–477.

McCallum, H., 2000. Population parameters: estimation for ecological models. Methods in Ecology, Blackwell. Oxford.

McCarthy, M.A., Possingham, H.P., 2007. Active adaptive management for conservation. Conservation Biology 21: 956–963 AUG.

McNamara, J., Webb, J., Collins, E., Szekely, T., Houston, A., 1997. A general technique for computing evolutionarily stable strategies based on errors in decision-making. Journal of Theoretical Biology 189: 211–225.

Meysman, F.J.R., Middelburg, J.J., Heip, C.H.R., 2006. Bioturbation: a fresh look at Darwin's last idea. Trends in Ecology and Evolution 21: 688–695.

Mooij W.M., Boersma, M., 1996. An Object Oriented Simulation Framework for Individual-Based Simulations (OSIRIS): Daphnia population dynamics as an example. Ecological Modelling 93: 87–97.

Nicholson, A.J., 1933. The balance of animal populations. Journal of Animal Ecology 2: 131–178.

Nicholson, A.J., Bailey, V.A., 1935. The balance of animal populations. Proceedings of the Royal Society of London 1: 551–598.

Okubo, A., 1980. Diffusion and ecological problems: Mathematical models. Springer-Verlag, Berlin.

Okubo, A., Levin, S.A., (eds.) 2001. Diffusion and ecological problems: Modern perspectives. Interdisciplinary Applied Mathematics 14.

Pacala S.W., 1986. Neighborhood models of plant population dynamics. 2. Multi-species models of annuals. Theoretical Population Biology 29: 262–292.

Petzoldt, T., 2003. R as a Simulation Platform in Ecological Modelling. R News, vol 3/3, pp 8–16. (http://www.r-project.org/).

Petzoldt, T., Rinke, K., 2007. simecol: An object-oriented framework for ecological modelling in R. Journal of Statistical Software, 22(9).

Powell, E., 1989. Oxygen, sulfide and diffusion: Why thiobiotic meiofauna must be sulfide-insensitive first-order respirers. Journal of Marine Research 47: 887–932.

Press W.H., Flannery B.P., Teukolsky S.A., Vetterling W.T., 1992. Numerical recipes. The Art of Scientific Computing. Second edition. Cambridge University Press, Cambridge.

Price, W.L., 1977. A controlled random search procedure for global optimization. The Computer Journal, 20: 367–370.

Proye, A., Gattuso, J-P., Epitalon, J-M., Gentili, B., Orr, J., Soetaert, K., 2007. seacarb: an R-package to calculate parameters of the seawater carbonate system. R package version 1.2.1. http://www.obs-vlfr.fr/~gattuso/seacarb.php

R Development Core Team (2007). R: A language and environment for statistical computing. R Foundation for Statistical Computing, Vienna, Austria. ISBN 3-900051-07-0, URL http://www.R-project.org.

Renshaw, E., 1991. Modelling biological populations in space in time. Cambridge studies in mathematical biology 11. Cabridge University Press, Cambridge.

Ricker, W.E., 1954. Stock and recruitment. Journal of the Fisheries Research Board of Canada 11: 559–623.

Riley, G.A., 1946. Factors controlling phytoplankton populations on Georges Bank. Journal of Marine Research 6: 54–73.

Rogers, D., 1972. Random search and insect population models. Journal of Animal Ecology 41: 369–383.

Sarukhán, J., Gadgil, M., 1974. Studies on plant demography: Ranunculus repens L., R. Bulbosus L. and R. Acris L. III. A mathematical model incorporating multiple modes of reproduction. Journal of Ecology 62: 921–936.

Sayles, F.L., Martin, W.R., Deuser, W.G., 2002. Response of benthic oxygen demand to particulate organic carbon supply in the deep sea near Bermuda. Nature 371: 686–689.

Schaefer, M.B., 1954. Some aspects of the dynamics of populations important to the management of commercial marine fisheries. Bulletin of the Inter-American Tropical Tuna Commission 1: 25–56.

Schaefer, M.B., 1967. Fishery dynamics and the present status of the yellowfin tuna population of the Eastern Pacific Ocean. Bulletin of the Inter=Americl Tophical Tuna Commission (12)3.

Scheffer, M., 2001. Climatic warming causes regime shifts in lake food webs. Limnology and Oceanography 46: 1780–1783.

Schink, D.R., Guinasso, N.L., Fanning, K.A., 1975. Processes affecting the concentration of silica at the sediment-water interface of the Atlantic Ocean. Journal of Geophysical Research, 80: 3013–3031.

Schwinning, S., Weiner, J., 1998. Mechanisms determining the degree of size asymmetry in competition among plants. Oecologia 113: 447–455.

Setzer, R.W, 2001. The odesolve Package: Solvers for Ordinary Differential Equations. R package version 0.1–1.

Shuter, B., 1979. Model of physiological adaptation in unicellular algae. Journal of Theoretical Biology 79: 519–552.

Silvertown, J., Holtier, S., Johnson, J., Dale, P., 1992. Cellular automaton models of interspecific competition for space – the effect of pattern on process. Journal of Ecology 80: 527–534.

Smith, V.L., 1969. On models of commercial fishing. Journal of Political Economy 63: 116–124.

Soetaert, K., Herman P.M.J., 1994. One foot in the grave: zooplankton drift into the Westerschelde estuary (The Netherlands). Marine Ecology Progress Series 105: 19–29.

Soetaert, K., Herman, P.M.J. 2008. ecol Mod: "A practical guide to ecological modelling – using R as a simulation platform". R package version 1.1.

Soetaert, K., Herman P.M.J., Middelburg J.J, Heip C., deStigter H.S., van Weering T.C.E., Epping, E., Helder W., 1996a. Modelling 210Pb-derived mixing activity in ocean margin sediments: Diffusive versus nonlocal mixing. Journal of Marine Research 54:1207–1227.

Soetaert, K., Herman P.M.J., Middelburg J., 1996b. A model of early diagenetic processes from the shelf to abyssal depths. Geochimica Et Cosmochimica Acta, 60(6): 1019–1040.

Soetaert, K., 2008. rootSolve: Nonlinear root finding and steady-state analysis. R package version 1.1.

Soetaert, K., Petzoldt, T., Setzer R.W., 2008. Solving differential equations in R: package deSolve.

Steele, J.H., Henderson, E.W., 1992. The role of predation in plankton models. Journal of Plankton Research 14:157–172.

Stubben, C.J., Milligan, B.G., 2007. Estimating and analysing demographic models using the popbio package in R. Journal of Statistical Software 22(2).

Tarantola, A., 2005. Inverse problem theory and methods for model parameter estimation. SIAM, Philadelphia.

Thomann, R.V., Mueller, J.A., 1987. Principles of surface water quality modelling and control. Harper and Row, New York.

Van De Koppel, J., Herman, P.M.J., Thoolen, P., Heip, C.H.R., 2001. Do alternate stable states occur in natural ecosystems? Evidence from a tidal flat. Ecology 82: 3449–3461.

Vandenborght, J.P., Billen, G., 1975. Vertical distribution of nitrate concentration in interstitial water of marine sediments with nitrification and denitrification. Limnology and Oceanography, 20: 953–961.

Van der Meer, J., 2006. An introduction to Dynamic Energy Budget (DEB) models with special emphasis on parameter estimation. Journal of Sea Research 56: 85–102.

Verhulst, P.F., 1838. Notice sur la loi que la population pursuit dans son accroissement. Correspondance Mathématique Et Physique 10: 113–121.

Vos, M., Flik, B.J.G., Vijverberg, J., Ringelberg, J., Mooij, W.M., 2002. From inducible defences to population dynamics: modelling refuge use and life history changes in Daphnia. Oikos 99: 386–396.

Volterra, V., 1926. Variazioni e fluttuazioni del numero d'individui in specie animali conviventi. Mem. R. Accad. Naz. dei Lincei. Ser. VI, vol. 2.

Wang, Y., Van Cappellen, P., 1996. A multi-component reaction-transport model of early diagenesis: Application to redox cycling in coastal marine sediments. Geochimica Et Cosmochimica Acta 60: 2993–3014.

Werner, P.A., Caswell, H., 1977. Population growth rates and age versus stage-distribution models for teasel (Dipsacus sylvestri Huds.). Ecology 58: 1103–1111.

Wiggins, S., 1990. Introduction to applied nonlinear dynamical systems and chaos. Springer-Verlag, New-York.

Wilson, W., 2000. Simulating ecological and evolutionary systems in C. Cambridge University Press, Cambridge.

Wolfram, S. 1983. Statistical mechanics of cellular automata. Reviews of Modern Physics 55: 601–644.

Wroblewski, J.S., 1977. A model of phytoplankton plume formation during variable Oregon upwelling. Journal of Marine Research 35: 357–394.

Wu, H., Sharpe, P.J.H., Walker, J., Penridge, L.K., 1985. Ecological field theory: a spatial analysis of resource interference among plants. Ecological Modelling 29: 215–243.

Wyszomirski, T., 1986. Growth, competition and skewness in a population of one-dimensional individuals. Ekologia Polska 43: 615–641.

Yodzis, P., 1989. Introduction to theoretical ecology. Harper and Row, New York.

Index